Lecture Notes in Computer Science 10517

More information about this series at http://www.springer.com/series/7411

Antonio Puliafito · Dario Bruneo
Salvatore Distefano · Francesco Longo (Eds.)

Ad-hoc, Mobile, and Wireless Networks

16th International Conference on Ad Hoc Networks and Wireless, ADHOC-NOW 2017
Messina, Italy, September 20–22, 2017
Proceedings

Editors
Antonio Puliafito
University of Messina
Messina
Italy

Dario Bruneo
University of Messina
Messina
Italy

Salvatore Distefano
University of Messina
Messina
Italy

and

Kazan Federal University
Kazan
Russia

Francesco Longo
University of Messina
Messina
Italy

ISSN 0302-9743 ISSN 1611-3349 (electronic)
Lecture Notes in Computer Science
ISBN 978-3-319-67909-9 ISBN 978-3-319-67910-5 (eBook)
DOI 10.1007/978-3-319-67910-5

Library of Congress Control Number: 2017953427

LNCS Sublibrary: SL5 – Computer Communication Networks and Telecommunications

Printed on acid-free paper

This Springer imprint is published by Springer Nature
The registered company is Springer International Publishing AG
The registered company address is: Gewerbestrasse 11, 6330 Cham, Switzerland

Preface

The International Conference on Ad-Hoc Networks and Wireless (ADHOC-NOW) is one of the most popular series of events dedicated to research on ad-hoc, mobile, and wireless sensor networks and computing. Since its inception in 2002, the conference has been held 15 times in six different countries and the 16th edition in 2017 was held in Messina, Italy, during September 20–22.

We wish to thank all the authors who submitted their work. This year, ADHOC-NOW received 55 submissions and 22 papers were accepted for presentation as full contributions after a rigorous review process involving the Technical Program Committee (TPC) members and the TPC chairs. Moreover, due to the high quality of the received submissions, nine more papers were accepted as short contributions.

The ADHOC-NOW 2017 program was organized in eight sessions grouping the contributions into the following topics: Internet of Things, security, smart cities, ad-hoc networks, implementation and validation, wireless sensor networks, data management, and wireless systems. In each of these sessions, new ideas and directions were discussed among attendees from both academia and industry thus providing an in-depth and stimulating view on the new frontiers in the field of mobile, ad hoc, and wireless computing.

The conference was also enriched by three distinguished keynote speakers, completing a high-level scientific program.

We would like to thank all the people involved in ADHOC-NOW 2017. First of all, we are grateful to the TPC members and the external reviewers for their help in providing detailed reviews of the submissions, to Maddalena Nurchis, our proceedings chair, to Riccardo Petrolo, our Web chair, to Riccardo Di Pietro and Giovanni Merlino, our publicity chairs, and to Marco Scarpa and Giuseppe Tricomi, our submission and registration chairs. A special thanks to Rossana Morana, Alfonso Panarello, and Maurizio Giacobbe for their valuable support in the local organization and arrangements of the event. We also thank Springer's team for their great support throughout the whole process, from submission to production.

Finally, the organization was made possible through the strong support of our sponsors: the University of Messina and its spinoff SmartMe.io. A special thank you to them.

September 2017

Antonio Puliafito
Dario Bruneo
Salvatore Distefano
Francesco Longo

Organization

ADHOC-NOW 2017 was organized by the University of Messina, Italy.

Executive Committee

General Chair

Antonio Puliafito	University of Messina, Italy

Technical Program Committee Chairs

Dario Bruneo	University of Messina, Italy
Salvatore Distefano	University of Kazan, Russia
Francesco Longo	University of Messina, Italy

Publicity Arrangements Chairs

Giovanni Merlino	University of Messina, Italy
Riccardo Di Pietro	University of Messina, Italy

Proceedings Chair

Maddalena Nurchis	University of Pompeu Fabra, Barcelona, Spain

Submissions and Registrations Chairs

Marco Scarpa	University of Messina, Italy
Giuseppe Tricomi	University of Messina, Italy

Local Arrangements Chairs

Maurizio Giacobbe	University of Messina, Italy
Alfonso Panarello	University of Messina, Italy
Rossana Morana	University of Messina, Italy

Web Chair

Riccardo Petrolo	Rice University, USA

Technical Program Committee

Assis Flavio	UFBA – Federal University of Bahia, Brazil
Barcelo-Ordinas Jose M.	UPC, Spain
Calafate Carlos	Universitat Politècnica de València, Spain
Cano Juan-Carlos	UPV, Spain
Dautov Rustem	Kazan Federal University, Russia
Mario Di Mauro	University of Salerno, Italy
El Barrak Soumaya	Abdelmalek Essaâdi University, Morocco
Falcon Rafael	University of Ottawa, Canada
Fischer Stefan	Universität zu Lübeck, Germany
Fortino Giancarlo	University of Calabria, Italy
Giacobbe Maurizio	University of Messina, Italy
Karyotis Vasileios	National Technical University of Athens, Greece
Klasing Ralf	CNRS and University of Bordeaux, France
Martinez Francisco J.	University of Zaragoza, Spain
Mezei Ivan	University of Novi Sad, Serbia
Mingozzi Enzo	University of Pisa, Italy
Natalizio Enrico	Université de Technologie de Compiègne, France
Papagianni Chrysa	NTUA, Greece
Papavassiliou Symeon	NTUA, Greece
Fabio Postiglione	University of Salerno, Italy
Carlo Puliafito	University of Pisa, Italy
Scarpa Marco	University of Messina, Italy
Stea Giovanni	University of Pisa, Italy
Syrotiuk Violet	Arizona State University, USA
Turau Volker	Hamburg University of Technology, Germany
Wrona Konrad	NCI Agency, The Netherlands
Xin Qin	University of the Faroe Islands, Denmark
Alessio Botta	Università degli Studi di Napoli Federico II, Italy
Davide Cerotti	Università del Piemonte Orientale, Italy
Jacek Cichon	Wroclaw University of Technology, Poland
Rustem Dautov	KFU, Russia
Alessandra De Paola	Università di Palermo, Italy
Rasit Eskicioglu	University of Manitoba, Canada
Massimo Ficco	Università degli Studi della Campania, Italy
Nidhi Kushwaha	IIIT-A, India
Pierre Leone	Université de Genève, Switzerland
Weifa Liang	Research School of Computer Science, Australia
Evgeni Magid	KFU, Russia
Giovanni Merlino	Università degli Studi di Messina, Italy
Marc Mosko	Park.com, USA
Michael Oche	Kampala International University, Uganda
Eleni Stai	NTUA, Greece
Max Talanov	KFU, Russia
Orazio Tomarchio	Università di Catania, Italy

Giuseppe Tricomi	Università di Messina, Italy
Salvatore Venticinque	Università degli Studi della Campania, Italy
Weigang Wu	Sun Yat-sen University, China
Riccardo Di Pietro	Università di Messina, Italy
Antonino Galletta	Università di Messina, Italy
Alfonso Panarello	Università di Messina, Italy
Carmelo Romeo	Università di Messina, Italy
Nachiket Tapas	IIIT Naya Raipur, India
Fabio Verboso	Università di Messina, Italy
Benenson Zinaida	FAU, Germany
R. Brust Matthias	University of Luxembourg, Luxembourg
Gozalvez Javier	Miguel Hernandez University of Elche, Spain
Mazzara Manuel	Innopolis University, Russia
Pfisterer Dennis	University of Lübeck, Germany
Rodrigues Joel	National Institute of Telecommunications (Inatel), Brazil
Omprakash Vyas	IIIT Naya Raipur, India

Contents

Internet of Things

SustainaBLE: A Power-Aware Algorithm for Greener Industrial IoT Networks . . . 3
Celia Garrido-Hidalgo, Diego Hortelano, Teresa Olivares, Vicente Lopez-Camacho, M. Carmen Ruiz, and Victor Brea

The 3D Redeployment of Nodes in Wireless Sensor Networks with Real Testbed Prototyping . . . 18
Sami Mnasri, Adrien Van Den Bossche, Nejah Nasri, and Thierry Val

Semantic Resource Management of Federated IoT Testbeds . . . 25
Marios Avgeris, Nikos Kalatzis, Dimitrios Dechouniotis, Ioanna Roussaki, and Symeon Papavassiliou

Targeted Content Delivery to IoT Devices Using Bloom Filters . . . 39
Rustem Dautov and Salvatore Distefano

Security

Trust Based Monitoring Approach for Mobile Ad Hoc Networks . . . 55
Nadia Battat, Abdallah Makhoul, Hamamache Kheddouci, Sabrina Medjahed, and Nadia Aitouazzoug

An Implementation and Evaluation of the Security Features of RPL . . . 63
Pericle Perazzo, Carlo Vallati, Antonio Arena, Giuseppe Anastasi, and Gianluca Dini

A Ticket-Based Authentication Scheme for VANETs Preserving Privacy . . . 77
Ons Chikhaoui, Aida Ben Chehida, Ryma Abassi, and Sihem Guemara El Fatmi

A Trust Based Communication Scheme for Safety Messages Exchange in VANETs . . . 92
Ryma Abassi and Sihem Guemara El Fatmi

Smart City

Mobility as the Main Enabler of Opportunistic Data Dissemination in Urban Scenarios . . . 107
Jorge Herrera-Tapia, Anna Förster, Enrique Hernández-Orallo, Asanga Udugama, Andrés Tomas, and Pietro Manzoni

Analysis and Classification of the Vehicular Traffic Distribution in an Urban Area 121
Jorge Luis Zambrano-Martinez, Carlos T. Calafate, David Soler, Juan-Carlos Cano, and Pietro Manzoni

User-Space Network Tunneling Under a Mobile Platform: A Case Study for Android Environments 135
Dario Bruneo, Salvatore Distefano, Kostya Esmukov, Francesco Longo, Giovanni Merlino, and Antonio Puliafito

Mobile Crowd Sensing as an Enabler for People as a Service Mobile Computing 144
Paolo Bellavista, Javier Berrocal, Antonio Corradi, and Luca Foschini

Ad-hoc Networks

SVM-MUSIC Algorithm for Spectrum Sensing in Cognitive Radio Ad-Hoc Networks 161
Soumaya El Barrak, Abdelouahid Lyhyaoui, Amina El Gonnouni, Antonio Puliafito, and Salvatore Serrano

Optimization of a Modular Ad Hoc Land Wireless System via Distributed Joint Source-Network Coding for Correlated Sensors 171
Dina Chaal, Asaad Chahboun, Frédéric Lehmann, and Abdelouahid Lyhyaoui

AdhocInfra Toggle: Opportunistic Auto-configuration of Wireless Interface for Maintaining Data Sessions in WiFi Networks 184
Anurag Sewak, Prakhar Mehrotra, Bhaskar Jha, Mayank Pandey, and Manoj Madhava Gore

Simulation of AdHoc Networks Including Clustering and Mobility 199
J.R. Emiliano Leite, Edson L. Ursini, and Paulo S. Martins

Implementations and Validations

Highlighting Some Shortcomings of the CoCoA+ Congestion Control Algorithm. 213
Simone Bolettieri, Carlo Vallati, Giacomo Tanganelli, and Enzo Mingozzi

Experimental Evaluation of Non-coherent MIMO Grassmannian Signaling Schemes 221
Jacobo Fanjul, Jesús Ibáñez, Ignacio Santamaria, and Carlos Loucera

WEVA: A Complete Solution for Industrial Internet of Things 231
Giuseppe Campobello, Marco Castano, Agata Fucile, and Antonino Segreto

Validating Contact Times Extracted from Mobility Traces 239
Liu Sang, Vishnupriya Kuppusamy, Anna Förster, Asanga Udugama, and Ju Liu

Wireless Sensor Networks

Optimising Wireless Sensor Network Link Quality Through Power Control with Non-convex Utilities Using Game Theory 255
Evangelos D. Spyrou and Dimitrios K. Mitrakos

Routing Protocol Enhancement for Mobility Support in Wireless Sensor Networks 262
Jinpeng Wang, Gérard Chalhoub, Hamadoun Tall, and Michel Misson

Correlation-Free MultiPath Routing for Multimedia Traffic in Wireless Sensor Networks 276
Dhouha Ghrab, Imen Jemili, Abdelfettah Belghith, and Mohamed Mosbah

Impact of Simulation Environment in Performance Evaluation of Protocols for WSNs 290
Affoua Therese Aby, Marie-Françoise Servajean, Nadir Hakem, and Michel Misson

Data Management

A Real-Time Query Processing System for WSN 307
Abderrahmen Belfkih, Claude Duvallet, Bruno Sadeg, and Laurent Amanton

Centralized and Distributed Architectures: Approximation of the Response Time in a Video Surveillance System of Road Traffic by Logarithm, Power and Linear Functions 314
Papa Samour Diop, Ahmath Bamba Mbacke, and Gervais Mendy

Secure Storage as a Service in Multi-Cloud Environment 328
Riccardo Di Pietro, Marco Scarpa, Maurizio Giacobbe, and Antonio Puliafito

Policy Management and Enforcement Using OWL and SWRL for the Internet of Things 342
Rustem Dautov, Symeon Veloudis, Iraklis Paraskakis, and Salvatore Distefano

Wireless Systems

$BSSA_{CH}$: A Big Slot Scheduling Algorithm with Channel Hopping for Dynamic Wireless Sensor Networks 359
Chi Trung Ngo, Quy Lam Hoang, and Hoon Oh

A Hybrid Ant-Genetic Algorithm to Solve a Real Deployment Problem: A Case Study with Experimental Validation 367
Sami Mnasri, Nejah Nasri, Adrien Van Den Bossche, and Thierry Val

Interference Analysis for Asynchronous OFDM in Multi-user Cognitive Radio Networks with Nonlinear Distortions 382
Hanen Lajnef, Maha Cherif Dakhli, Moez Hizem, and Ridha Bouallegue

Author Index 395

Internet of Things

SustainaBLE: A Power-Aware Algorithm for Greener Industrial IoT Networks

Celia Garrido-Hidalgo[1(✉)], Diego Hortelano[1], Teresa Olivares[2], Vicente Lopez-Camacho[1], M. Carmen Ruiz[2], and Victor Brea[1]

[1] Albacete Research Institute of Informatics, University of Castilla-La Mancha, Albacete, Spain
{celia.garrido,diego.hortelano,vicente.lcamacho,victor.brea}@uclm.es

[2] Faculty of Computer Science Engineering, University of Castilla-La Mancha, Albacete, Spain
{teresa.olivares,mcarmen.ruiz}@uclm.es

Abstract. Industry 4.0 is based on interoperability among smart factories, products and services embedded in the Industrial Internet of Things (IIoT), which provides huge opportunities for sustainable manufacturing using ubiquitous information and ICT infrastructure. This paper has a twofold goal, to propose an IoT architecture linking physical devices with operators to afford efficient communication (green-BY concept) and to provide energy-saving oriented techniques to encourage the deployment of cost-effective networks, which enhances the lifespan of IoT devices (green-IN concept). Namely, we propose a flexible sleep-based algorithm called *SustainaBLE*, suitable for a wide range of domains using Bluetooth Low Energy (BLE). A real case studio has been carried out to provide power consumption measurements using *SustainaBLE* proposal, achieving one-year autonomy compared to one-day autonomy with BLE *Connection* mode. This achievement encourages us to broaden our experimental study considering more nodes, different topologies and additional standards for contributing to the future Industrial IoT networks.

Keywords: SustainaBLE · Industrial IoT · BLE algorithm · IoT architecture

1 Introduction

A brand-new Industrial Revolution, the so-called Industry 4.0, is expected to take place during the following years, achieving its full development by 2020. In this new paradigm not only machines but products, operators and conveyors cooperate according to Industrial IoT (IIoT) [1,2], leading the Fourth Industrial Revolution. The incoming communication networks are cognitive tools for sustainable manufacturing processes. Considering the main research areas of Industry 4.0, in [3] is defined how a Cyber-Physical Systems (CPS) should be: different mechatronic components based on sensors and actuators embedded in complete systems for collecting data in physical processes. CPS will be intelligently linked

A. Puliafito et al. (Eds.): ADHOC-NOW 2017, LNCS 10517, pp. 3–17, 2017.
DOI: 10.1007/978-3-319-67910-5_1

via cloud and the use of smart interfaces for interacting with operators [4] will be mandatory. The emergence of new CPS architectures contributes to structured networks where communication allows an optimal resource allocation in cognitive factories. In this work, we propose an architectural approach based on the reference models for Industrial IoT networks [5,6] and a power-aware algorithm, *SustainaBLE*, for encouraging the future industry transformation.

Industry 4.0 provides great opportunities to exploit the three dimensions of sustainability: economic, social and environmental. This paper encourages a greener IoT by approaching concepts for Information and Communications Technology (ICT). Our architecture not only reduces power consumption (which is referred as green-BY ICT) but also allows the development of power-aware algorithms to enhance lifespan of smart devices (green-IN ICT) [7,8].

The novelty of this paper lies on the introduction of green-ICT techniques into industrial environments for deploying cooperative networks. For this, it stands for the use of BLE [9] as emergent technology to provide a common infrastructure for IoT and IIoT. The major impact of BLE on ultra low cost IoT networks [10,11] encourage us to investigate alternative ways of using this standard for achieving more sustainable Cyber-Physical Systems (CPS). Two different algorithms are proposed in this paper for adapting BLE duty cycle towards meaningful energy savings, according to strict IIoT requirements. Following this idea, a BLE-based architectural prototype has been deployed in a real scenario emulating a smart factory for testing how heterogeneous devices cooperate according to our power-aware proposal.

This paper is structured as follows: Sect. 2 introduces related works; Sect. 3 proposes our reference architecture; Sect. 4 provides a description of our power-aware algorithm proposal; Sect. 5 shows the real consumption study carried out and, finally, Sect. 6 summarizes our conclusions and suggestions for future works.

2 Related Works

Greener ICT are being increasingly demanded to promote a higher throughput. Although some papers provide studies concerning CPS [12–14] they show no clear evidence of sustainability. This work encourages green networking as the major research area towards sustainable factories for [15] Industry 4.0. Furthermore, different techniques and algorithms have been released aiming to provide a greater efficiency by the use of as less resources as possible [16]. A. Dementyev et al. propose in [17] an approach based on cyclic sleep intervals. However, although real power consumption measurements are provided, there is no evidence of real environment or application. Conversely, [18] provides an approach for managing BLE devices by the performance of adaptive sleep intervals based on workload and battery level. Although it concludes that a 30% improvement with respect to fixed sleep intervals is obtained, it must be highlighted that a one-second fixed sleep interval is assumed for carrying out these results. Actually, such a reduced sleep interval is not a viable sampling rate for indoor monitoring systems, where typically magnitudes (e.g. humidity, temperature, etc.) might not change during short periods of time.

Some theoretical dynamic models are being developed for Industry 4.0 collaborative CPS [19]. The proliferation of architectural models leads to the establishment of a reference model for carrying out real experiments. Two well-known institutions have released architectural cooperative approaches to define the development of partial contents for Industry 4.0, identifying a set of norms or standards according to determine future challenges: *Plattform Industrie 4.0* [5] and the *Industrial Internet Consortium* [6].

Plattform Industrie 4.0 released the Reference Architecture Model for Industrie 4.0 (RAMI4.0), a three-dimensional approach covering the whole value chain through several hierarchy levels as well as IT layers. Conversely, the *Industrial Internet Consortium* is the international organization which introduced the Industrial Internet Reference Architecture (IIRA) to match smart manufacturing with Industrie 4.0. In the following section our architectural model is introduced, which is based on the previous reference models for the IIoT.

3 Architecture

This sections introduces an IIoT architecture considering a BLE-based mesh network composed of wearables, beacons, and static nodes, where a nodes are

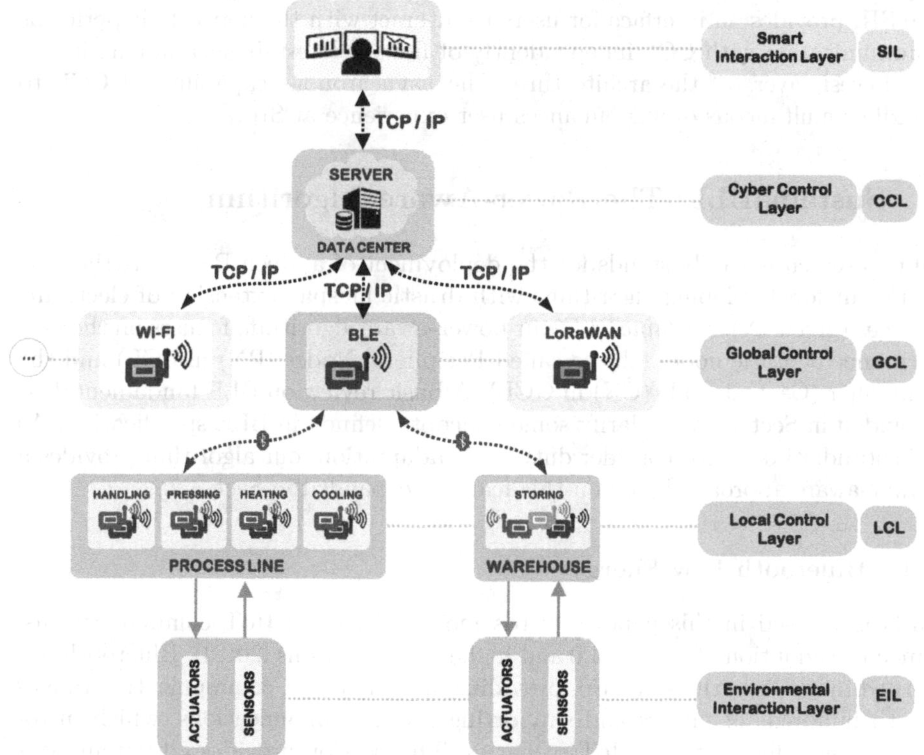

Fig. 1. Architecture overview for Industrial IoT networks

connected to a centralized system (controller). The use of BLE mesh overcomes an existing coverage gap, while our centralized system manages information from mesh sub-networks. An outline of the layered architecture proposed for the case studio is shown in Fig. 1, including some representative stages for industrial scenarios. The principal layers of this architecture are: *Environmental Interaction Layer* (EIL), *Local Control Layer* (LCL), *Global Control Layer* (GCL), *Cyber Control Layer* (CCL) and *Smart Interaction layer* (SIL).

The aim of EIL is to provide LCL with gathered information from the physical world and to manage actuators. This layer processes, encodes and sends data via BLE to GCL in order to make the extraction of specific features fairly automatic for GCL. A set of static nodes from LCL are responsible for gathering BLE mesh sub-network information and sending it via BLE to GCL. Regardless of the origin of commands they are hierarchically allocated so that they reach their destination sector. GCL is responsible for gathering data from LCL and transmitting it via TCP/IP to CCL. Focusing on BLE technology, the controller has been programmed with a non-blocking event-driven runtime environment so as to optimize communication among layers improving scalability.

The cloud-based cognitive role is performed by CCL, which provides the network with a data persistence service. It is able to make decisions considering stored information, and uses a WebSocket server to achieve an improved user experience in terms of managing the network and monitoring data. Eventually, the SIL provides an interface for users to interact with the network, it performs communication with CCL independently of flow rates established among it and the lowest layers of the architecture. The asynchronous capability of CCL to handle simultaneous events enhances user experience at SIL.

4 SustainaBLE: The Power-Aware Algorithm

This research not only stands for the deployment of a green-BY industrial network but for developing algorithms with drastic lifespan extension of electronic devices (green-IN). *SustainaBLE*, our power-aware algorithm, focuses on the performance of static nodes (the so-called Peripheral Nodes (PN) in LCL) and the controller (Central Node (CN) in GCL). A basic review on BLE fundamentals is provided in Sect. 5. A to clarify some concepts defined in BLE specification. As this standard does not consider duty cycle adaptation, our algorithm provides a power-aware approach based on this feature, expanded in Sect. 4.2.

4.1 Bluetooth Low Energy

BLE is focused in this paper towards more efficient in IIoT communications. Since the adoption of version 4.0 and higher specifications [20,21], Bluetooth has turned into a versatile standard, providing sustainability, communication ranges up to hundreds of meters and supporting mesh communications, which introduces user collaboration in IoT scenarios. This section provides a brief analysis of BLE for a better understanding of *SustainaBLE* algorithm.

According to BLE standard, a *Master Node* (MN) is a CN which is able to acquire data from several PN and response, whilst a *Slave Node* (SLN) allows its MN to read and write specific PN registers (the so-called *characteristics*, grouped into *services*). According to BLE specification some *characteristics* have *notification* or *indication* properties, which allow MN to be notified in case information is received. In both cases, MN and SLN remain connected and keep transmitting packets to maintain communication. *Scanner* and *Advertiser* roles are connectionless ways of exchanging data. A *Scanner Node* (SCN) scans the network looking for advertisement packets, whilst an *Advertiser Node* (AN) is a PN able to send data to be received by a CN acting as SCN. However, these roles do not allow information flow in both directions; only from PN to CN.

4.2 Overview

This paper stands for a fractioned duty cycle for PN, where active and sleep modes are considered. Different configurations for our algorithm proposal, *SustainaBLE*, are described in this section according to special requirements. In the following lines, we define the two operating modes to be compared: *Connection* and *SustainaBLE* (our proposal). Eventually, regarding the sleeping interval to be performed by each node, *Non-Advertiser while Sleeping* (NAS) and *Advertiser while Sleeping* (AS) modes are defined.

Connection is defined in BLE standard. It refers to a MN-SLN connection where information arrival is notified. In this mode CN and PN keep always connected, which implies the shortest time for information flow. Conversely, *SustainaBLE*, our power-aware algorithm proposal, allows PN to sleep and minimise power consumption. There are two different methods to wake up a device, depending on whether the interruption is programmed via RTC (*Fixed Sleep Interval*, FSI) or is caused by a sensor (*Variable Sleep Interval*, VSI). Both FSI and VSI operating modes are defined as default configuration for each node within our network under *SustainaBLE* algorithm. Conversely, VSI operating mode allows a PN to wake up as a result of interrupts carried out by certain sensor readings or connection events (in case BLE module is turned on).

Regardless of interruptions, two operating modes are defined in *SustainaBLE* depending on the state of BLE radio while sleeping: NAS and AS. NAS is set up as default configuration, in which the BLE module will remain turned off by the time PN starts sleeping, drastically reducing power consumption. AS mode, at the expense of higher consumption, maintains the BLE radio module active while PN is sleeping, allowing actuators to immediately execute user requests.

Table 1 shows a summary of the previous operating modes: if an ACK packet is required when a data packet is received (ACK); if a PN turns into sleep mode (Sleep); if there is a possibility for PN to send data (PN_{send}) and to receive orders (PN_{rec}) in any time; necessary time to send data to CN (T_{send}) and to receive data from CN (T_{rec}); the maximum PNs per CN (PN per CN).

***SustainaBLE* in Peripheral Nodes (PN).** *SustainaBLE* (Fig. 2) is presented in this paper as a power-aware algorithm for PN by means of saving the greatest

Table 1. Comparison of *SustainaBLE* proposal and BLE standard modes

	BLE Standard			SustainaBLE Algorithm			
	Master-Slave			Fixed Sleep Interval (FSI)		Variable Sleep Interval (VSI)	
	Notification	Indication	Read-Write characteristics	NAS[a]	AS[b]	NAS[a]	AS[b]
ACK	✗	✓	✓	✓	✓	✓	✓
PN_{send}	✓	✓	✗	✗	✗	✓	✓
PN_{rec}	✓	✓	✓	✗	✓	✗	✓
Sleep	✗	✗	✗	Fixed	Variable	Fixed	Variable
T_{send}	<0.4 s	<0.4 s	<0.4 s	Fixed	Fixed	<0.7 s	<0.7 s
T_{rec}	<0.4 s	<0.4 s	<0.4 s	Fixed	<0.7 s	Fixed	<0.7 s
PN[c] per CN[d]	5–8	5–8	5– 8	-	-	-	-

[a] Non-Advertiser while Sleeping
[b] Advertiser while Sleeping
[c] Peripheral Node
[d] Central Node

amount of energy in PN-CN communication. PN is firstly performing a fixed sleep-mode period until a programmed interruption event takes place, however, BLE radio module can be either switched on or off (NAS and AS configurations). Then the node wakes up and waits for the CN to request a connection. Once connection takes place, PN sends gathered data (own and mesh-based), and asks for incoming commands to be performed. After managing commands received from GCL, the node turns into its initial low power consumption state.

Three different interruption sources have been defined, according to Fig. 2: *Cyclic*, *Sensor-based* and *Connection-based* interruptions. A *cyclic interruption* arises due periodic alarms configured in PN timers (mainly the Real Time Clock) and, after the node wakes up, it will turn on its sensors to collect data. *Sensor interruptions* are defined to manage unexpected situations where the PN will wake up as a result of changes in the output of certain sensors. Finally, *connection-based interruption* affords cases where a PN turns into sleep mode keeping its BLE module switched on, waiting for a connection event interruption.

PN will turn into slave role and CN into MN by the time a connection event is noticed to start data transmission. Thus, regarding autonomous decisions of MN or operator requests, objects are directly managed by MN with an ACK-based handshaking mechanism. In any case, when actions are performed SLN remains waiting until a disconnection event is noticed, moment in which it will turn to sleep mode. However, not only fixed interruption sources have been considered but unexpected ones as well. According to Fig. 2, any sensor can be configured to trigger an specific interruption by the time an unexpected change takes place. Furthermore, every PN is user-configurable in order to offer the possibility of keeping into sleep mode while its BLE radio module is still working (AS mode).

***SustainaBLE* from Central Node (CN).** No synchronization among peripherals is expected, which means that connection events could happen at any time and, in order to save as much energy as possible, they must be immediately

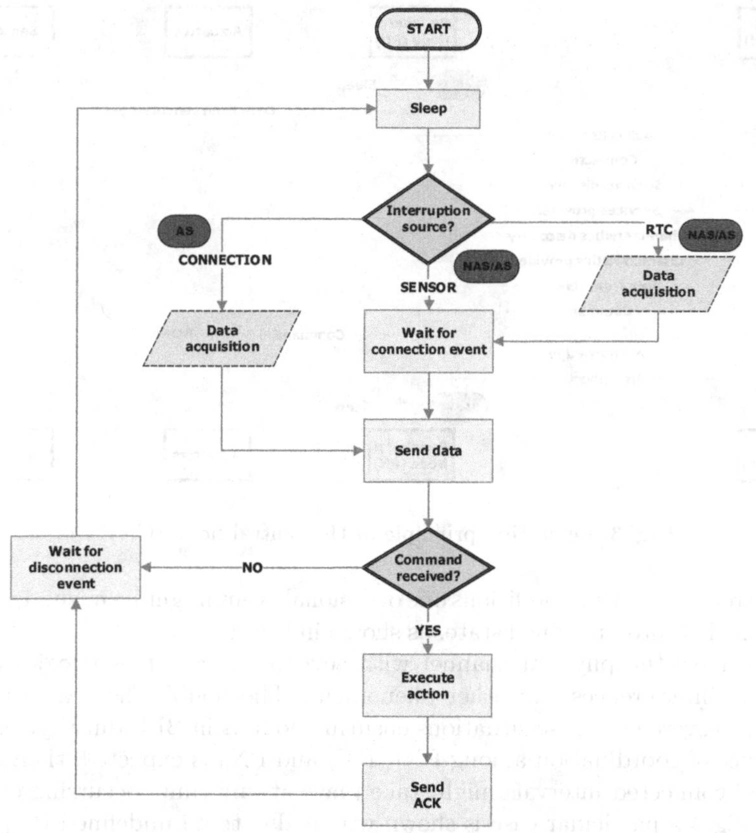

Fig. 2. SustainaBLE algorithm defined for Peripheral Nodes (NAS and AS)

handled by CN. Figure 3 shows a sequence diagram in which representative events taking place during a connection with any PN are represented.

By the moment CN receives RSSI from PN (just for NAS-defined PN nodes), it sends PN a connection request to be authenticated. Once the connection is handled, CN (MN from now on) must discover BLE *services* and *characteristics* to acquire data previously collected from sensor readings. For *service discovery* MN sends a request to the slave in order to discover its specific *services.* Then, MN remains waiting until it notices that *services* are discoverable. Once services are discovered, for each one of them MN has to send a request in order to discover every *characteristic* belonging to each *service.* MN will read data uploading it to CCL and, if no commands are sent from CCL, it will remain into standby state until an ACK is received. Eventually, MN triggers the disconnection event and PN remains in a low-energy mode period until a new interruption takes place.

Double-Check Algorithm. A complementary algorithm is proposed to enhance communication. Although CN manages both connection and disconnection events,

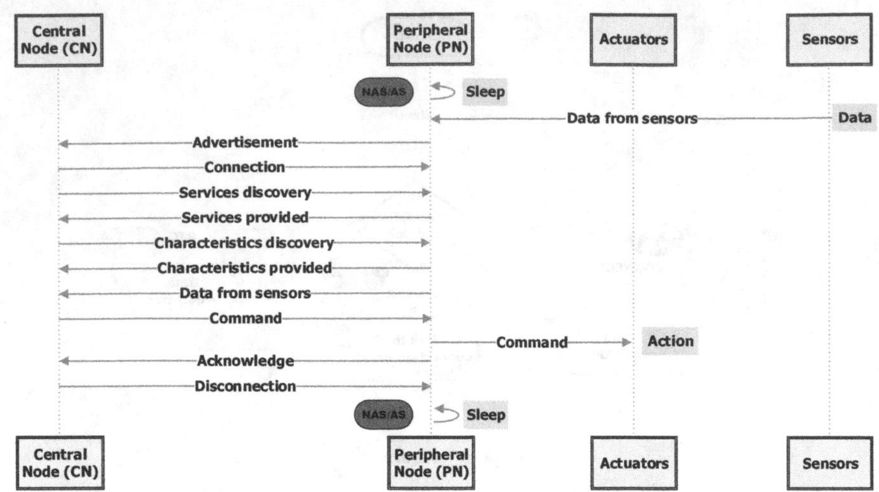

Fig. 3. Operating principle of the central node (CN)

under extreme operating conditions one occasional event might be misunderstood, driving one PN into undefined state, as shown in Fig. 4.

BLE shares the physical channel with several wireless technologies and is exposed to interferences and other phenomena. The double-check algorithm is proposed to overcome these situations ensuring no fails in BLE duty cycle. Since no evidence of coordination among PN_1, PN_2 and PN_3 is expected, their advertising and connected intervals might take place at any time, occurring overlapping. In Fig. 4 a particular case is shown where, due to an undefined PN_1 state, CN triggered the synchronization event. As a result, CN waited during a predefined time interval and attempted disconnection from PN_1 to ensure it started performing its sleep period.

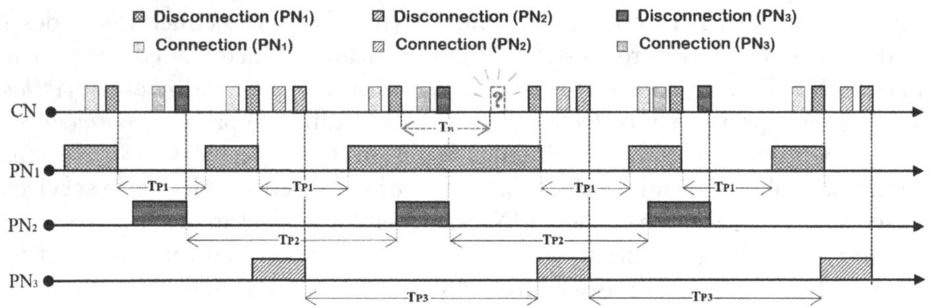

Fig. 4. Double-check mechanism performed by the central node

5 Case Study

This section offers a review based on a real case study carried out by the use of different platforms taken as reference model by means of verifying green-IN contributions. Several power consumption measurements were carried out during the study, being the most relevant ones presented in this paper. The sensors to be attached to PN (and so the operation of each node, *connection* or *SustainaBLE*) depend on the necessities of the magnitude to be sensed. For a basis of 10 PN deployed in a typical industrial environment the roles were established as follows:

- *Connection mode.* Since accelerometers or ultrasonic sensors are necessary and require real-time interaction, 2 *connection*-based nodes were defined.
- AS *SustainaBLE* algorithm. Thermal cameras, inductive proximity sensors or extensiometric gauges are commonly used in industrial environments, and must be connected to a PN which has the ability to be woken up by CN at any time. Thus, 4 AS *SustainaBLE* nodes were defined in our case study.
- NAS *SustainaBLE* algorithm. Temperature, humidity, gas, vibration and luminosity sensors are usually deployed to gather context information, which do not require real-time operation. Those sensors were used in 4 PN performing NAS *SustainaBLE* so that power consumption was drastically minimised.

5.1 Experimental Results

This case study is based on real current measurements, provided and analysed in order not to neglect real factors such as current peaks taking place during wake-up events. With this, we aimed to evaluate our green contribution. Namely, INA219 [22] operational amplifier was used to determine, with 0.1 Ω accuracy, load current consumption. LightBlue Bean [23] platform was selected as prototyping microcontroller-based board (LCL), which was programmed under both *connection* mode and *SustainaBLE* algorithm. Temperature-humidity sensor board DHT-22 [24] and luminance sensor board TSL-2561 [25] were attached in both cases to provide the same increase in current consumption due to EIL and provide measurements on how efficient our algorithm is with respect to BLE *connection* mode. Eventually, as BLE controller, Raspberry Pi 3 [26] platform performed GCL role.

We deployed an initial testbed in the Albacete Research Institute of Informatics [27] including 10 PN and 1 CN, which did not imply any kind of communication or coverage constrain [28]. Then, our proposal was tested from SIL to EIL and vice versa and no constraints were noticed regarding interaction among layers in our IoT architecture, integrating it on our own BLE-mesh network [29].

Measurements and Analysis. A five-second sleep duty cycle was programmed to extrapolate further on these results to greater intervals and reach firm conclusions. Figure 5a and b show real current consumption measurements of LightBlue Bean PN. For making our results generalizable, the algorithms selected were FSI *connection* and *SustainaBLE* NAS.

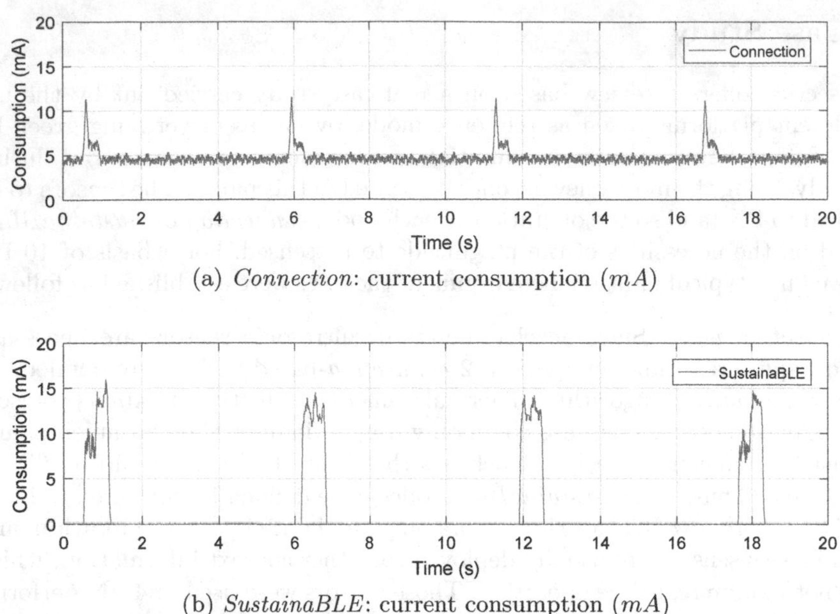

(a) *Connection*: current consumption (mA)

(b) *SustainaBLE*: current consumption (mA)

Fig. 5. Instantaneous current consumption measurements (PN)

According to BLE specification, advertising intervals range from 20 ms to 10 s (higher advertising frequency by means of a greater consumption). To enhance user experience, the minimum advertising interval, 20 ms, was set during the case study and the measurements shown in Fig. 5a and b. For a better quantification of the current consumed, Table 2 shows real average values obtained by measuring LightBlue Bean consumption configured under different intervals.

Table 2. Average consumption per duty cycle (mA)

Advertising Interval	Active Advertising (mA)	Sleeping	
		Radio OFF (mA)	Radio ON (mA)
20 ms	10.5848	≈ 0	1.2043
1 s	10.1973	≈ 0	0.8621
3 s	10.1116	≈ 0	0.2340
6 s	9.5790	≈ 0	0.1611
9 s	9.3328	≈ 0	0.1428

Extrapolation. In order to extrapolate the five-second-interval measured conditions to a wide range of advertising values we calculated the weighted average. As the down-time (either idle or sleep) increases the time spent to send data remains constant, reason why the consumption relation is drastically enhanced and the duty cycle is reduced. We reached a representation extrapolating this five-second-interval data to the range shown in Fig. 6, [0, 20] s. Since we obtained the break-event point at exactly 0.77 s, our proposal becomes profitable for sleeping intervals greater than, approximately, one second.

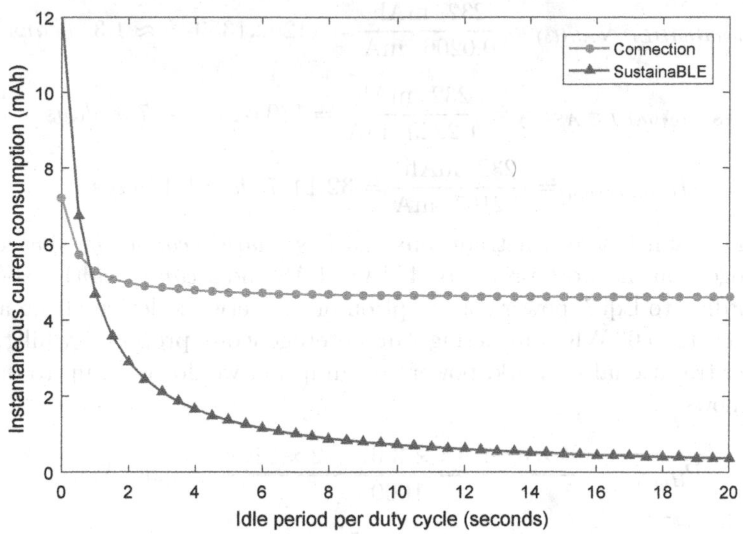

Fig. 6. Comparison: break-even point of *SustainaBLE* proposal

The five-second interval was extrapolated to a five-minute interval, which is sufficient for magnitudes sensed by *SustainaBLE*-based nodes such as humidity, temperature or luminosity. Power consumption values shown in Eqs. (1) and (2) were obtained for each one of the proposed algorithms regarding a fixed advertising interval. For *connection* evaluation, we considered an scenario where the PN is continuously transmitting data, which is a requirement for accelerometers, gyroscopes and real time data-gathering sensors. Under such conditions, (3) shows the equivalent power consumption with regard to previous cases.

$$I_{SustainaBLE(NAS)}(t) = 0.0206 \text{ mA} \rightarrow P_{SustainaBLE,NAS} = 0.06 \text{ mWh} \tag{1}$$

$$I_{SustainaBLE(AS)}(t) = 1.2229 \text{ mA} \rightarrow P_{SustainaBLE,AS} = 3.67 \text{ mWh} \tag{2}$$

$$I_{Connection}(t) = 7.2167 \text{ mA} \rightarrow P_{Connection} = 21.65 \text{ mWh} \tag{3}$$

Evaluation. According to the previous measurements, lifespan of devices was determined for both *SustainaBLE* (NAS and AS) and *connection* algorithms. The capacity of the testing cell was 232 mAh (CR2032 [30]). According to Eqs. 4, 5 and 6 the proposed *SustainaBLE* algorithm achieved, under the same testing hardware, a consumption relation of 350 with respect to *connection.*

$$t_{SustainaBLE,NAS(ls)} = \frac{232 \ \ \text{mAh}}{0.0206 \ \ \text{mA}} = 11262.1359 \ \ h \approx 1.3 \ \ years \tag{4}$$

$$t_{SustainaBLE,AS(ls)} = \frac{232 \ \ \text{mAh}}{1.2229 \ \ \text{mA}} = 189.613 \ \ h \approx 7.9 \ \ days \tag{5}$$

$$t_{Connection} = \frac{232 \ \ \text{mAh}}{7.2167 \ \ \text{mA}} = 32.1477 \ \ h \approx 1.4 \ \ days \tag{6}$$

A comparison between a traditional BLE scenario (*connection*-based) and our heterogeneous network proposal (4 NAS, 4 AS and 2 *connection*) was carried out. According to Eq. 7, power consumption of the network derived from all PNs would reach to 0.06 Wh considering our heterogeneous proposal, whilst, for a *connection* traditional network, power consumption would reach up to 0.22 Wh as Eq. 8 shows.

$$P_{HetNet} = \frac{4 \times 0.06 + 4 \times 3.67 + 2 \times 21.65}{1000} = 0.06 \ \ \text{Wh} \tag{7}$$

$$P_{Conn} = \frac{10 \times 21.65}{1000} = 0.22 \ \ \text{Wh} \tag{8}$$

Considering our heterogeneous proposal during a full-time basis of 10-hour performance power consumption reached 0.582 W (almost 4 times lower than the typical consumption of 2.17 W that would be required in the case of a full *Connection* network). Figure 7 shows a summary of the methodology applied during the case studio, including the most significant results.

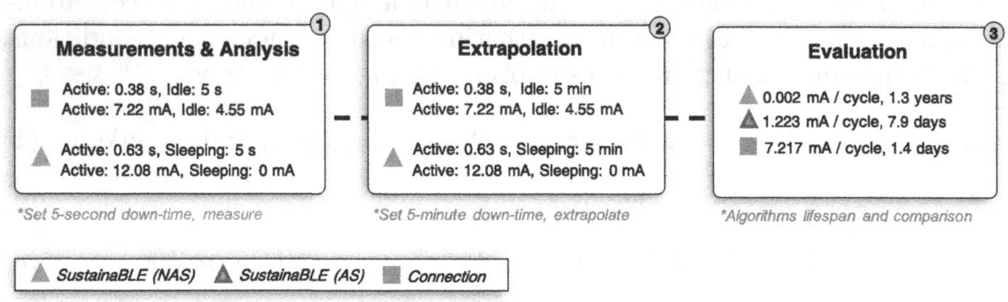

Fig. 7. Experimental methodology for the experiments and conclusion

6 Conclusion and Future Works

This research work is based on a generic IIoT architecture which establishes the basic hierarchies among its layers, especially suitable for Industry 4.0 environments composed of heterogeneous devices expected to cooperate. A power-aware BLE algorithm is proposed (*SustainaBLE*) and compared with the approach provided by BLE specification (*connection*). Both algorithms were tested in a real environment establishing how many nodes were able to operate under our *SustainaBLE* proposal for an industrial scenario composed of 10 PN. Our green-IN contribution was evaluated by measuring real instantaneous current consumed by a specific PN for either *connection* and *SustainaBLE* algorithm and extrapolating to the 10-PN heterogeneous network proposed. The most remarkable contributions of this research paper are summarised in the following lines:

- *Architecture.* This work is based on an architectural model for IIoT networks which was tested in a real environment. This architecture is focused on BLE technology since it provides a wide coverage range by means of such a reduced power consumption, which contributes to the development of a greener IoT.
- *Synchronization algorithm.* A double-check synchronization algorithm is introduced in this paper to ensure cooperation within the network, avoiding failed transmissions.
- *Power-aware Algorithm.* Considering real measurements, the so-called *SustainaBLE* algorithm proposed is profitable for sleep intervals greater than one second. Nevertheless, strict timing issues of certain sensors in our emulated system determined that, at least, two nodes of the network should operate under *connection* at the expense of higher consumption. Taking into account our heterogeneous network proposal:
  - This work contributes to a green-BY approach on how to afford a sustainable development for Industry 4.0. Power consumption derived from the whole network is almost four times lower than a *connection* network model, where nodes transfer information indefinitely. The waste of four times more resources than required may not be alarming, but taking into account the massive emerging revolution which will take place in the following years, industrial infrastructures must be prepared to deploy CPS governed by IoT technologies with negligible environmental impact.
  - Our algorithm stands for the use of BLE-based devices to promote green-IN initiatives oriented to low power IoT networks. Real measurements provided in this paper show how the autonomy of the network has been enhanced from one-day to one-year lifespans (under the same hardware conditions) without worsening user experience.

The incoming Industrial Revolution is influenced by an emerging IoT where heterogeneous sets of devices and technologies are expected to collaborate for achieving global goals. For this, we propose as future work to use our generic architecture for integrating systems based on widely-extended standards such as Wi-Fi, LTE, Sigfox, ZigBee or LoRaWAN, not only developing heterogeneous algorithms but technologies according to industrial requirements. Eventually, the integration of these technologies should be tested under several analyses and experiments to proof their cooperation and sustainability.

Acknowledgment. This work has been supported by the JCCM under project POII-2014-010P and by the Spanish MEC under projects TIN 2015-66972-C5-2-R and TIN 2015-65845-C3-2-R.

References

1. IBM: Industry 4.0 and IoT. In: Proceedings of the Interconnect 2016, February 2016
2. Mui, C.: Thinking big about the industrial internet of things, March 2016. Forbes Journal Online
3. Helmrich, K.: On the way to industry 4.0-the digital enterprise. https://www.siemens.com
4. Stock, T., Seliger, G.: Opportunities of sustainable manufacturing in industry 4.0. In: Proceedings of Global Conference on Sustainable Manufacturing, September 2016
5. Plattform Industrie 4.0: Industrie 4.0 Plattform. https://www.plattform-i40.de/I40
6. Industrial Internet Consortium: Industrial Internet Consortium. http://www.iiconsortium.org/
7. Murugesan, S., Laplante, P.: IT for a greener planet. J. IT Prof. **13**(1), 16–8 (2011)
8. Kern, E., Dick, M., Naumann, S., Guldner. A., Johann, T.: Software and green software engineering-definitions, measurements and quality aspects in green software engineering. In: Proceedings of First International Conference on Information and Communication Technologies for Sustainability, February 2013
9. Bluetooth SIG.: Bluetooth Technology Basis: Bluetooth Low Energy. https://www.bluetooth.com/
10. Liu, J., Chen, C., Ma, Y., Xu, Y.: Energy Analysis of Device Discovery for Bluetooth Low Energy (2013)
11. Galinin, O., Mikhayhov, K., Andreev, S., Turlikov, A., Koucheryavy, Y.: Smart home gateway system over bluetooth low energy with wireless energy transfer capability. EURASIP J. Wirel. Commun. Netw. **2015**(1), 1–18 (2015)
12. Hehenberger, P., Vogel-Heuser, B., Bradley, D., Eynard, B., Tomiyama, T., Achiche, S.: Design, modelling, simulation and integration of cyber physical systems: methods and applications. Comput. Ind. **82**, 273–289 (2016)
13. Mosterman, P.J., Zander, J.: Industry 4.0 as a cyber-physical system study. Softw. Syst. Model. **15**(17), 17–29 (2016)
14. Wang, S., Wan, J., Zhang, D., Li, D., Zhang, C.: Towards smart factory for industry 4.0: a self-organized multi-agent system with big data base feedback and coordination. J Comput. Netw. **101**, 158–168 (2016)
15. Khan, S., Mauri, J.L.: Green Networking and Communications: ICT for Sustainability. CRC Press, Boca Raton (2013)
16. Swain, F.: Are We Headed for Wireless Chaos. http://www.bbc.com/future/story/20131014-are-we-headed-for-wireless-chaos
17. Dementyev, A., Hodges, S., Taylor, S., Smith, J.: Power Consumption Analysis of Bluetooth Low Energy, ZigBee, and ANT Sensor Nodes in a Cyclic Sleep Scenario, April 2013

18. Collotta, M., Pau, G.: Bluetooth for internet of things: a fuzzy approach to improve power management in smart homes. Comput. Electr. Eng. **44**, 137–152 (2015)
19. Ivanov, D., Sokolov, B., Ivanova, M.: Schedule coordination in cyber-physical supply networks. IFAC-PapersOnLine (2016)
20. Bluetooth SIG: Specification of the Bluetooth System. Covered Core Package version: 4.1. https://www.bluetooth.com/specifications/adopted-specifications
21. Bluetooth SIG: Specification of the Bluetooth System. Covered Core Package version: 4.2. https://www.bluetooth.com/specifications/adopted-specifications
22. Texas Instruments: INA219 Technical Datasheet Document.http://www.ti.com/product/INA216
23. Punch Through: The LightBlue Bean Family. https://punchthrough.com/bean/
24. Sparkfun Electronics: DHT-22 Technical Datasheet Document. https://www.sparkfun.com/datasheets/Sensors/Temperature/DHT22.pdf
25. Texas Instruments: TSL-2561 Technical Datasheet Document. http://pdf1.alldatasheet.es/datasheet-pdf/view/203054/TAOS/TSL2561
26. Raspberry Pi Foundation: Raspbian Jessie Operative System. https://www.raspberrypi.org/downloads/raspbian/
27. University of Castilla-La Mancha: Albacete research Institute of Informatics. http://www.i3a.uclm.es/i3a_w/
28. Garrido, C., Olivares, T., Lopez, V., Ruiz, M.C.: Architecture Proposal for Heterogeneous, BLE-based Sensor and Actuator Networks for Easy Management of Smart Homes, April 2016
29. Hortelano, D., Olivares, T., Ruiz, M.C., Garrido, C., Lopez, V.: From sensor networks to internet of things. Bluetooth low energy, a standard for this evolution. Sensors **17**(2), 372 (2017)
30. Energizer: CR2032 Lithium Coin Technical Datasheet. http://data.energizer.com/PDFs/cr2032.pdf

The 3D Redeployment of Nodes in Wireless Sensor Networks with Real Testbed Prototyping

Sami Mnasri[1(✉)], Adrien Van Den Bossche[1], Nejah Nasri[2], and Thierry Val[1]

[1] University of Toulouse, UT2J, CNRS-IRIT-IRT, Toulouse, France
{Sami.Mnasri,Vandenbo,val}@irit.fr
[2] ENIS, LETI, University of Sfax, Sfax, Tunisia
nejah.nasri@isecs.rnu.tn

Abstract. In wireless sensor networks (WSNs), prototyping systems facilitate the realization of real node deployment, enabling to test new algorithms, protocols, and networking solutions. This paper investigates the 3D indoor redeployment problem in WSNs by finding the positions where nodes should be added in order to improve an initial deployment while optimizing different objectives. For this purpose, an approach based on a recent evolutionary optimization algorithm (NSGA-III) is used. The latter algorithm is hybridized with a strategy of incorporating of the user preferences (PI-EMO-VF). The major contributions of this work are as follows: testing the NSGA-III efficiency in the case of real world problems, comparing it with another recent many-objective algorithm (MOEA/DD), and incorporating the concept of preferences of users into NSGA-III. The real experiments performed on our testbeds indicate that the results given by the proposed algorithm are better than those given by other recent optimization algorithms such as MOEA/DD.

Keywords: Testbed · Prototyping · 3D indoor deployment · DL-IoT · NSGA-III · Preference · Many-objective optimization

1 The Problem of 3D Indoor Redeployment in WSNs

The redeployment of nodes in WSNs consists in identifying the number and positions of the nodes to be added to an initial configuration of nodes. It is a process that greatly influences the performance of a network. In this paper, we study the problem of redeploying nodes in three-dimensional spaces which represent the real topology of the RoI (region of interest) better than two-dimensional spaces. The improvement of the initial 3D indoor deployment is achieved by the addition of a set of nodes in a deterministic way while optimizing different objectives, such as localization, coverage, link quality, connectivity, or the lifetime of the network.

In fact, the recourse to use real experiments rather than simulations is due to the increasing simplicity of prototyping of communication devices and access to hardware. The major contribution of using real experiments than simulations is the realism of the obtained results. In this regard, the deployment of nodes with real prototyping that incorporates human experience is a more interesting to researchers than approaches

A. Puliafito et al. (Eds.): ADHOC-NOW 2017, LNCS 10517, pp. 18–24, 2017.
DOI: 10.1007/978-3-319-67910-5_2

which are based on theoretical assumptions, simulations or formal calculations. Indeed, the objective of our work is to finely represent the real world using real physical nodes (36 TeensyWiNo nodes). These nodes are deployed using our rapid prototyping platform. Details of the implementation of this platform are presented in the experimental section.

In order to obtain the best distribution of the nodes in the RoI, while satisfying the mentioned objectives, we propose an approach which is based on an evolutionary genetic optimization algorithm, the NSGA-III [1]. Indeed, genetic algorithms (AGs), of which NSGA-III is a recent variant, are based on a set of individuals constituting the population. Each individual in this population has a genotype that encodes a candidate solution for a particular problem. After the determination of the initial population, a set of mechanisms such as recombination, mutation, and selection ensure the convergence of the population towards a better solution.

The rest of the paper is composed of the following sections: The new NSGA-III hybrid algorithm is detailed in Sect. 2. Then, our experiments on real testbeds are presented in Sect. 3. Finally, a conclusion and different perspectives are identified in Sect. 4.

2 A Many-Objective Hybrid Algorithm (PI-NSGAIII-VF) for the 3D Indoor Redeployment Problem

2.1 The NSGA-III Algorithm

NSGA-III is a new many-objective evolutionary optimization algorithm (MaOA) which is designed to solve many-objective problems (MaOPs). Based on the concept of reference points, NSGA-III is introduced as an extension of the NSGA-II algorithm [2] to resolve to problem of bad performance of the latter when the number of objectives exceeds three. As in the case of the MOEA/D algorithm [3], the NSGA-III is based on the concept of generating weight vectors to identify the reference points which are disseminated in the objective space.

In this regard, Algorithm 1 shows one generation of the NSGA-III algorithm. Indeed, for each generation of a solution, the values of the objective function are normalized to a binary value. Thereafter, each solution is associated with a reference point according to the perpendicular distance of each reference point to the reference line. This ensures having a uniform distribution of the reference points in the normalized hyper-plane.

Therefore, a hybrid population composed of the parent and an offspring is obtained. Subsequently, this population is divided into a set of non-domination levels using a non-dominated sorting procedure. The solutions constituting the first level serve as the next parents so on and so forth. Next, the solutions of the last acceptable level are selected using a niche preservation operator. Finally, a maximum number of iterations serves as the termination condition of the algorithm. In the case of many-objectives problems, NSGA-III has been shown to perform better than MOEA/D and NSGA-II for several theoretical test problems [1]. Remain the NSGA-III test on a real problem, compare it with other many-objective algorithms like MOEA/DD and incorporate it with a user preference procedure.

```
Algorithm 1 . One generation of the NSGA-III
Input: P0 (Initial Population), NPop (size of population),
       t (iteration) = 0, Itmax (Max iteration).
Output : Pt
While t<Itmax do
    Create Offspring Qt
    Mutation and recombination on Qt
    Set Rt=Pt∪Qt
    Apply non-dominated sorting on Rt and find F1, F2, ...
    St={}, i=1;
    While |St|≤ NPop do
        St=St∪Fi
        i=i+1
    End While
    If |St| = NPop do
        Pt+1=St
    Else
      Pt+1= ∪_{j=1}^{l-1} Fj
      Normalize St using min and intercept points of each objective
      Associate each member of St to a reference point
      Choose NPop−|Pt+1| members from Fl by niche preserving operator
    End if
t= t+1;
End While
```

2.2 Incorporation of Preferences

When solving a many-objective problem, MaOAs often provide a set of non-dominated solutions that are close as possible to the Pareto front (PF). In the case of real-world problems, which are known by a large number of objectives, the size of the population and the number of needed solutions exponentially depend on this number of objectives [4]. On the other hand, the user, often called the decision-maker (DM), usually only needs a very small number of non-dominated solutions. Thus, it is preferable concentrate the search on a set of specific regions guided by the preferences of the user. Based on the integration time of the preferences, we can classify the preference procedures into three main classes [5]: A priori, interactive and a posteriori ones. The most relevant class is the second one, because it allows the progressive engagement of the preferences of the DM and the readjustment of its decisions as the algorithm generations evolve. Nevertheless, few research papers propose solutions for the interactive integration of the DM preferences. Our paper proposes a new hybridization scheme that incorporates NSGA-III into an interactive preference algorithm called PI-EMO-VF [6] where the DM is asked to interactively integrate its preferences in order to guide the searches towards a specific subset of the PF. Indeed, incorporated into any evolutionary multi-objective optimization (EMO) algorithm, PI-EMO-VF is a generic procedure that uses an approximate value function which is progressively generated. Indeed, after each small number of generations (iterations) of the EMO algorithm used (NSGA-III in our case), a set of non-dominated uniformly distributed solutions is determined, and the DM proposes its preferences (in general, an information about the relationship between the solutions). The DM can either classify all solutions from the best to the worst, or provide partial preference information. The given information helps to construct an increasing polynomial value function which determines the stop condition. More details on the PI-EMO-VF procedure are given in [6].

3 Experiments with Real Prototyping

A testbed can be any platform that allows to test and experiment new network protocols, models, or technologies in order to prove its effectiveness in real-world environments. Indeed, formal analyzes and simulations cannot exactly reproduce the technical and physical characteristics of the real environments. Also, the current tendency is to experiment protocols and algorithms with real environments [7]. Therefore, the experiments carried out on our testbeds allow reducing the gap between the theory and the practice in the WSN deployment problems.

3.1 Parameters of Experiments

The following parameters were used in our experiments:

- Number of nodes: 36 (30 fixed nodes, 6 added ones)
- Antenna model: transceiver RFM22
- Distribution of nodes: 200 * 200 m^2
- Bit rate: 256 kbps
- Transmission power: 100 mW
- Indoor transmission range: 7 m
- Modem configuration: 12 # GFSK_Rb2Fd5
- Reception gain: 50 mA
- Indoor Sensing Range: 8 m
- Frequency: 434.79 MHz
- Modulation model: 125 Kbit/s GFSK
- Number of messages: 1000
- Average of runs: 20 experiments
- Message-wait: 5
- Tx power: 7 (which is the maximum of the standard RFM22)
- Length of the message: 16

3.2 Description of the Prototyping Platform

In our experiments, the deployed nodes are TeensyWiNo ones. They are based on the WiNoRF22 nodes, and equipped with several types of sensors (pressure, temperature, acceleration, brightness, etc.). These WiNo nodes have the advantage of allowing the developer to access to low layers. Indeed, this facilitates control of the average access time and the sleep one, the restricted memory and the CPU time. This control is necessary in order to ensure compliance with the real-time constraints and the management of drastic energy-saving policies. Moreover, these nodes with their effective material energy features (several months of operation using two AAA batteries), allow hosting protocols with high temporal constraints. Figure 1 illustrates an example of the deployed WiNo nodes.

Fig. 1. The used Teensy WiNo nodes

The compatibility of these nodes with the Arduino environment enables researchers to easily integrate software bricks (processing algorithms, prototyping solutions, etc.) and hardware ones (sensors, actuators, interaction devices, etc.) which enables the feedback from users.

3.3 Results

Conduct of the Experiments. 30 fixed nodes are initially deployed in known positions that are chosen according to the application needs of the users. Six added nodes, having positions that are identified by the tested optimization algorithms. The experimental scenario is as follows: Initially, all nodes are flashed. Afterwards, they are sent the initial configuration parameters (transmission power, etc.). Then a randomly selected node sends a message to the other nodes of the network and records the RSSI (received signal strength indicator) and FER (frame error rate) measurements detected by these nodes and the received measurements (RSSI and FER) of the nodes receiving the signal. After a predefined waiting time, the sender ends the process. Afterwards, the transmitter is changed in such a way that each time a receiver node becomes a transmitter. These steps are repeated until 36 experiments (one send and 35 receptions in each experiment) are performed. The obtained values representing the relations between all the nodes are collected and recorded in two connectivity matrices (an RSSI matrix and an FER one). These two matrices give also information about the neighbors of each node. Indeed, we consider a node i as a neighbor to another node j if the following two conditions are satisfied: (a) The average of the transmitted RSSI between the two nodes (from i to j and vice versa) is higher than a predefined threshold (set to 100 in our experiments); and (b): The average FER value is less than a predefined threshold (set to 1/10 in our experiments).

To assess the impact of the new chosen positions of the added nodes on the performance of the network, this process is repeated several times (20 times in our experiments, given the variation of RSSI and FER rates and the stochastic nature of the evolutionary algorithms). The performance of the proposed algorithm (PI-NSGAIII-VF) is compared to the MOEA/DD [8] which is another recent competitive MaOA.

Evaluation of the RSSI Rates. The evaluation of a set of the considered objectives in our study (such as localization, link quality, and connectivity) is achieved using the RSSI rate measurement. Indeed, in our experiments, the localization is based on a protocol that combines the RSSI rates and the Distance-VectorHop protocol. Therefore, the localization is proportionally dependent on the RSSI rate. Figure 2 illustrates the average RSSI rate between each node and the other ones. The RSSI is represented by a value between 0 and 256, which is convertible to a dBm measure.

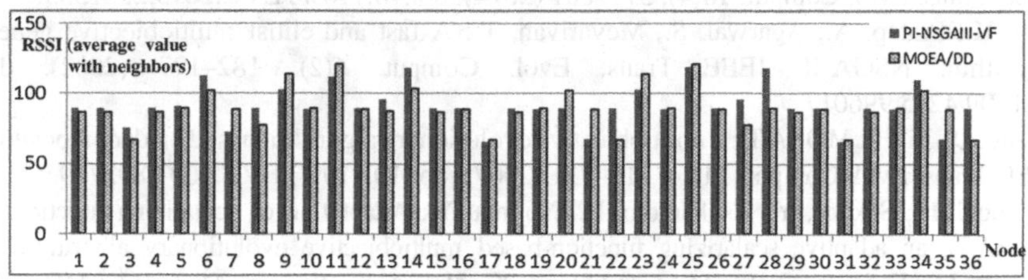

Fig. 2. Average rates of RSSI between nodes

Evaluation of FER Rates. Similarly, the measurement of the FER rate is used to assess the coverage and the quality of link between nodes. Indeed, the coverage is proportionally dependent on the FER rate. Figure 3 illustrates the average rate of FER between each node and the other ones.

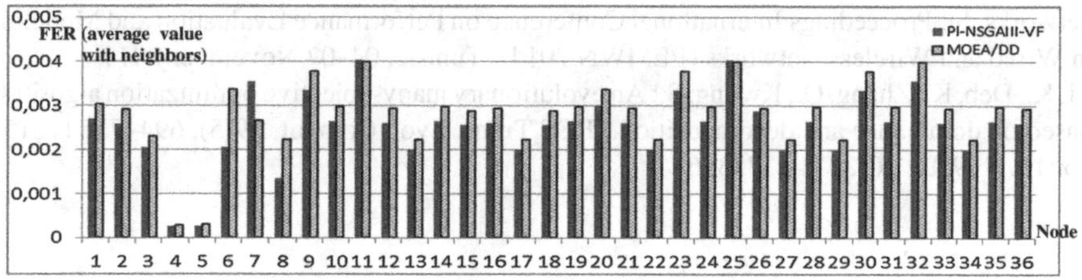

Fig. 3. Average rates of FER between nodes

4 Conclusion and Perspectives

This paper aims to resolve the problem of 3D indoor redeployment in WSNs. It presents a real world deployment which is based on a real prototyping with WiNo nodes. To find the best node positions, we use a hybrid algorithm, called PI-NSGA-III-VF, which integrates the evolutionary optimization algorithm NSGA-III into the PI-EMO-VF preference procedure. This allows testing the performance of the NSGA-III with a real world problem, and evaluating its behavior by hybridizing it with the PI-EMO-VF. Different extensions of this work can be envisaged. Among others, supporting different transmission protocols by developing them on the OpenWiNo library which lacks the implementation of several standard protocols. Moreover, considering several other constraints that makes experiments more realistic, such as the mobility of the deployed nodes. Also, taking into account the existence of obstacles in the deployment area.

References

1. Deb, K., Jain, H.: An evolutionary many-objective optimization algorithm using reference-point-based nondominated sorting approach, part I: solving problems with box constraints. IEEE Trans. Evol. Comput. **18**(4), 577–601 (2014). doi:10.1109/TEVC.2013.2281535
2. Deb, K., Pratap, A., Agarwal, S., Meyarivan, T.: A fast and elitist multiobjective genetic algorithm: NSGA-II. IEEE Trans. Evol. Comput. **6**(2), 182–197 (2002). doi: 10.1109/4235.996017
3. Zhang, Q., Li, H.: MOEA/D: a multiobjective evolutionary algorithm based on decomposition. IEEE Trans. Evol. Comput. **11**(6), 712–731 (2007). doi:10.1109/TEVC.2007.892759
4. Ishibuchi, H., Sakane, Y., Tsukamoto, N., Nojima, Y.: Adaptation of scalarizing functions in MOEA/D: an adaptive scalarizing function-based multiobjective evolutionary algorithm. In: Ehrgott, M., Fonseca, Carlos M., Gandibleux, X., Hao, J.-K., Sevaux, M. (eds.) EMO 2009. LNCS, vol. 5467, pp. 438–452. Springer, Heidelberg (2009). doi:10.1007/978-3-642-01020-0_35
5. Jaimes, A.L., Montaño, A.A., Coello, C.A.C.: Preference incorporation to solve many-objective airfoil design problems. In: IEEE Congress of Evolutionary Computation (CEC2011), pp. 1605–1612. New Orleans, LA (2011). doi:10.1109/CEC.2011.5949807
6. Deb, K., Sinha, A., Korhonen, P., Wallenius, J.: An interactive evolutionary multi-objective optimization method based on progressively approximated value functions. IEEE Trans. Evol. Comput. **14**(5), 723–739 (2010). doi:10.1109/TEVC.2010.2064323
7. Mnasri, S., Nasri, N., Val, T.: An overview of the deployment paradigms in the wireless sensor networks. In: Proceedings International Conference on Performance Evaluation and Modeling in Wired and Wireless Networks (PEMWN 2014), Tunisie, 04–07 November 2014
8. Li, K., Deb, K., Zhang, Q., Kwong, S.: An evolutionary many-objective optimization algorithm based on dominance and decomposition. IEEE Trans. Evol. Comput. **19**(5), 694–716 (2015). doi:10.1109/TEVC.2014.2373386

Semantic Resource Management of Federated IoT Testbeds

Marios Avgeris(✉), Nikos Kalatzis, Dimitrios Dechouniotis, Ioanna Roussaki, and Symeon Papavassiliou

School of Electrical and Computing Engineering,
National Technical University of Athens, Athens, Greece
{mavgeris,ddechou}@netmode.ntua.gr, {nikosk,ioanna.roussaki}@cn.ntua.gr, papavass@mail.ntua.gr

Abstract. Testbeds and experimental network facilities accelerate the expansion of disruptive Internet services and support their evolution. The integration of IoT technologies in the context of Unmanned Vehicles (UxVs) and their deployment in federated, real–world testbeds introduce various challenging research issues. This paper presents the *Semantic Aggregate Manager (SAM)* that exploits semantic technologies for modeling and managing resources of federated IoT Testbeds. SAM introduces new semantics–based features tailored to the needs of *IoT enabled UxVs*, but on the same time allows the compatibility with existing legacy, "de facto" standardised protocols, currently utilized by multiple federated testbed management systems. The proposed framework is currently being deployed in order to be evaluated in real–world testbeds across several sites in Europe.

Keywords: IoT · Semantics · Unmanned vehicles · Interoperability · Federated testbeds

1 Introduction

Internet of Things (IoT) aims "to connect everything and everyone, everywhere to everything and everyone else" [9]. It enables innovative applications for daily life activities, such as transportation, education, healthcare, city's administration, natural or human–made disasters etc. Despite that IoT paradigm is on the way to dominate service delivery, interoperability is necessary to bridge the diverse technologies of sensors, actuators and communication hardware. Novel management frameworks for supporting the entire lifecycle of IoT applications in an automated manner are essential towards resource description and discovery, reservation and provisioning, security, experimental control and monitoring.

IoT orchestration and management can leverage already mature technologies, such as cloud computing and federated heterogeneous testbed facilities. Over the past few years, the Global Environment for Network Innovation (GENI) [3] in USA, the Future Internet Research and Experimentation (FIRE) [4] project in

A. Puliafito et al. (Eds.): ADHOC-NOW 2017, LNCS 10517, pp. 25–38, 2017.
DOI: 10.1007/978-3-319-67910-5_3

EU and its successor FIRE+ have provided federated testbeds and platforms that bridge the gap between research and large–scale experimentation, through experimentally driven advanced research on new networking and service architectures. These initiatives focus on the management and the orchestration of heterogeneous resources and support the experimental lifecycle in areas of sensor networks, 5G/wireless broadband, Cloud Computing, Software Defined Networking, wired networks, IoT etc.

An important challenge that arises in this context is the exchange of information about the provided resources with their types and characteristics. Existing works rest upon certain interfaces and syntactic data models with arbitrary extensions and identifiers, which aggravate the management of heterogeneous resources across autonomous testbeds. To tackle this issue, it is proposed that management of the resources be based on their semantics, i.e. their underlying meaning and relations, while specific descriptions, data models and necessary API interactions are abstracted. In other words, heterogeneous resources are described in a formalized manner to build a basis for their management.

This paper presents a Semantic Aggregate Manager (SAM) for federated IoT testbeds that focuses on:

- Management of resources' reservation lifecycle (discovery, booking, provision, release).
- Management of federated testbeds' inventory from testbed administrator perspective (addition of resources, update of their description, removal from testbed).
- Semantic modeling of resources.
- Interoperability and compatibility with external tools and testbeds through the Slice–based Federation Architecture (SFA) legacy protocol [2].

In our case, SAM is deployed on a unmanned vehicles (UxVs) testbed federation. Given that UxVs can be equipped with a wide variety of sensors, they can be considered as a sub–category of IoT devices with application in surveillance, physical disasters, environmental pollution etc. UxVs are distinguished into the following categories according to the terrain of action: unmanned aerial vehicles (UaVs); unmanned ground vehicles (UgVs); unmanned surface (water) vehicles (UsVs); autonomous underwater vehicles (AuVs). Depending on the application environment, UxVs can be equipped with a plethora of devices and sensors (e.g. equipment for spectral imaging and photogrammetry, sound sensors, temperature sensors, accelerometers, wireless interfaces, etc.). Experimenters or other SAM service consumers must be able to search and discover UxVs based on their field of operation (e.g. air, ground, surface, underwater) but also based on the equipment that is appropriate for the desired operation. Ontologies described in detail in the following sections, are utilized by SAM in order to describe all kinds of UxVs and their attached sensors. Already well–defined ontologies are adopted to support the entire reservation lifecycle.

The rest of the paper is structured as follows. Section 2 discusses related work. Section 3 contains an analytical description of semantic modeling for IoT equipped UxVs and Sect. 4 presents the Aggregate Manager's architecture.

Section 5 describes the deployment of the proposed framework in a federated testbed environment that involves various types of UxVs across various EU regions while conclusions and future work are drawn in Sect. 6.

2 Related Work

To the best of our knowledge, there exists no study in bibliography that proposes a similar management approach on reservation lifecycle of sensor equipped UxVs, especially in the context federated IoT testbeds. In the first part of this section, an analysis of existing semantic descriptions for UxV mission control are presented, followed by the current status of semantic data modeling of federated testbeds' resources. In the second part, the present state of federated testbeds' management is discussed.

2.1 Semantic Descriptions of UxVs and Federated Testbeds

Schumann et al. [11] introduced an ontology for simulating and recreating civil UaV lifecycle. Such lifecycle consists of missions, targets and segments' information derived from Geographical Information Systems (GIS) maps. In [12,13], the authors proposed and integrated an ontology to improve the navigation planning capabilities and performance of UgVs. The suggested ontology and automatic reasoning are used to determine the damage consequences after collisions between UgVs and several kinds of objects under different driving scenarios. In [14], the authors presented a hierarchical semantic–based framework that provides the core architecture for knowledge representation of service oriented agents in AuVs. It focused on the problem of fault tolerant adaptive mission planning in order to improve interoperability, situation–awareness and independence of operation of the embedded agent for autonomous platforms. SWARMs ontology [15] is a network of domain–specific ontologies including the mission and planning, the robotic vehicle, the communication and networking and the environment recognition and sensing ontology that are interrelated by one core ontology. Furthermore, this ontology annotates context uncertainty based on the Multi–Entity Bayesian Network (MEBN) theory. This results in formal abstraction of information and uncertainty reasoning.

Over the past few years, there have been significant efforts on semantic information modeling of federated resource management. Based on Network Description Language (NDL), the Network Mark–Up Language (NML), and OGF standard [16] was introduced to describe and define computer networks as generally as possible allowing future extensions/customization for emerging network architectures and novel use cases. The NOVI [17] project analyzed challenges for semantically describing federated virtualized infrastructures. NDL–OWL [18] redefines NDL as an OWL DL ontology. Created within the GENI initiative, it extends NDL, introducing the notion of resources and their lifecycle into the ontology by creating the request, advertisement and manifest models. The OWL encoded Open–Multinet (OMN) ontology suite [19] is built upon many of the existing ontologies and facilitates the management of federated infrastructures with heterogeneous resources through the combination of semantically enriched information models. Similarly, in [20], authors presented a high–level ontology and initial integration concepts for management of distributed federation testbeds.

2.2 Testbed Management Frameworks and Testbed Federation

Within the FIRE community, the SFA protocol is the agreed standard to federate testbeds that belong to different administrative domains. For this purpose, a number of different Aggregate Manager (AM) implementations are available, that fall under two basic categories [19]: resource specific AM implementations (e.g., Flowvisor OpenFlow Aggregate Manager – FOAM) and generic wrappers or lightweight AMs like the SFAWrap[1], the NITOS broker [1], FITeagle[2], AMsoil[3] and the GENI Control Framework (GCF). Considering only the second category, we look into implementations either that have been deployed to more than one Future Internet (FI) testbeds or provide functionality directly related to our objectives.

Python–based SFAWrap can be used to wrap an existing testbed management framework, providing support for SFA AM API (v2/v3) [3]. SFAWrap provides a testbed driver, with appropriate calls to the underlying testbed management software, in order to enable SFA support for the testbed. The main components of SFAWrap are: the Registry, the AM and the Slice Manager. SFAWrap supports reservation of resources as long as the underlying testbed management framework supports such an operation, while it uses XML–based RSpecs [5] for describing available resources, requests, and reservations.

The Federated Infrastructure Resource Management Architecture (FIRMA) [8] is a Java–based semantic testbed management framework that supports SFA AM API v3, REST API, FRCP for the provisioning and control of resources. The architecture currently relies on the OMN [19] semantic model. FIRMA is designed to be reusable by supporting arbitrary protocols and heterogeneous resources. In order to achieve these goals, the framework consists of multiple decoupled microservices that exchange information using OMN. Its reservation module supports booking of resources. FITeagle, the management framework for the FuSeCo testbed is based on FIRMA. However, the latest version of FITeagle is only currently being deployed at FuSeCo testbed.

The Ruby–based testbed management framework detailed in [1] (aka NITOS broker) proposes an architecture for AMs that separates the actual management of the resources from its core components, introducing an extensible way of supporting versatile type of resources, exposing multiple APIs (SFA AM v2 and v3, REST and FRCP). The framework is divided in several layers. Each layer includes a set of fundamental components that have a role in a specific area of the

[1] http://sfawrap.info/.
[2] http://fiteagle.github.io.
[3] http://fp7-ofelia.github.io/ocf/.

facility management (e.g., scheduling resources for provisioning). The AM supports booking of resources and uses XML–based RSpecs for describing available resources, requests, and reservations and the corresponding leases extension.

2.3 SAM Contribution and Innovation

In comparison with the aforementioned studies on UxV semantic descriptions and federated testbed management, the present study introduces the following contributions; semantic modeling of UxV resources and attributes; management of UxV reservation lifecycle; concurrent support of semantic and SFA–based resource descriptions as a means of communication. These will accelerate the innovations on the Future Internet area and enable the testing and deployment of new IoT and UxV services.

3 Semantic Data Modeling

As mentioned in the previous section, there currently exists no work regarding the semantic description of the features of UxVs and their attached sensors. In a federated environment, an in–depth description of UxVs and their equipped sensors would support users in all phases of an experiment, i.e. resource discovery, reservation and construction of an execution scenario. Given the plethora of UxVs, one emerging issue is the description of these offerings. SFA, the standard API for testbed federation, uses XML–based RSpecs with arbitrary extensions to describe, discover, provision and release resources. However, such tree–based data models, lack consistency, standardized vocabularies as well as semantic meanings, therefore impede interoperability within a federation [19,21]. For this reason, SAM reuses and extends already well–defined standard semantic models for representing and linking federated UxV resources. Additionally, the deployment of a semantic repository for testbeds and resources enables experimenters to discover and reserve resources more easily.

The OMN ontology suite, being developed within the W3C Federated Infrastructures Community Group[4], combined with the respective management mechanisms fulfill a set of requirements for achieving semantic and syntactic interoperability among federated heterogeneous infrastructures. It has already been implemented and successfully evaluated for domains referring to wireless (e.g., 802.11) and wired (e.g., Cloud) resources. In the SAM framework context, the OMN ontology suite is adopted and extended towards the semantic description of federated vehicular, aerial and maritime environments. As a result, two new domain specific ontologies referring to UxV resources and their embedded sensors, named OMN UxV and OMN Sensor ontology respectively, are introduced and aligned with the OMN upper ontology, as illustrated in Fig. 1.

The developed OMN UxV ontology is as generic as possible and aims to describe any UxV testbed's requirements. It focuses on the semantic modeling

[4] https://www.w3.org/community/omn/.

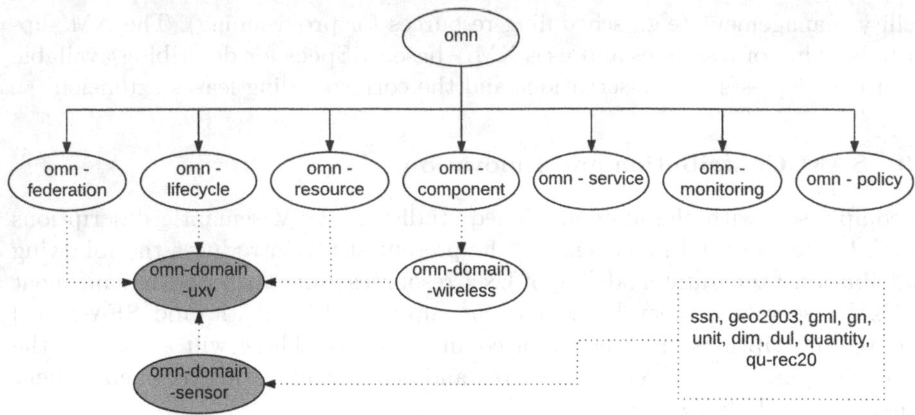

Fig. 1. Extending omn for UxV & sensor resources

of the resources, testbeds and experimenters, as well as the reservation lifecycle (discovery, reservation and release). It consists of 59 classes, which are defined directly in the ontology or imported by others, 41 object properties that represent the relationships between the classes and 55 data properties describing classes attributes. The *Uxv* class represents any kind of UxV and is descendant of the *omn:Resource* class of the upper omn–resource ontology. Data properties like *Battery, Max Take–off Weight, Speed, Endurance* etc. provide all the essential UxV characteristics. Every testbed is defined by the *Testbed* class and is descendant of the *Infrastructure* class of omn–federation ontology. It is linked with the *UxV* and *foaf:Person* classes via the *:hasResource* and *:isTestbedOf* object properties respectively.

The testbed's information is represented by data properties such as *name, description, countryCode, geo:alt, geo:long, geo:lat*, which specifically provide geographical position of each testbed's area using the GeoRSS Feature Model and ontology [22]. The *foaf:Person* class defines every concept regarding the testbed experimenters and administrator and is imported from the FoaF ontology[5].

Several classes define essential concepts of UxVs attributes. For instance, the *Interface, Network Controller, Channel* classes describe the characteristics of the communication interfaces of UxVs. The *HealthStatus* class represents the current state of each UxV, while the *ConfigParameters* and *ExperimentResourceConfig* classes define the concepts of the configuration parameters of the unmanned vehicles and the experimental scenarios. As mentioned earlier, the OMN UxV ontology supports the reservation lifecycle of UxVs. For this purpose, many concepts from the omn–lifecycle ontology are imported, which describe the entire lifecycle of resource management in federated testbeds. The reservation of a UxV is modeled by the *omn:Reservation* and *omn–lifecycle:Lease* classes while the state of a reservation is defined by the *omn–lifecycle:ReservationState* and

[5] http://www.foaf-project.org/.

its subclasses, named *Allocated, Cancelled, Pending, Provisioned, Unallocated.* The object properties *omn:hasReservation* and *omn:isReservationOf* express the relationships between them.

The OMN Sensor ontology reuses already defined models from relevant ontologies on sensors and measurements. Each UxV is equipped with several sensors and some of them are able to measure different phenomena simultaneously. For this reason, each UxV has a root multi–sensor system that contains all underlying individual and multiple sensors, as defined in [23]. The concepts of sensors and multi–sensor systems are defined by the *ssn:SensingDevice* and *ssn:System* classes respectively. The relationships between a UxV and its multi–sensor system are expressed by the *hasSensorSystem* and *isSensorSystemOf* object properties. The *ssn:hasSubSystem* object property links every multi–sensor system with the underlying individual sensors. On the other hand, data properties provide information about single sensors and multi–sensor systems. Every sensor is associated with a measuring property and one or more measuring units, e.g. *quantity:tempareture, unit:kelvin* and *unit:degreeCelcius.* These concepts are denoted by the *qu:QuantityKind* and *qu:Unit* classes which model a large number of physical quantities (i.e. mass, pressure, velocity, electrical current etc.) and their corresponding units of measurement, and are imported from the W3C ontology for quantity kinds and units [24]. Finally the "Feature of Interest" concept is an abstraction of real world phenomena, which includes air, ground and water concepts and is defined by the *ssn:FeatureOfInterest* class. Every physical quantity can be property of one or more of the three "Feature of Interest" concepts. This relationship is expressed by the *ssn:isPropertyOf* and *ssn:hasProperty* object properties.

4 Architecture

SAM's architecture follows design principles that were formed after considering all the desired features that a management framework should provide. Initially based on the NITOS broker architecture [1], the framework was modified and expanded to accommodate the experimentation workflow based on semantically enriched resource descriptions, in parallel with SFA–based [2] legacy resource descriptions. As depicted in Fig. 2, the resulted framework consists of distinguished fundamental architectural components, each of which contributes to a specific area of the reservation lifecycle management.

4.1 Communication Layer and Access Control

The framework's ambidexterity in terms of communication interfaces (APIs) is provided by the implementation of a semantic aware REST API alongside an XML–RPC API supporting SFA federation. Security issues are tackled with the employment of an Access Control module that contains API–specific submodules for assessing the incoming requests.

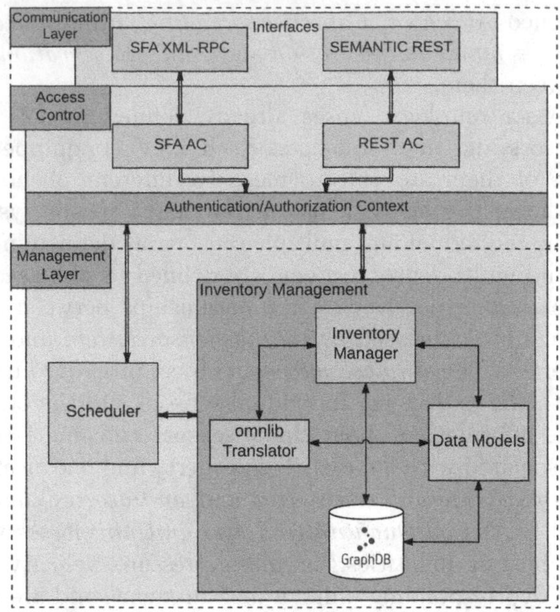

Fig. 2. SAM architecture

REST API. The REST API is tailored to support the discovery, reservation, provision and release functionality of UxV resources. It leverages the semantic, OMN–based resource descriptions stored in the local RDF [7] triplestore, to provide experimenters with semantically enriched information regarding the resources managed by the respective testbed. Thus, the users are able to allocate and provision resources that correspond to their experiments' specifications, as well as release them when no longer used. Complementary to this functionality, this API exposes the essential administrative management methods; namely, UxV resource descriptions retrieval, creation, update and deletion are supported.

XML–RPC API. Alternately, the XML–RPC API is exposed by the Aggregate Manager in order for the framework to be interoperable with existing SFA enabled provisioning tools (e.g. SFA Command–Line Interface (SFI), jFED [6] and Omni[6]) and to allow federation with existing testbed management platforms that confront with the SFA and the GENI API v3 [3] specification. This SFA endpoint is practically an implementation of the GENI AM API v3 which is used both in GENI and FIRE [4] testbeds as a way of federation. Intra federation communication is achieved through the exchange of standardized RSpec XML documents which follow agreed schemas to represent resources. It is worth mentioning that these

[6] http://trac.gpolab.bbn.com/gcf/wiki/Omni.

documents are products of a translation process between semantic and XML data formats, taking place in a separate module, as discussed in later sections. Backwards compatibility with GENI AM API v2 [3] is also present.

Access Control. Security is of paramount importance to any testbed management framework and SAM is no exception. In order to provide a filtering intermediary between the internal system and the outside world, the Access Control (AC) API–specific submodules are responsible to selectively restrict access to the deeper architecture layers. Experimenters' privileges are derived through client side X.509 certificates, which accompany each request to the aforementioned interfaces, and then the AC module decides whether to invoke the corresponding functionality, based on the Authentication/Authorization Context.

4.2 Management Layer

The Management Layer is where all the related policies and resource management functionality is concentrated, including request processing, event scheduling, storage transactions orchestration, as well as data format conversion.

Inventory Manager. Requests regarding resource discovery, booking and reservation, resource provisioning and release (i.e. user tasks), as well as resource description retrieval, creation, update and deletion (i.e. administrative tasks) are forwarded to the Inventory Manager via the Communication Layer. This component facilitates the orchestration and coordination of the actions required to fulfill the aforementioned requests. These actions include, but are not limited to, forwarding the received GENI RSpec to the Translator and receiving the respective semantic description (and vice versa), consulting the Scheduler about the feasibility of booking the requested resources, manipulating proper objects which achieve compliance with the established Data Models and storing/retrieving them to/from the RDF triplestore, formulating responses and directing them back to the Communication Layer. Throughout the whole process, the manager constantly addresses the AC module in order to access policy–sensitive content and perform policy–sensitive tasks.

Scheduler. The Scheduler component is the provider of the main functionality of the framework, since decisions regarding resource reservation take place there, based on their availability. More specifically, when a request for booking a resource is received in the Communication Layer and forwarded to the Scheduler component via the Inventory Manager, it first compares the requested booking's start and expiration time with those of the existing, active reservations. Afterwards, it decides based on possible time slot conflicts and while taking into account the authorization context, whether to fulfill the request or reject it. While by default a simple First-Come–First-Served (FCFS) policy is applied to the requests for resource reservation, with only minor modifications, testbeds administrators are able to define their own resource allocation policies.

omnlib Translator. Given the fact that semantically enriched resource descriptions and legacy RSpec XML documents are simultaneously supported as a means of communication, an efficient mechanism is needed to facilitate a seamless transition between the two formats. Within the OMN suite, the omnlib Translator already provides this functionality. While this component was initially implemented to help developers work with existing OMN related ontologies, within the SAM context it was extended to support the OMN UxV ontology. In particular, this integration provides support for translating the SFA RSpecs into RDF–based graphs and vice versa. The main advantage of this approach is the automation and speed up of non RDF data conversion, while ensuring that the quality of the generated RDF data corresponds to its counterpart representation in the legacy system. This architectural decision contributes to one of the core innovations of the SAM framework; prevention of stored data redundancy. Resource descriptions need only be stored once in a Semantic Graph Database and converted respectively as requested.

Data Models and RDF Triplestore. As a corollary to the above–mentioned functionalities, a way to store and access the RDF data in an agile, software–integrated way is necessary. To this end, the Spira[7] Ruby framework has been utilized, which provides the capability of programmatically manipulating the data stored in the RDF triplestore as model objects, while ensuring data integrity. Harnessing this, every OMN UxV ontology class has been modeled as a Spira Class and each included property as a Spira Class property, hence enabling the instantiation and manipulation of RDF data as Ruby Objects, leveraging the ORM technique.

In the architecture's southbound lies an RDF persistent triplestore responsible for storing the resource descriptions. For this purpose, a GraphDB[8] semantic graph database, compliant with W3C Standards, has been employed. Inside it, data are modeled in a way that allows interlinking and querying entities, while profiling the relationships between them.

5 Proof of Concept

The first version of the described SAM architecture has already been implemented and the source code along with the implementation and installation details are publicly available[9], alongside the developed ontologies[10]. In order to evaluate the overall testbed management architecture, SAM instances have been deployed to real UxV testbeds provided within the H2020 RAWFIE project [10]. Under the FIRE+ initiative, RAWFIE (Road-, Air-, and Water- based Future Internet Experimentation) aims at providing research facilities for IoT devices.

[7] https://rubygems.org/gems/spira.
[8] http://graphdb.ontotext.com/graphdb/.
[9] https://github.com/nikoskal/samant.
[10] http://samant.lab.netmode.ntua.gr/documents.

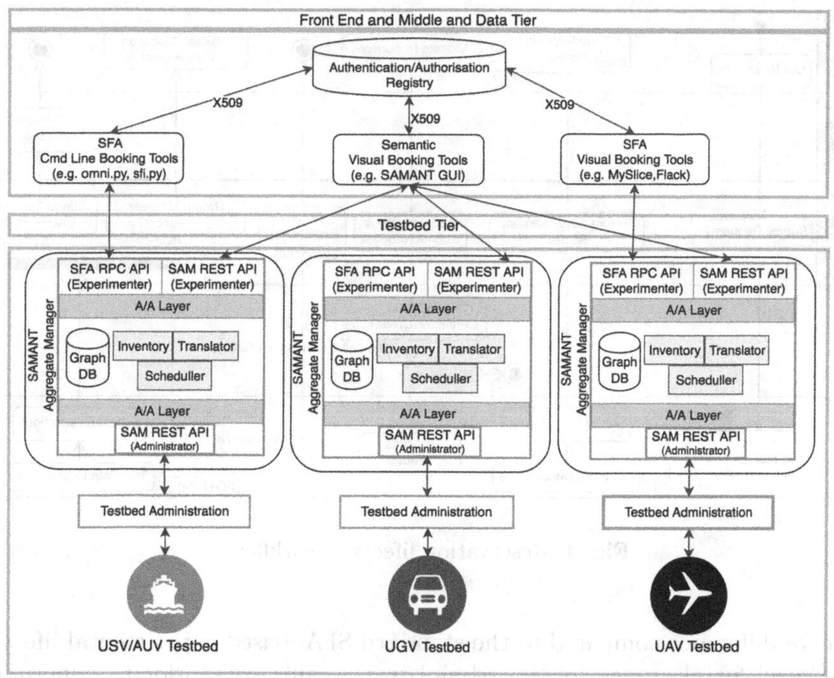

Fig. 3. RAWFIE deployment architecture

The project is introducing a unique platform that enables research experimentation across geographically distributed UxV testbeds. This platform supports experimenters with smart tools to conduct and monitor experiments in the domains of IoT, networking, sensing and satellite navigation. To this end, SAM provides the necessary tools and software enhancements for discovering, reserving and releasing experimental resources that belong to RAWFIE federation testbeds (e.g. UxVs) and/or to wider existing federations (e.g. FIRE, GENI).

The RAWFIE platform provides an umbrella that enables virtualization in accessing UxVs resources lying in different testbeds and in disparate geographical locations. Currently, there are at least three RAWFIE testbeds facilities: HAI Tanagra Greece (UaVs); HMOD Skaramagkas Greece (UsVs); PEGASE Pourrieres France (UsVs), while it is expected more to be added in the near future. As depicted in Fig. 3, an instance of the SAM framework is deployed in each testbed in order to handle the reservation lifecycle of the respective resources. Each instance maintains semantic descriptions, based on the OMN UxV and OMN Sensor ontologies, of testbed resources to a GraphDB database which is deployed locally in the testbeds' facilities. It is worth mentioning that the management of UxV testbeds imposes additional requirements and constraints compared with the more "traditional" computing and networking testbed facilities, as the UxV experimentation is subject to additional regulations and standards.

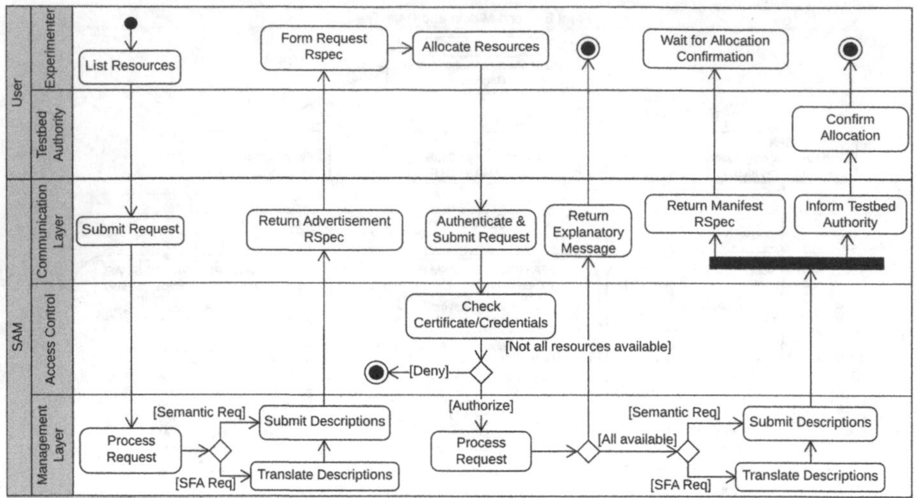

Fig. 4. Reservation lifecycle workflow

One core difference compared to the standard SFA–based experimental lifecycle is the need for clearance by the administrative authority prior to experiment execution. This imposes a major modification as it is not possible for direct resource booking and experiment launching which is a core characteristic of SFA standard. In order to overcome this discrepancy, but to also facilitate SFA compatibility, the SAM framework keeps the requested experimental resources in a "Pending Allocation" state - which is part of the defined SFA GENI v3 operational states. When the request is approved by the testbed authority an additional SAM REST API call sets the experimental resources to "Allocated" (ready to be utilized by the experimenter) state. Figure 4 visualizes a basic reservation scenario via a UML activity diagram. The process is initialized with the experimenter requesting a list with the testbed's available resources for a specified time slot. SAM responds with the respective resource descriptions, either in SFA or semantic data format, as requested. Then the experimenter forms an allocation request, containing a subset of these resources, suitable for the experiment. After authenticating the experimenter, SAM checks again for the availability of the requested resources and informs the experimenter and the testbed authority accordingly. The whole process terminates with the administrative confirmation. Several in–depth details are omitted for the sake of simplicity.

In practice, the HMOD testbed's GraphDB instance contains information about the locally available resources; 10 UsVs, equipped with infra–red camera, day HD camera, gyroscope, magnetometer, accelerometer/gyroscope, barometer and GPS available for vision, air and water monitoring purposes. More specifically, 4 vision UsVs with HD camera, 3 air monitoring UsVs equipped with CO, CO_2 and temperature sensors and 3 water monitoring UsVs with conductivity, pH and temperature sensors are available.

6 Conclusions and Future Work

The SAM framework aims to equip federated IoT testbeds with open–source tools that allow experimenters to discover and select available testbeds and their resources in order to build experiments. To allow maximum expressiveness, a generic semantic model describing UxV resources and their embedded sensors is developed, as an extension of already well defined ontologies. This study presents detailed descriptions of the respective individual components that comprise the architecture of SAM. The proposed framework has been implemented and currently is being deployed in various federated testbeds. This will allow to further test the proposed functionality and to extract the appropriate conclusions with regard to future refinements.

In the future, we intend to extend our work with an engine offering reasoning capabilities, able to extract implicit information based on generic inference rules. This will allow experimenters to perform combinatorial queries.

Acknowledgments. This work has been partially supported by the European Commission, Horizon 2020 Framework Programme for research and innovation under grant agreement no. 645220.

References

1. Stavropoulos, D., Dadoukis, A., Rakotoarivelo, T., Ott, M., Korakis, T., Tassiulas, L.: Design, architecture and implementation of a resource discovery, reservation and provisioning framework for testbeds. In: 2015 13th International Symposium on Modeling and Optimization in Mobile, Ad Hoc, and Wireless Networks (WiOpt), pp. 48–53. IEEE (2015)
2. Peterson, L., Ricci, R., Falk, A., Chase, J.: Slice-based federation architecture. Ad Hoc Des. Doc. **2008**, 1 (2010)
3. Berman, M., Chase, J.S., Landweber, L., Nakao, A., Ott, M., Raychaudhuri, D., Seskar, I.: GENI: a federated testbed for innovative network experiments. Comput. Netw. **61**, 5–23 (2014)
4. Gavras, A., Karila, A., Fdida, S., May, M., Potts, M.: Future internet research and experimentation: the FIRE initiative. ACM SIGCOMM Comput. Commun. Rev. **37**(3), 89–92 (2007)
5. Faber, T., Ricci, R.: Resource description in GENI: RSpec model. In: Presentation Given at the Second GENI Engineering Conference, March 2008
6. Vermeulen, B., Van de Meerssche, W., Walcarius, T.: jFed toolkit, Fed4Fire, Federation. In: GENI Engineering Conference, vol. 19 (2014)
7. Brickley, D., Guha, R.V.: RDF vocabulary description language 1.0: RDF schema (2004)
8. Willner, A., Magedanz, T.: A future Internet resource management architecture. In: 2014 26th International Teletraffic Congress (ITC), pp. 1–4. IEEE
9. Atzori, L., Iera, A., Morabito, G.: The Internet of Things: a survey. Comput. Netw. **54**(15), 2787–2805 (2010)
10. H2020 RAWFIE project, Road-, Air- and Water- based Future Internet Experimentation. http://www.rawfie.eu/

11. Schumann, B., Scanlan, J., Fangohr, H., Ferraro, M.: A generic unifying ontology for civil unmanned aerial vehicle missions. In: 12th AIAA Aviation Technology, Integration, and Operations (ATIO) Conference and 14th AIAA/ISSMO Multidisciplinary Analysis and Optimization Conference, p. 5504 (2012)
12. Schlenoff, C., Balakirsky, S., Uschold, M., Provine, R., Smith, S.: Using ontologies to aid navigation planning in autonomous vehicles. Knowl. Eng. Rev. **18**(03), 243–255 (2003)
13. Provine, R., Schlenoff, C., Balakirsky, S., Smith, S., Uschold, M.: Ontology-based methods for enhancing autonomous vehicle path planning. Robot. Auton. Syst. **49**(1), 123–133 (2004)
14. Patron, P., Miguelanez, E., Cartwright, J., Petillot, Y.R.: Semantic knowledge-based representation for improving situation awareness in service oriented agents of autonomous underwater vehicles. In: OCEANS 2008, pp. 1–9. IEEE (2008)
15. Li, X., Bilbao, S., Martín-Wanton, T., Bastos, J., Rodriguez, J.: SWARMs ontology: a common information model for the cooperation of underwater robots. Sensors **17**(3), 569 (2017)
16. van der Ham, J., Dijkstra, F., Lapacz, R., Zurawski, J.: Network markup language base schema version 1. In: Grid Forum Documents (2013)
17. van der Ham, J., Steger, J., Laki, S., Kryftis, Y., Maglaris, V., de Laat, C.: The NOVI information models. Fut. Gener. Comput. Syst. **42**, 64–73 (2015)
18. Xin, Y., Baldin, I., Chase, J., Ogan, K.: Leveraging semantic web technologies for managing resources in a multi-domain infrastructure-as-a-service environment. arXiv preprint arXiv:1403.0949 (2014)
19. Willner, A., Papagianni, C., Giatili, M., Grosso, P., Morsey, M., Al-Hazmi, Y., Baldin, I.: The open-multinet upper ontology towards the semantic-based management of federated infrastructures. EAI Endorsed Trans. Scalable Inf. Syst. **2**(7), 1–10 (2015)
20. Willner, A., Loughnane, R., Magedanz, T.: Federated infrastructure discovery and description language. In: 2015 IEEE International Conference on Cloud Engineering (IC2E), pp. 465–471. IEEE
21. Morsey, M., Willner, A., Loughnane, R., Giatili, M., Papagianni, C., Baldin, I., Al-Hazmi, Y.: DBcloud: semantic dataset for the cloud. In: 2016 IEEE Conference on Computer Communications Workshops (INFOCOM WKSHPS), pp. 207–212. IEEE (2016)
22. Lieberman, J., Signh, R., Goad, C.: W3C Geospatial Vocabulary (2007). https://www.w3.org/2005/Incubator/geo/XGR-geo-20071023/
23. Compton, M., Barnaghi, P., Bermudez, L., GarciA-Castro, R., Corcho, O., Cox, S., Huang, V.: The SSN ontology of the W3C semantic sensor network incubator group. Web Semant. Sci. Serv. Agents World Wide Web **17**, 25–32 (2012)
24. Lefort, L.: Ontology for quantity kinds and units: units and quantities definitions. W3 Semantic Sensor Network Incubator Activity (2005)

Targeted Content Delivery to IoT Devices Using Bloom Filters

Rustem Dautov[1(✉)] and Salvatore Distefano[1,2]

[1] Higher Institute of Information Technology and Information Systems (ITIS), Kazan Federal University (KFU), Kazan, Russia
{rdautov,s_distefano}@it.kfu.ru
[2] University of Messina, Messina, Italy
sdistefano@unime.it

Abstract. The increasing number of smart interactive devices connected to the network opens new business opportunities for digital content and advertisement providers, interested in reaching out to new customer audiences. To this end, they employ various device discovery and data collection techniques to gather user- and device-specific information in order to build a user profile and deliver targeted content accordingly. However, the extreme (and constantly growing) number of smart devices, dynamically connecting to and disconnecting from a network in the IoT scenario, renders existing routing techniques, such as multicasting and broadcasting, unscalable, especially when using the IPv6 128-bit addresses. Moreover, these existing solutions can hardly provide information about technical capabilities of end devices. To address this limitation, this paper discusses the potential of implementing the IoT device discovery for device-specific content delivery, based on device properties, such as screen size and resolution, network connectivity, presence of speakers, supported languages, etc., and presents an approach to enable property-based access to IoT nodes using Bloom filters. The proposed approach demonstrates space- and network-efficient characteristics, as well as provides an opportunity to perform device discovery at various granularity levels.

Keywords: Internet of Things · Edge computing · Content delivery · Device discovery · IPv6 · Bloom filter

1 Introduction

According to recent statistics [1], there are already about 10 billion connected smart objects – a constantly growing number, which has already exceeded the human population, and is expected to reach hundreds of billions in 10 years from now. This opens up several issues, targeted by joint research efforts of both academia and industry, collaboratively working towards fulfilling the vision of the ubiquitous Internet of Things (IoT). Among these issues, networking aspects, concerning how to deal with this increasing population of Internet-connected

A. Puliafito et al. (Eds.): ADHOC-NOW 2017, LNCS 10517, pp. 39–52, 2017.
DOI: 10.1007/978-3-319-67910-5_4

devices, are recognised among the most critical to enable the IoT vision. As IPv4 IP addresses are running out, keeping track of such a huge number of addresses is not trivial, even if in the next 5–6 years IPv6 will be eventually fully adopted. IP addresses, routing tables and all related mechanisms should be adapted to this wider context and higher order of magnitude. It could be even necessary to reshape and tailor them towards new solutions, since the existing ones not necessarily scale out to the IoT paradigm.

The IoT implies that a user (or an application) has to deal with considerably long network addresses to uniquely identify and refer to an overwhelming amount of network-connected devices. This exponential growth is expected to introduce new challenges to traditional computer network protocols, such as, for example, *(i)* efficient access to a huge number of devices; *(ii)* security and privacy; *(iii)* interoperability and standardisation; *(iv)* efficient energy consumption.

One of the most important challenge in the IoT is device discovery [5] – i.e. searching for useful devices able to accomplish a task or matching specific query parameters. Currently, there are two established ways of performing device discovery: *multicasting* and *broadcasting*. A group of devices can be discovered by multicast routing by, for example, adopting multi-hop routing. The problem, however, is that multicast addresses cannot be aggregated in routing tables (as opposed to unicast addresses, which have a single entry on the same subnet in the routing tables of the intermediate routers. For multicast addresses aggregation is not possible, since any two adjacent multicast addresses from the address space reserved for multicasting will correspond to multicast groups that might have completely different members, located in different parts of the network, so that the corresponding multicast trees will have different shapes as well. In this case, each multicast group requires a separate entry in the routing tables of the intermediate routers [12], which renders this approach hardly scalable.

On the other hand, broadcast routing does not have similar limitations – this, however, comes at a cost of increased network latency and waiting time. Due to the fact that, if we want to address some devices using a broadcast call, we have to send a request to all of them and wait for their replies. This results in an increased amount of traffic, proportional to the number of nodes in the network. As a result, broadcasting is not scalable either, and thus can hardly address the device discovery requirements, introduced by large IoT networks.

Moreover, given the extreme amounts of *heterogeneous* devices constituting the IoT ecosystem, ranging in their sensing/actuation, computing, storage and networking capabilities, timely and accurate device discovery and access, as well as content delivery, based on some specific parameters such as device type, network connection, status, powering options, etc. are also seen as a pressing and challenging issue. The latter feature of targeted content delivery is becoming specifically important with the rise of the IoT as a global commercialisation, advertisement and entertainment channel, through which more and more enterprises aim to deliver content tailored to end users and reach out to their potential customers. More specifically, as more and more connected objects are equipped with display devices – e.g. vehicle on-board ‘infotainment’ systems, smart kitchen appliances, various kinds of interactive surfaces (boards, walls, screens, etc.), not to mention the more traditional mobile and portable devices – they all are nowadays seen as potential consumers of some targeted and differentiated multimedia content. This current trend is supported by various techniques, serving to collect some information about users and their contexts and provide this information to content providers. Existing techniques for targeted content delivery typically rely on collecting user-specific information either explicitly (i.e. end users agree to send their behavioural information to the content provider), or implicitly (i.e. behavioural information is indirectly inferred from a browsing history, cookies, search queries, etc.). The collection of potentially large user behaviour data sets from end user devices is typically implemented as a background process (i.e. transparent to the user), thus possibly increasing network latency and creating network overheads [11].

However, the existing routing mechanisms, as described above, are only able to discover and access connected devices using their network IP addresses and ports, which do not take into account the properties of the device, and, therefore, have little value, as far as targeted content delivery is concerned. In this light, a potential way to enable device discovery and access in the IoT, taking into account the size and complexity of underlying networks, could be to include additional search parameters in the routing procedure so as to support targeted device discovery and limit the search space. More specifically, a potential solution would be able address IoT network devices not only through their IP addresses, but also through a combination of device properties, such as *device type, location and timezone, network connection and bandwidth, screen type and resolution, presence of speakers, supported languages*, etc. From this perspective, an envisaged solution is expected to implement some kind of selective routing algorithm, which would facilitate time- and network-efficient device discovery in the IoT context based on device properties and functionalities, and consequent delivery of tailored content. For instance, this will allow identifying specific edge devices (e.g. equipped with high resolution screens and broadband network connection) and deliver corresponding full-size rich video-content.

Taking into considerations these desired features of a potential solution, this paper presents an approach to facilitate property-based device discovery in complex IoT networks using counting Bloom filters to enable tailored device-specific content delivery. As it will be further explained below, the proposed approach adopts the content provider's perspective and benefits from the space-efficient way of storing information about devices and their properties, as well as fast calculation times when deciding whether a matching device is present in the network. Moreover, with property-based search using Bloom filters, it becomes possible to perform device discovery at various granularity levels, and, as a result, enable even more targeted content delivery.

Accordingly, the rest of the paper is organised as follows. Section 2 contains relevant background information and briefs the reader on the theoretical underpinnings of Bloom filters, as well as discusses some related research works. Section 3 presents and explains the actual approach. A sample use case scenario in the IoT context including preliminary evaluation on false positive results is discussed in Sect. 4. Section 5 summarises the main potential benefits of the approach while Sect. 6 concludes the paper.

2 Background

2.1 Bloom Filters and Counting Bloom Filters

In 1970, Burton Howard Bloom [2] proposed the technique for applications where the amount of source data would require an impractically large amount of memory if 'conventional' error-free hashing techniques were applied. A Bloom filter is a space-efficient probabilistic data structure that is used to test whether an element is a member of a set. False positive matches are possible, but false negatives are not – in other words, a query returns either 'possibly in set' or 'definitely not in set'. Elements can be added to the set, but not removed (though this can be addressed with a 'counting' filter); the more elements that are added to the set, the larger the probability of false positives. Typically, fewer than 10 bits per element are required for a 1% false positive probability, independent of the size or number of elements in the set [3].

More formally, a Bloom filter represents a set $\mathcal{S}$ of m elements using an array of n bits $\mathbf{B} = (B[1], ..., B[n])$ initially set to 0. The filter uses a set of k independent hash functions $\mathcal{H} = \{h_1, ..., h_k\}$ with range $\{1, ..., n\}$ uniformly mapping each element of $\mathcal{S}$ to random position over the $\mathbf{B}$ array as shown in Fig. 1.

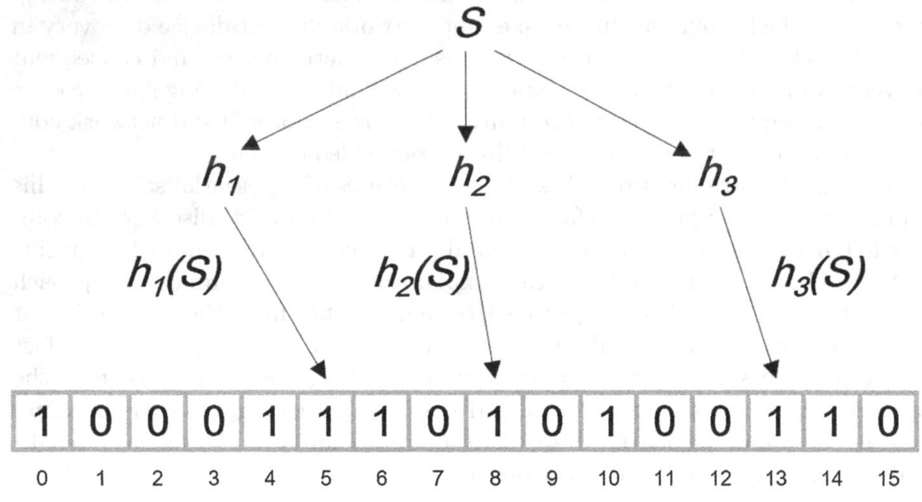

Fig. 1. An illustration of a Bloom filter.

More specifically, for each element $s \in S$, the bits $B[h_i(s)]$ are set to 1 $\forall i \mid 1 \leq i \leq k$. A bit can be set to 1 multiple times either through different hash functions for the same element s or different elements of S. An answer to the query 'Is $b \in S$?' is true if all $h_i(b)$ are set to 1; otherwise, if at least one bit is 0, b is not in S.

While the Bloom filter has many advantages, such as fast access time and a relatively small size (a few bytes per element at most), it does come with an obvious drawback: a possibility of false positive on membership checks. A false positive occurs when the hashes from an element not in the Bloom filter overlap with a combination of hashes from elements that are in the Bloom Filter. The probability of a false positive for an element not in the set can be derived. Given a size m of a bit vector, a number of elements n present in a Bloom Filter, and k hash functions, a probability of a false positive P_{fp} is as follows [14]:

$$P_{fp} = \frac{m!}{m^{k(n+1)}} \sum_{i=1}^{m} \sum_{j=1}^{1} (-1)^{i-j} \frac{j^{kn} i^{k}}{(m-i)!j!(i-j)!} \tag{1}$$

Deleting elements from a Bloom filter cannot be done simply by changing ones back to zeros, as a single bit may correspond to multiple elements. To enable deletion of elements, the so-called counting Bloom filter uses an array of n counters instead of bits (Fig. 2). These counters are able 'track' the number of elements currently hashed to that location [8]. Deletions can now be safely done by decrementing the relevant counters. A standard Bloom filter can be derived from a counting Bloom filter by setting all non-zero counters to 1. It is worth noting that counters must be chosen large enough to avoid overflow – for most applications, four bits are sufficient.

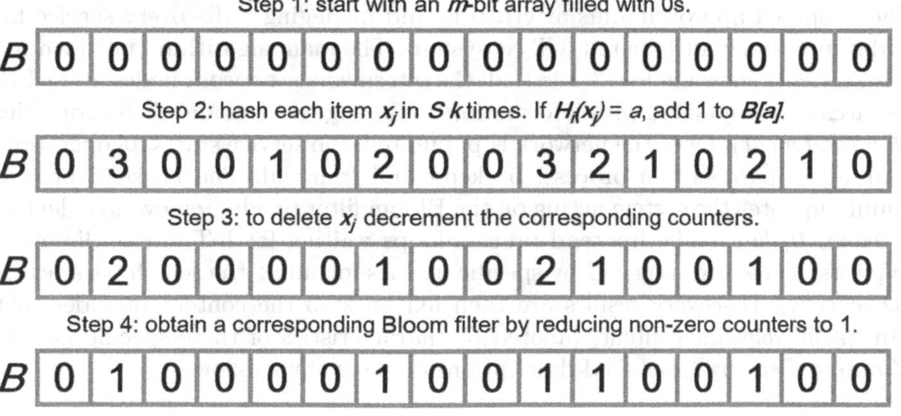
Step 1: start with an m-bit array filled with 0s.
B 0 0 0 0 0 0 0 0 0 0 0 0 0 0 0 0
Step 2: hash each item x_j in S k times. If $H_i(x_j)$ = a, add 1 to B[a].
B 0 3 0 0 1 0 2 0 0 3 2 1 0 2 1 0
Step 3: to delete x_j decrement the corresponding counters.
B 0 2 0 0 0 0 1 0 0 2 1 0 0 1 0 0
Step 4: obtain a corresponding Bloom filter by reducing non-zero counters to 1.
B 0 1 0 0 0 0 1 0 0 1 1 0 0 1 0 0

Fig. 2. An illustration of a counting Bloom filter.

2.2 Related Work

Bloom filters have been originally introduced to improve data management performance, quickly becoming popular in a variety of databases and storage systems [2]. Then, they have been widely used in distributed systems [15] across several different application domains. Nowadays, Bloom filters are becoming a consolidated and stable technology that can be found in different solutions such as the Squid Web proxy for the cache management[1] and the SPIN model checker [6]. In recent years, a Bloom filter experienced an increased interest in a networking domain and security [4]. One of the first applications of Bloom filters for security purposes traces back to Spafford's work on storing dictionary of weak passwords [14]. They are also widely applied in intrusion detection, virus and spam detection, access control [9], and IP traceback [13] systems.

Recently, a fertile field of application for Bloom filters is the IoT, due to the issues discussed above. Some attempts in this direction are available in literature. Among them, [10] adopts Bloom filters in context-aware addressing and routing for the IoT. It is a position paper, which primarily aims at demonstrating the feasibility and main benefits of a parameter-based approach to access IoT devices using Bloom filters. The authors also report on a simple, proof-of-concept implementation, which was developed to this purpose. The existing approaches, however, do not seem to consider and explore the potential of applying counting Bloom filters to address situations, commonly present in the IoT context, when devices disconnect from the network, and thus information about them is expected to be updated accordingly.

3 Proposed Approach

The proposed approach aims at creating and managing a discovery service for nodes in a widely distributed IoT ecosystem. The main algorithm implementing this service is shown in Fig. 3. First, all the intermediate network nodes in the IoT hierarchy are populated with information about edge devices and their properties (*Establishment*). Once the network is populated, the service starts its operation, running in parallel two process: backend and frontend. The backend process mainly updates the system acting on the Bloom filter to add/remove IoT devices (*System Update*). The frontend process loops waiting for IoT device discovery requests. These are triggered by specific queries including the search parameters (*Discovery*). Discovery results are then fed back to the content provider that can 'push' relevant content, tailored to characteristics of these specific devices (*Content Delivery*). Detailed descriptions of these steps follow.

3.1 Service Establishment

The proposed approach relies on representing an IoT device in a network hierarchy as a tuple $D = (ID, Prop)$, where ID is a unique identifier of this device,

[1] http://www.squid-cache.org/.

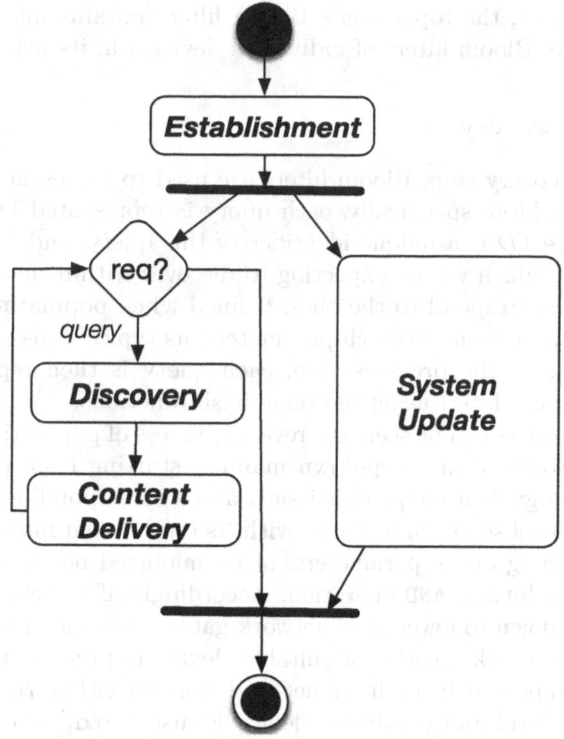

Fig. 3. Bloom filter-based discovery service algorithm.

and *Prop* is a set of properties of this device, such as, for example, *type* (e.g. smartphone, tablet, embedded 'infotainment' terminal, etc.), *power supply* (e.g. solar panel, battery, power cord, etc.), *network connectivity* (e.g. wireless, mobile, Ethernet cable, etc.), *operating system and available software, location, timezone, language*, and so on. All these properties, as will be demonstrated below, serve to enable fine-grained search and discovery of network IoT devices. Using suitable hash functions, each set of properties *Prop* is converted into a corresponding Bloom filter array.

First, a Bloom filter, representing device properties, is created on each edge device. Next, the network hierarchy is 'populated' by these newly-generated Bloom filters. This process is done in a bottom-up manner, such that an edge device provides its immediate subnet gateway with its Bloom filter representation. The gateway, in its turn, 'populates' its own Bloom filter with arrays coming from all the edge devices in its subnet by doing the bitwise `OR` operation. This process is then iteratively repeated up until the very top of the given network. As a result, the top server's Bloom filter contains information about all edge nodes (i.e. Bloom filters of individual devices) in its network.[2]

3.2 Device Discovery

At the device discovery step, Bloom filters are used to represent corresponding discovery queries. More specifically, each query is represented by a tuple $Q = (ID, Prop)$, where ID is a unique identifier of this query, and $Prop$ is a set of device properties, which we are expecting to discover within the given network. These properties correspond to the ones, defined when populating the network hierarchy, and can be seen as search parameters, as typically used in traditional searching. Similar to the previous step, each query is then represented by a corresponding Bloom filter, using the same hash functions.

Discovering devices can be seen as a reverse process of populating the network hierarchy. It is executed in a top-down manner, starting from the very top of the network topology. Once a query is issued and a corresponding Bloom filter is created, the top-level server first checks with its own Bloom filter whether there is a device, matching query parameters, in its managed network. This is done by performing the bitwise `AND` operation. Accordingly, if the evaluation is true, the query is sent down to lower-level network gateways, which perform a similar `AND` evaluation to check whether a suitable device is present in their subnet. This process is repeated in each subnet and iterates either *(i)* until reaching the very bottom level and a suitable device is discovered, or *(ii)* until one of the intermediate network nodes replies that no matching device is present in its subnet.

3.3 Content Delivery

Finally, once matching devices and their IP addresses have been identified, the content provider can now 'push' corresponding content, matching and fully leveraging the end user device's characteristics. It is worth noting that in the context of the presented research, the provider acts as the initiator of the content delivery, and is therefore needs to know in advance the specifics of the target customers. This is different from the more traditional content delivery, where required meta-information is sent to the server along with user requests (i.e. content delivery is initiated by users).

3.4 System Update

False positive results, despite being a shortcoming of Bloom filters in general and of the proposed approach in particular, may potentially be useful as well.

[2] It is worth noting that this network population process might be continuously repeated, so as to update the bits related to the actual network connection of a device.

More specifically, they can be used to update Bloom filters throughout the network hierarchy, whenever a device unexpectedly disconnects from the network. In these circumstances, the upper-level network gateway is typically unaware that a specific device is no longer present in its network, whereas its counting Bloom filter indicates the opposite. To handle such situations, it is possible to decrement corresponding bits in the filter, whenever the gateway transferred a device discovery query down to nodes in its subnet, but never received a reply from a matching device. Removing elements, however, needs to be taken with care, as in this case false positive results may also occur due to the nature of Bloom filters, not just because a device has disconnected from the network.

4 From Theory to Practice

In order to explain how the proposed Bloom filter-based IoT discovery service works, without loss of generality, a simple example in this context is considered more thoroughly below.

4.1 A Sample Scenario

In this example, target devices are equipped with high-resolution screens and speakers, and connected to the Internet via a high-speed Ethernet network. This way, it is safe to stream the 'richest' content (i.e. high-definition advertisements with sound) to the user, assuming that the device will be able to download the content at a high speed with no charges, and to playback the content at the best quality with sound. Accordingly, it is assumed that each device (and its properties) is represented by a Bloom filter containing multiple bits, among which only six are relevant in the given scenario. Namely, these bits represent the following properties of an edge device:

1. High-resolution display
2. Low-resolution display
3. Equipped with speakers
4. Connected via an Ethernet network
5. Connected via a mobile network
6. Connected via a wireless network

It is also assumed that there are three subnets in the network, each containing three devices, as illustrated by the diagram in Fig. 4. The three network gateways contain combined Bloom filters of their respective subnets, and the server contains the overall Bloom filter representation of the network. According to the goal of this scenario, a device discovery query is represented by the following Bloom filter $BF = (1, 0, 1, 1, 0, 0)$.

At the first step, the server evaluates the query against its own Bloom filter, and decides that there is indeed a matching device present somewhere down the network. Next, the query is propagated down to three subnets. The respective gateways start evaluating the query against their own Bloom filters. As it is seen from the diagram, *Subnet A* contains some devices with high-resolution screens and speakers, none of which, however, is connected via an Ethernet cable. *Subnet B*, on contrary, contains an Ethernet-connected and speakers-enabled devices, but does not have a device with a high-resolution display. Accordingly, for both subnets, the query evaluation returns *false*, and, as a result, the network call is not propagated down the first two subnets. Finally, *Gateway C*, by evaluating the query, understands that there is a matching device in its subnet and sends the query to all three nodes. Two of these nodes do not match the search parameters at all, whereas only the last one – *Device C3* – perfectly matches the all three criteria. By evaluating the incoming query, it realises that it matches query parameters, and replies back with its ID and network location. The reply is then sent back to the server through intermediate hops. Having identified target devices for the richest content, the content provider can then identify devices with lower capabilities (e.g. with low-resolution displays or without speakers) in a similar manner, so as to push appropriate content.

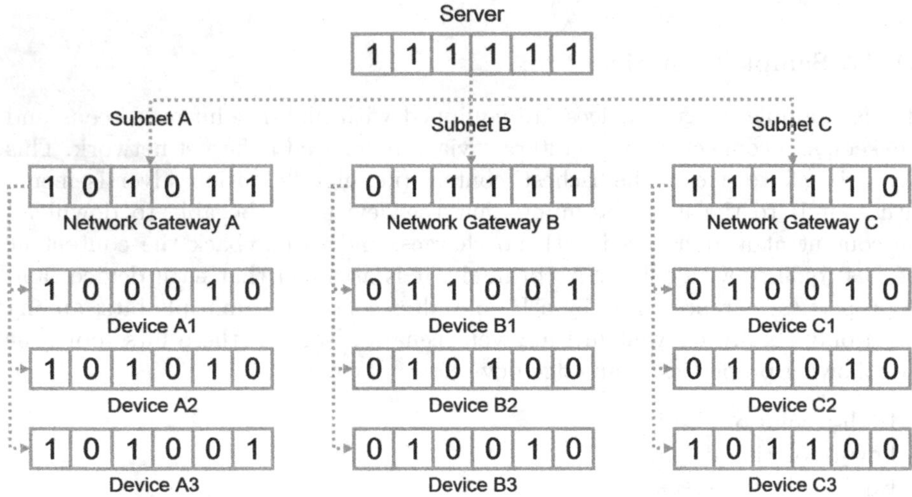

Fig. 4. Sample IoT hierarchy to demonstrate property-based device discovery using Bloom filters.

4.2 Preliminary Evaluation

As explained above, the presented approach is based on property-based search for IoT devices, which enables flexible, fine-grained discovery of IoT nodes. This means that it is possible to discover devices by specifying one or more properties. Specifying a large number of parameters in the query will impact on the search results, reducing the number of matching devices. On the other hand, if few parameters are specified, the search space is expected to be wider, since more devices might satisfy the search parameters. For example, searching for a device with speakers will yield 4 results, also including 2 matching devices, which are additionally enabled with wireless connectivity.

There is a potential shortcoming to be taken into account, however. As explained in Sect. 3.4, Bloom filters might be particularly prone to false positive results when performing query evaluation, especially when the size of the Bloom filter bit array is relatively small. In these circumstances, a small-size search query, containing just a single property, might yield a positive result, because a specific bit in the Bloom filter was previously set to a value, belonging to some other device's hash value.

In this light, it becomes particularly important to find the right balance between the preciseness of the search query (i.e. number of properties specified) and the false positive rate when matching the query against a Bloom filter. To demonstrate this dependency, a series of experiments were conducted with a goal to see how the false positive rate changes with respect to the number of device properties, specified in the search query.

In the experiments, we referred to the above described example considering 6 properties and exploiting the latest version of MurmurHash3[3], a widely used 32-bit hash function implementation [7], in development. Being able to accommodate 2^{32} bits of information (i.e. more than 4 billion unique values), the resulting Bloom filter was populated with 500 million values. Then, we started evaluating this Bloom filter against two types of queries. In the first case, the Bloom filter has been split into as many parts as the properties specified in the query, one BF per property. In the second case, all properties were specified within a single Bloom filter. In both cases, the number of properties ranged from 1 to 6 as in the above specified example where a 1-bit representation for each property has been adopted, while in the experiments a larger number of bits ($\lfloor 2^{32}/6 \rfloor$) per property has been used. We therefore performed 6 different pairs of tests, by increasing the number of properties (and split Bloom filters), comparing the split and single Bloom filters against this property-based benchmark.

Table 1. The false positive rate (percentage) from experiments on split and a single Bloom filters.

# of properties	False positive rate (%)	
	Split BF	Single BF
1	2.407	2.407
2	5.131	2.419
3	8.246	2.523
4	11.268	2.557
5	15.328	2.628
6	18.451	2.687

[3] https://github.com/aappleby/smhasher/blob/master/src/MurmurHash3.cpp.

As shown in Table 1, the evaluation of more than 900,000 queries (explicitly known not to belong to the given Bloom filter) indicates that there is a considerable increase in the number of yielded false positive results when split Bloom filters are used, as opposed to a single Bloom filter, for all the query properties. This is due to the fact that a single Bloom filter representing properties altogether allows to exploit a larger space than the one obtained by splitting the domain into different subdomains as in split Bloom filters, thus resulting in a higher precision.

5 Discussion

This paper presented a property-based approach to discover IoT nodes using Bloom filters. Potential benefits of the proposed solution are the following:

Flexible device discovery at different levels of granularity: as opposed to the traditional access to edge nodes in IoT environments, where IP addresses need to be known in advance, the proposed approach enables searching for devices based on their properties, such as type, network connection, screen resolution, etc. This kind of property-based device discovery can be performed at various levels of granularity – i.e. coarse-grained (e.g. discover any kind of device with a display) or fine-grained (e.g. discover a low-resolution embedded device, connected by an Ethernet cable, and powered by a cord). This flexibility has the potential to contribute to creation of a wide range of IoT systems, where the network topology is not static, but rather devices are constantly joining and leaving the network.

Network efficiency: a Bloom filter (as suggested by its name itself) serves to filter incoming search queries to avoid redundant broadcast calls through the whole network. As demonstrated in Sect. 3, if an intermediate node understands that there is no matching device within its subnet, it does not allow the query to go down that specific subnet, thus *(i)* decreasing the amount of time needed to discover a device, and *(ii)* minimising the amount of redundant network traffic and improving network latency. Moreover, the query evaluation procedure – i.e. performing the bitwise `AND` operation on two bit arrays – is a time-efficient operation with minimum impact on the overall device discovery process. In the presence of hundreds and thousands of edge devices and intermediate network nodes, both network- and time-efficiency is seen as considerable benefits when discovering devices (even taking into consideration the false positive rate).

Space efficiency: as described in Sect. 2, Bloom filters are space-efficient data structures, requiring minimum amount of memory. Even when using counting Bloom filters (i.e. each counter within a Bloom filter occupies 4 bits), resulting arrays typically do not exceed 4 MB of storage space – a highly-relevant feature in the context of IoT environments, where individual nodes are not necessarily equipped with mass storage facilities.

High accuracy: despite potential false positive results that may occur during device discovery at intermediate network nodes, the overall accuracy is not affected, since a final decision whether a matching device is present in the network or not is taken by edge devices themselves. Only if an edge device's Bloom filter matches an incoming query, a corresponding acknowledgement is sent back to the server. Otherwise, it is assumed that no matching devices were identified.

Scalability and extensibility: thanks to the ability to store large amounts of hashed values and high calculation speed, a Bloom filter can be easily updated with new elements without affecting the overall performance. This is especially important in the context of the IoT networks, which are already constituted by millions of devices, and keep on exponentially growing in size and complexity.

6 Conclusions

The presented solution enables flexible property-based device discovery in the context of complex IoT networks using counting Bloom filters to enable content delivery, tailored to specific end devices. As opposed to the IPv6 routing, which requires 128 bits to encode an address, the proposed approach benefits from the space-efficient way of representing and storing data in a Bloom filter. This also contributes to decreased traffic and network latency, as the device discovery duration depends on how narrow-focused the search query is – i.e. the less devices matching the query, the less network traffic generated, since using a Bloom filter to decide whether a device belongs to a subnet branch or not can 'cut off' the entire branch before actually checking it. As a result, this considerably reduces the amount of network traffic, especially when compared to broadcast and multicast routing techniques. Moreover, by knowing the specific device capabilities in advance, the content provider is able to tailor the delivered content towards them, thus making sure that the content is fully supported by devices. This way, no unnecessarily heavy and rich content is pushed over the network.

Acknowledgements. The work presented in this paper was partially supported by the ERASMUS+Key Action 2 (Strategic Partnership) project IOT-OPEN.EU (Innovative Open Education on IoT: improving higher education for European digital global competitiveness), reference no. 2016-1-PL01-KA203-026471. The European Commission support for the production of this publication does not constitute endorsement of the contents which reflects the views only of the authors, and the Commission cannot be held responsible for any use which may be made of the information contained therein.

References

1. Google IPv6 statistics (2017). https://www.google.com/intl/en/ipv6/statistics.html. Accessed 14 July 2017
2. Bloom, B.H.: Space/time trade-offs in hash coding with allowable errors. Commun. ACM **13**(7), 422–426 (1970)
3. Bonomi, F., Mitzenmacher, M., Panigrahy, R., Singh, S., Varghese, G.: An improved construction for counting Bloom filters. In: Azar, Y., Erlebach, T. (eds.) ESA 2006. LNCS, vol. 4168, pp. 684–695. Springer, Heidelberg (2006). doi:10.1007/11841036_61
4. Broder, A., Mitzenmacher, M.: Network applications of Bloom filters: a survey. Internet Math. **1**(4), 485–509 (2004)
5. Ccori, P.C., De Biase, L.C.C., Zuffo, M.K., da Silva, F.S.C.: Device discovery strategies for the IoT. In: Proceedings of 2016 IEEE International Symposium on Consumer Electronics (ISCE), pp. 97–98. IEEE (2016)
6. Dillinger, P.C., Manolios, P.: Bloom filters in probabilistic verification. In: Hu, A.J., Martin, A.K. (eds.) FMCAD 2004. LNCS, vol. 3312, pp. 367–381. Springer, Heidelberg (2004). doi:10.1007/978-3-540-30494-4_26
7. Estébanez, C., Saez, Y., Recio, G., Isasi, P.: Performance of the most common non-cryptographic hash functions. Softw. Pract. Exp. **44**(6), 681–698 (2014)
8. Fan, L., Cao, P., Almeida, J., Broder, A.Z.: Summary cache: a scalable wide-area web cache sharing protocol. IEEE/ACM Trans. Netw. (TON) **8**(3), 281–293 (2000)
9. Foley, S.N.: A Bloom filter based model for decentralized authorization. Int. J. Intell. Syst. **28**(6), 565–582 (2013)
10. Kalmar, A., Vida, R., Maliosz, M.: Context-aware addressing in the internet of things using Bloom filters. In: Proceedings of 2013 IEEE 4th International Conference on Cognitive Infocommunications (CogInfoCom), pp. 487–492. IEEE (2013)
11. Nychis, G., Licata, D.R.: The impact of background network traffic on foreground network traffic. In: The Proceeding of the IEEE Global Telecommunications Conference (GLOBECOM), pp. 1–16 (2001)
12. Sebestyen, G., Hangan, A.: Bloom filters for information retrieval in the context of IoT. In: Proceedings of 2016 IEEE International Conference on Automation, Quality and Testing, Robotics (AQTR), pp. 1–6. IEEE (2016)
13. Snoeren, A.C., Partridge, C., Sanchez, L.A., Jones, C.E., Tchakountio, F., Kent, S.T., Strayer, W.T.: Hash-based IP traceback. In: ACM SIGCOMM Computer Communication Review, vol. 31, pp. 3–14. ACM (2001)
14. Spafford, E.H.: Preventing weak password choices. Technical report 91–028, Department of Computer Science, Purdue University (1991)
15. Tarkoma, S., Rothenberg, C.E., Lagerspetz, E., et al.: Theory and practice of Bloom filters for distributed systems. IEEE Commun. Surv. Tutor. **14**(1), 131–155 (2012)

Security

Trust Based Monitoring Approach for Mobile Ad Hoc Networks

Nadia Battat[1], Abdallah Makhoul[2(✉)], Hamamache Kheddouci[3], Sabrina Medjahed[1], and Nadia Aitouazzoug[1]

[1] LIMED Laboratory, University of Bejaia, Algeria, France
nadiabattat@yahoo.fr

[2] FEMTO-ST Institute, Univ. Bourgogne Franche-Comté, CNRS, Belfort, France
abdallah.makhoul@univ-fcomte.fr

[3] LIRIS Laboratory, University of Lyon 1, Villeurbanne, France
hamamache.kheddouci@univ-lyon1.fr

Abstract. Mobile ad-hoc networks (MANET) are vulnerable to many types of attacks. Monitoring MANET is then essential to ensure high level performance. Many challenges arise in the MANET self-monitoring. Namely, the limited storage and energy resources of mobile nodes, the high topological dynamism and the unpredictable behaviors, etc. In this paper we propose a new self monitoring scheme that comprises a new multi criteria monitors' election method while integrating a new trust based cooperation technique. This scheme does, not only, elect the trustworthy monitors having a large capacity, but it also can guarantee the continuous participants' control in order to measure their sincerity. We validate our approach through several simulations. The experimental results indicate that the proposed scheme outperforms the cluster-based and CDS-based architectures in terms of the number of exchanged messages, excluded regular monitors and that of detected irregular monitors.

1 Introduction

Self monitoring of MANET consists in evaluating the operational state of its mobile devices, the links between them as well as its quality of service. This is achieved by a subset of mobile nodes (called monitors) which are elected according to several predefined parameters [5]. Each monitor performs its assigned tasks and in the same time is responsible for controlling a subset of mobile nodes in its area called the monitored nodes. In their turn, the monitored nodes are responsible for enforcing the policies they receive from their monitors, collecting the requested information and delivering them to their corresponding monitors. The monitoring evaluation can be performed by analyzing and processing the local collected data by the nodes as well as the information received from their neighbors. To realise high level monitoring, it is vital that each participant (monitored node or monitor) contributes correctly to the election of monitors and the monitoring process. However, this leads to consume more computational and energy

A. Puliafito et al. (Eds.): ADHOC-NOW 2017, LNCS 10517, pp. 55–62, 2017.
DOI: 10.1007/978-3-319-67910-5_5

resources. Actually, not all nodes participate in this process. Selfish nodes can use the resources of the others without participating in the monitoring functions. Malicious nodes can falsify the collected data, modify the distributed policies and make illegal and inappropriate decisions. In order to force mobile nodes to obey the monitoring approach and cooperate with each other, we propose in this paper a new monitor electing method. It is based on two main factors: truthfulness and capability.

In literature, several monitoring approaches were proposed for MANET [5]. These approaches can be classified as follows: ***Unique criterion based election approaches:*** These approaches [5,6] use only one criterion such as Lowest-ID or Highest-Degree to elect monitors in that they are easily achievable. However, these algorithms do not take into account all MANET characteristics and the resources level of the elected nodes. This can lead to reapply the election process which reduces the lifetime of monitoring cycle and increases the network overhead in addition to the consumed energy. Moreover, they do not balance the monitoring tasks uniformly among all the nodes. This can result in electing the same node as monitor frequently. ***Multi criteria based election approaches:*** These ones [4,7] use a diversity of criteria to elect monitors. They aim to increase the lifetime of the monitoring cycle by electing the most cost-efficient nodes as monitors. In this paper, we present a new scheme to guarantee an efficient monitoring in multi-hop mobile ad hoc networks. Furthermore, we define a set of rules in order to detect the malicious and selfish behaviors. We study the performance analysis and evaluation of the proposed architecture through simulations. The obtained results show that the proposed scheme can significantly reduce the overhead and maintain a high level of detection.

2 MANET Self Monitoring Scheme

2.1 Messages Structure

We propose some modifications on the original hello message by adding the following fields: *Weight* ($W(n_i) \in [0,1]$): this field contains the weight of the node n_i that is initialized to 0. Its value is estimated in Sect. 2.3; *Trust value* $T(n_i) \in [0,1]$: this field represents the trust value of the node n_i that is initialized to 0.5; *Energy level* $E(n_i)$: this field indicates the remained energy level of the node n_i; *Role*: this field defines the role of the node n_i: monitor, delegated monitor or ordinary node. $NeighborsList$: this field contains the IDs of the node neighbors and their estimated trust values.

2.2 Trust Computation Method

Each node must observe the behaviors of its neighbors to detect their malicious or selfish comportments. Consequently, it can observe and trace their behaviors by the continuous updates of the trust values. Initially, we assign to each node a trust value equal to 0.5[1]. Furthermore, we define the following rules for identifying the selfish or the malicious behaviors.

- The contribution level of each mobile node can provide falsified evaluation about its collaborations in order to raise its trust value. Therefore, a monitor can distribute a part or the full report to its controlled node to confirm its honesty. A monitored node can either select a route containing a maximum number of its neighbors for forwarding its data and/or the local report, or divide this quantity of data into N packets (N represents the number of its neighbors). Then it sends each one through each neighbor. When one neighbor drops packets and that this behavior is observed by a sender, the latter will decrease its trust value. A local analysis is needed to avoid monitors to act maliciously and to detect their selfish behaviors. Misbehaved or selfish nodes will be penalized by decreasing their trust values.
- As mobile nodes use limited storage capacities, they can discard not only their collected data but also data of other nodes, in order to exploit its resources for further interesting uses. Therefore, the monitored node (resp. the monitor) can periodically ask its monitors (resp. the data holder) to send a randomly selected piece of its collected data at a specific time. Once receiving this requested data, the monitored node (resp. the monitor) compares it to its stored data hunk and then increases or decreases the corresponding node trust value.
- The participants' contributions of mobile nodes can indicate the existence of malicious or selfish nodes. For instance, if a node exchanges its opinions on neighbors periodically and performs a local analysis without participating in forwarding data or data storage, it will be considered as malicious.
- A monitor can compare the received data within its radio range to detect the malicious or selfish behaviors of its neighbors. For instance, if more than one neighbor indicate that two nodes X and Y are neighbors and the neighbors' list of X does not contain any information about Y, a monitor can conclude that X is either selfish or malicious.

For updating the value of confidence, we use the activity rate (AR), which is calculated according to the number of positive realized tasks including the packet forwarding rate and the realized monitoring tasks. If we consider two nodes i and j, the node i calculates the $AR(j)$ as follows: the node i should record the number of positive interactions ($pos(i,j)$) with the node j, and the total number of interactions ($total(i,j)$), over a given interval of time, and then it calculates the activity rate as follows:

$$AR = pos(i,j)/total(i,j) \tag{1}$$

The trust value is estimated over time to reflect changes in the activity rate. Nevertheless, local estimation on each mobile node might not be enough to detect any node bad behavior. It should have information from other nodes.

[1] To not consider a node in advance as being selfish, malicious or confident.

Moreover, in some cases, a mobile node can monitor only the behavior of its direct neighbors. As a result, not all neighbors at $n - hops$ will honestly share the real values. Consequently, we propose that each node calculates the trust values based on the combination of direct and indirect estimations that derive from neighbors. Therefore, we consider also the two following cases: **(a)** A neighbor does not report his accurate trust value about the corresponding monitor (resp. monitored node) in case of hardware or software failures held by this node; **(b)** A neighbor can provide a false trust value about the corresponding monitor (resp. monitored node). It may provide a negative (or positive) value to misbehaved/trusted monitor (resp. monitored node): *false accusation attack* [3] (or *false praise attack* [1]).

Algorithm 1. Locally detection of regular, irregular and normal nodes

Constant $MaxNbrF = 3$;
$T(n_i)$: Trust value of the node n_i; $NbrF$: number of node' faults ;
A: Last activity that must be realized by the node n_i;
$The function K$: $K(A) = 1$, if A is correctly realized by n_i, otherwise $K(A) = 0$;
$The function B$: $B(n_i) = M$, if n_i acts maliciously *or* $B(n_i) = S$ if n_i acts selfishly;
SL: List of detected selfish nodes;
ML: List of detected malicious nodes;
$E(n_i)$: Energy level of the node n_i; Et: Necessary energy level for realizing A ;
$The function S$: $S(n_i) \in$ {Irregular, Regular, Normal}; Dt: Penalty period;

```
Begin
if (((n_i ∉ SL)and(n_i ∉ ML))or(S(n_i) = Normal)) then
  if (K(n_i) = 1) then
    if (T(n_i) ≺ 1) then
      Recompute T(n_i);
      if (T(n_i) ≻ 0.5)and(S(n_i) = Normal)) then
        S(n_i) = Regular;
  else
    if (T(n_i) ≻ 0) then
      Recompute T(n_i);
      NbrF = NbrF + 1 ;
  if (T(n_i) ≤ Bt) then
    if (((n_i ∉ SL)and(n_i ∉ ML))) then
      if (E(n_i) ≻ Et) then
        if (NbrF ≥ MaxNbrF) then
          S(n_i)= Irregular;
          Dt = CurrentTime + Tb ;
          if (B(n_i) = S) then
            add n_i to SL;
          else
            add n_i to ML;
End
```

After receiving the indirect estimations, a node i calculates the trust values $T(c)$ (its and that of its neighbors) using the following formula:

$$T(n_j) = (\sum_{k=1}^{n}(T(k,j)) + T(i,j))/(n+1) \qquad (2)$$

n is the nodes number having sent their trust values about the node j to the node i. $T(k,j)$ is node k trust value about node j.

When a node does not receive any trust value, it can rely either on its trust values or on the previously gathered ones. The trust value can be increased or decreased by a chosen changing step *Stp*, according to nodes' behaviors. We assume that the chosen changing step $Stp = 0.1$ [2,8]. Mobile node can behave selfishly or maliciously following to its features or according to the mobile environment characteristics. Nevertheless, environmental conditions can lead to intensively deteriorate trust values. Therefore, we propose to use the maximum authorized faults number *MaxNbrF* to avoid the inexactitude of trust value estimation. If a node does not participate in *three* successive activities while it has sufficient energy level to perform them and its trust value is equal or less than a predefined threshold $Bt = 0.3$, it will be irregular. A detected node will be added to the selfish or malicious nodes list (see Algorithm 1) according to its last behavior.

A mobile node will be considered as malicious if it: falsifies the monitoring policies; generates unnecessary traffic; advertises non-existing monitors; modifies the monitoring system; provides fake data; broadcasts a false alarm; contributes in some monitoring tasks only. On the other hand, it can be considered as selfish if it: refuses to participate in monitoring process; discards the collected data; drops the exchanged monitoring messages.

2.3 Monitors Election

Our approach is a multi criteria based election method. The network is logically divided into clusters with a single monitor (cluster-head). We assume that only regular nodes can participate in monitors election. Every regular node n_i, aware of its neighbors, performs the following steps:

1. it calculates its weight $W(n_i)$ which indicates its ability to serve as monitor as follows: $W(n_i) = COF1 * T(n_i) + RS(n_i)$, where, $COF1 \in [0,1]$ is the metric trust value coefficient. A monitor can consume more resources than a monitored node because of the monitoring tasks that must be performed. In fact, we also use the weighted parameter $RS(n_i)$ to elect monitors. This weighted parameter can be computed according to: the processing power[2]; the energy level; the storage capacity.
2. it forwards a hello message, containing its weight, its trust value, its neighbors and their estimated trust values list and its energy level, to its neighbors.

[2] Node with little processing power can slow the forwarding or analyzing of collected data.

3. it waits a time period for receiving messages from its neighbors.
4. it compares its weight with those of its neighbors. It becomes monitor, if it has the maximum weight.

A monitor informs its neighbors about its presence by sending hello message, while initializing the field *Role* to 1. Each neighbor selects the nearest monitor based on hop count.

2.4 Maintenance of the Monitoring Architecture

The proposed topologies for the monitoring approaches are usually based on the construction of cluster or CDS (Connected Dominating Set) [5]. Our proposed approach is also cluster-based where each participant (monitor or monitored node) is controlled by its regular neighbors. To detect any mobile node neighbors, periodic hello messages are exchanged. Once these messages are received, each mobile node can update the list of its neighbors, their weights, their trust values and their roles with minimum transmission overhead. When mobile nodes voluntarily/involuntarily disconnect or move, our approach faces these topological changes by applying the following policies. **(a)** When a new regular node joins a network, it exchanges with its neighbors its data, and then chooses the nearest monitor. **(b)** When a mobile node loses connectivity with its monitor, two cases can be considered: **1.** ***Voluntary disconnection of regular monitor:*** the regular monitor can select one of its regular neighbors having the maximum weight to replace it. Then, it informs its neighboring nodes and the other monitors about the new one. **2.** ***Involuntary disconnection of monitor:*** when a mobile node detects the sudden death of its monitor, it launches the election of new monitor.

3 Simulation Results

To study the effectiveness of our scheme, we compare it with the cluster-based and CDS-based architectures. We use the same metrics as our approach to elect cluster-heads and dominator nodes. We assume that $RS(n_i) = COF2 * EC(n_i)$ where $COF1 = COF2 = 0.5$ and $EC(n_i)$ indicates the remaining energy level of the node n_i. We also assume that the necessary energy for performing a given monitoring task, the mobile node trust value and remaining energy level are randomly selected from the range $[0, 1]$ following a uniform distribution. The settings of our simulations are as follows: $Duration = 5000\,\text{s}$, $Numberofnodes = 100$, $Territoryscale = 100\,\text{m}^2$, $Rangeofnode = 20$, the mobility model is random waypoint, $Pauseinterval = [0, 20]$ (s), $speedinterval = [0, 20]$ (m/s).

Figure 1 indicates the number of the exchanged messages in order to construct topologies through time. From the results, we can observe that our scheme outperforms the cluster and CDS based architectures by attaining low message overhead. This explains that only regular nodes can perform the monitoring plan. Figure 2 illustrates the evolution of the irregular monitors detection rate through

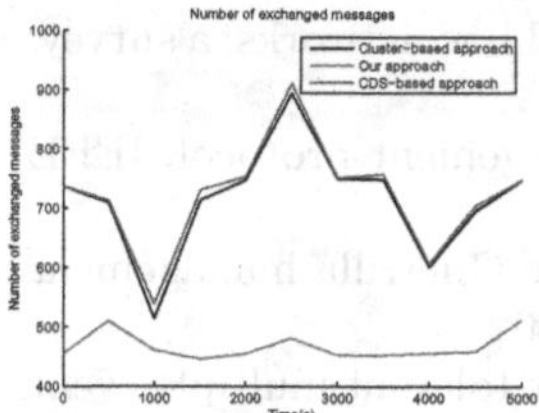

Fig. 1. Number of exchanged message

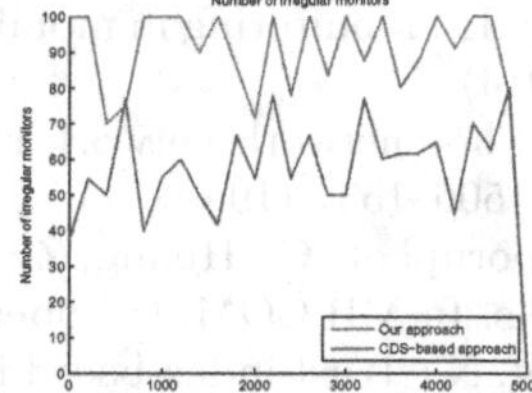

Fig. 2. Number of detected irregular monitors

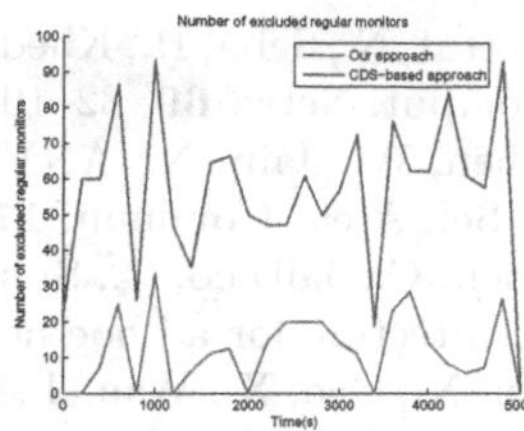

Fig. 3. Number of excluded regular monitors

time. The results indicate that an important detection rate of our scheme. This is interpreted by the fact that in CDS-based architecture, regular monitors can be isolated and consequently, cannot detect the malicious or selfish behaviors of irregular ones. Figure 3 shows that our scheme decreases the number of the excluded regular monitors compared to CDS-based architecture.

4 Conclusion and Perspectives

Monitoring the behavior of each mobile is an essential requirement for developing a robust and reliable monitoring approach. Therefore, we propose to elect only well behaving and honest nodes as monitors. We select monitors based on a weighing factor which uses the trust value. The latter is measured using the rate of contribution in monitoring process of the participants and their neighbors. We evaluated the performance of this scheme compared with the cluster-based and CDS-based architectures. The obtained results demonstrate the effectiveness of our scheme in terms of the numbers of the exchanged messages, the excluded regular monitors and the detected irregular monitors. The proposed scheme also decreases the maintenance time.

References

1. Alzaid, H., Alfaraj, M., Ries, S., Jsangand, A., Albabtain, M., Abuhaimed, A.: Reputation-based trust systems for wireless sensor networks: a comprehensive review. In: IFIPTM, pp. 66–82 (2013)
2. Awad, A.I., Hassanien, A.E., Baba, K. (eds.): Advanced in Security of Information and Communication Networks. Springer, Heidelberg (2013). doi:10.1007/978-3-642-40597-6
3. Banković, Z., Vallejo, J.C., Fraga, D., Moya, J.M.: Detecting bad-mouthing attacks on reputation systems using self-organizing maps. In: Herrero, Á., Corchado, E. (eds.) CISIS 2011. LNCS, vol. 6694, pp. 9–16. Springer, Heidelberg (2011). doi:10.1007/978-3-642-21323-6_2
4. Battat, N., Kheddouci, H.: HMAN: hierarchical monitoring for ad hoc network. In: Embedded and Ubiquitous Computing (2011)

5. Battat, N., Seba, H., Kheddouci, H.: Monitoring in mobile ad hoc networks: a survey. Comput. Netw. **69**, 82–100 (2014)
6. Chen, W., Jain, N.: ANMP: ad hoc network network management protocol. IEEE J. Sel. Areas Commun. **17**(8), 1506–1531 (1999)
7. Shen, C., Jaikaeo, C., Srisathapornphat, C., Huang, Z.: The Guerrilla management architecture for ad hoc networks. In MILCOM, October 2002
8. Xu, X., Gao, X., Wan, J., Xiong, N.: Trust index based fault tolerant multiple event localization algorithm for WSNs. Sensors **11**, 6555–6574 (2011)

An Implementation and Evaluation of the Security Features of RPL

Pericle Perazzo(✉), Carlo Vallati, Antonio Arena, Giuseppe Anastasi, and Gianluca Dini

Department of Information Engineering, University of Pisa, 56122 Pisa, Italy
{pericle.perazzo,carlo.vallati,antonio.arena,giuseppe.anastasi, gianluca.dini}@iet.unipi.it
http://www.dii.unipi.it

Abstract. Wireless Sensor and Actuator Networks (WSANs) will represent a key building block for the future Internet of Things, as a cheap and easily-deployable technology to connect smart devices on a large scale. In WSAN implementation, the Routing Protocol for Low-Power and Lossy Networks (RPL) has a crucial role as the standard IPv6-based routing protocol. The RPL specifications define a basic set of security features based on cryptography. Without these features, RPL would be vulnerable to simple yet disruptive routing attacks based on forgery of routing control messages. However, the impact of these features on the performances of the WSAN has not been investigated yet. The contribution of this paper is twofold: an implementation of the RPL security features for the Contiki operating system, which is, at the best of authors' knowledge, the first available, and an evaluation of their impact on the WSAN performances by means of simulations. We show that the protection against eavesdropping and forgery attacks has a modest impact on the performances, whereas the protection against replay attacks has a more considerable impact, especially on the network formation time which increases noticeably.

Keywords: Internet of things · Embedded systems · Secure routing · RPL

1 Introduction

Recent technology advancements are rapidly making real the Internet of Things (IoT), a future in which objects will be empowered with communication capabilities to enable seamless integration with information systems. In this future, such smart objects will penetrate the physical world around us, in some cases implementing remote monitoring and control capabilities, in others, offering enhanced features that exploit automation and self-coordination. IoT applications are expected to cover a wide range of domains such as smart home, smart city, e-health, and so on.

A. Puliafito et al. (Eds.): ADHOC-NOW 2017, LNCS 10517, pp. 63–76, 2017.
DOI: 10.1007/978-3-319-67910-5_6

Wireless Sensor and Actuator Networks (WSANs) will be a key enabler for all these IoT applications, because they allow for rapid and cost-effective installation of smart objects over large areas. The IPv6 Routing Protocol for Low-Power and Lossy Networks (RPL) [16], standardized in 2012 by the IETF ROLL working group, is considered at the present the most mature option to connect IPv6-enabled devices and form WSANs over lossy links with minimal overhead [5]. Considering the importance of the services delivered, protecting the routing functionalities from attacks will be a major challenge to prevent malicious attempts to disrupt IoT network operations [10]. A basic set of cryptographic security mechanisms to guarantee routing resilience to external attackers has been introduced by design in RPL specifications [16]. However, the impact of such mechanisms on the WSAN performances has not been investigated yet.

In this paper, we give a first evaluation of the impact of the RPL security mechanisms on the WSAN performances. We develop a standard-compliant implementation of the RPL security mechanisms for the Contiki operating system [3], and we evaluate it by means of simulations. We show that the RPL security mechanisms have a negligible impact on the performances in terms of network formation time and power consumption, if they do not have to defend against replay-based attacks. Otherwise, if also replay protection is needed, the impact on performances is more pronounced, especially on the network formation time which increases noticeably. To the best of our knowledge, this is the first implementation of the standard security mechanisms of RPL.

The remainder of the paper is organized as follows. In Sect. 2 we review related work. In Sect. 3 we offer a short introduction to the RPL specifications, including its security mechanisms. In Sect. 4 we describe our implementation of the security mechanisms of RPL. In Sect. 5 we evaluate the impact of security in RPL performances. Finally, Sect. 6 concludes the paper.

2 Related Work

Many research papers [1,4,6,8,12,14] studied possible attacks against RPL, and proposed countermeasures. Dvir et al. [4] took into consideration Rank attack and DODAG Version attack. Both these attacks can be considered as RPL-specific instances of the more general sinkhole attack [7], in which a malicious node attracts a large amount of traffic from surrounding nodes in order to eavesdrop or interrupt it. The authors presented a countermeasure to both attacks based on asymmetric cryptography. Perrey et al. [12] presented an improvement of such countermeasure which corrects some of its vulnerabilities, but requires round-trip protocols for path validation. Weekly and Pister [14] presented and evaluated the synergy between two countermeasures against sinkhole attacks in RPL: parent fail-over and rank authentication. Iuchi et al. [6] presented a countermeasure against Rank attack based on a particular next-hop selection policy. This countermeasure requires the nodes to choose sub-optimal routes. Le et al. [8] studied the impact of an attack in which a malicious node deviates from the normal behavior by selecting as next hop the worst neighbor instead of the best one. Airehrour et al. [1] proposed a countermeasure against blackhole attack, in which a malicious node drops all the traffic forwarded to it, and breaks the availability of large parts of the network. Their countermeasure requires every node to operate in promiscuous mode, and to receive and process also packets not destined to it.

All these countermeasures provide for partial security, since they defend against specific attacks only, namely Rank attacks [4,6,12,14], DODAG Version attacks [4,12], blackhole attacks [1], and attacks involving next-hop selection [8]. All these attacks can be avoided, at least in case of external adversaries, by simply using the standard RPL security mechanisms. Indeed, they impede a malicious entity to become part of the network and transmit routing control messages. In this paper, we evaluate the impact of RPL security mechanisms on the WSAN performances.

3 IPv6 Routing Protocol for Low-Power and Lossy Networks

The IPv6 Routing Protocol for Low-Power and Lossy Networks (RPL) [5,16] is an IPv6 distance-vector routing protocol focused on resource-constrained devices and lossy wireless environments. RPL assumes that the majority of the traffic is upstream, i.e., directed towards a single node acting as a border router. Downstream traffic, i.e., generated by the border router and directed towards other nodes, is considered to be sporadic, and node-to-node traffic to be rare. For this reason, RPL builds and maintains a logical topology for upstream data delivery, and downstream routes are established only when required. Specifically, the built topology is a *Destination Oriented Directed Acyclic Graph* (DODAG), in which every node has a set of neighbors (*parent set*), which are candidates for upstream data delivery. Among the nodes in the parent set, one node is selected as the *preferred parent*. The preferred parent is the node exploited for upstream data forwarding, whereas the other parents are kept as failover.

The DODAG is rooted in a single node, called *DODAG root*, to which all upstream data is directed. The DODAG root is also a border router for the other nodes. It is responsible for triggering the network formation through the emission of *DODAG Information Object* (DIO) messages. Initially, every non-root node listens for DIO messages. When a DIO is received, the node joins the network using the information included in the message. Right after having joined the network, the node starts emitting DIOs to advertise its presence and its distance to the root. During regular operations, the emission of DIO messages is regulated by the *Trickle* algorithm [9], which aims at reducing the power consumption of the nodes by minimizing the redundant messages and by adapting dynamically the transmission rate over time. The asynchronous emission of DIOs can be requested through *DODAG Information Solicitation* (DIS) messages, e.g., to accelerate the join process of a node during network formation or to recover from errors during regular network operations. All the RPL messages (DIO, DIS, etc.) are ICMPv6 messages of Type 155.

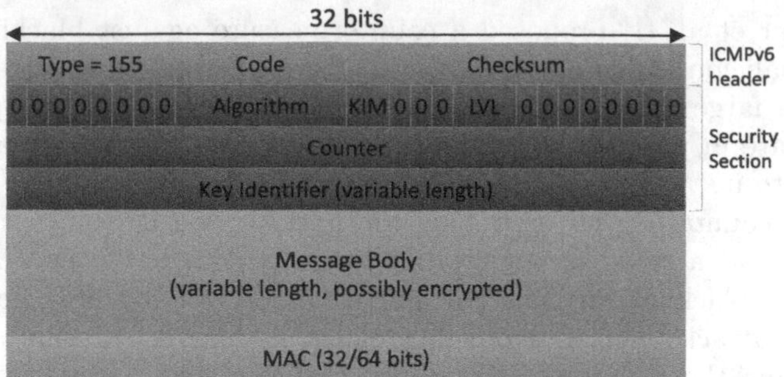

Fig. 1. Secured RPL message format.

3.1 Security Mechanisms

A DODAG can operate in one of the following security modes: unsecured mode, preinstalled mode, or authenticated mode. In the unsecured mode, the RPL messages are sent in the clear and without any security protection. In the preinstalled mode, the RPL messages are protected by cryptography-based security mechanisms using keys assumed to be already present in each node at boot time. In the authenticated mode, the RPL messages are protected in the same way, but the nodes receive keys from some key authority after having underwent an authentication process. The preinstalled and the authenticated modes differ only in the way the keys are deployed on the nodes, and they have in common all the other security mechanisms. In the scope of the present paper, we will refer to both preinstalled and authenticated modes with the general term "secured modes".

The RPL specifications define the following security services: (a) data confidentiality, (b) data authenticity, (c) replay protection. Data confidentiality assures that routing information is not disclosed to unauthorized entities. Data authenticity assures that routing control messages are not modified in transit. Replay protection assures that malicious duplicates of routing control messages are discarded.

If the DODAG operates in a secured mode, all the RPL messages are *secured.* A secured RPL message follow the general format shown in Fig. 1, in which the message body is preceded by a *Security Section.* The Code field of the ICMPv6 header determines the type of the RPL message: secured DIO, secured DIS, etc. The Algorithm field of the Security Section specifies the algorithm suite employed to authenticate and encrypt the message. With the current version of the specifications, only CCM (CTR with CBC-MAC) with AES-128 [15] is supported. CCM is a mode of operation for 128-bit block ciphers which can provide for both confidentiality and authenticity by combining the CTR (Counter) encryption mode of operation and CBC-MAC (Cipher Block Chaining Message Authentication Code). The LVL bits (Security Level) specify whether the message is only authenticated or both authenticated and encrypted, and the length of the MAC field. The Key Identifier field, whose format is specified by the

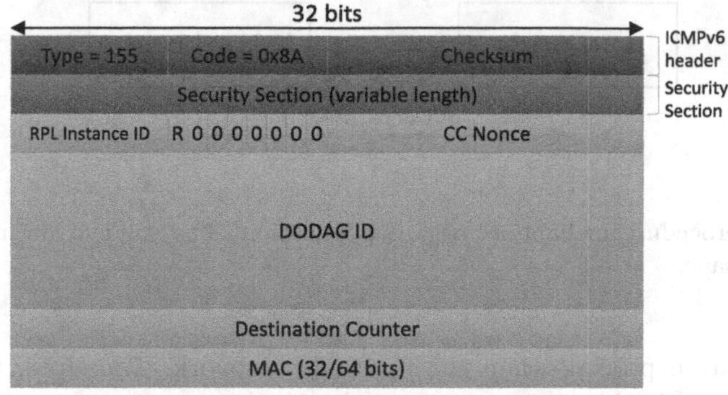

Fig. 2. Consistency Check (CC) message format.

KIM bits (Key Identifier Mode), identifies the employed cryptographic key. The Counter field is a value incremented at each sent RPL message. It is used both as input of CCM and to implement replay protection mechanisms.

To recover from situations in which a node loses its current Counter value, for example after a reboot, the RPL specifications foresee a *Consistency Check* (CC) message. The format of such a message is shown in Fig. 2. The CC message is used to inform a destination about the last valid value of its Counter field, and to issue generic challenge-response handshakes. The Destination Counter field contains the last valid value of the Counter field of the target node. The CC Nonce field is used as a proof of freshness within challenge-response handshakes, and the R bit specifies whether the message is a challenge (CC request) or a response (CC response). The RPL Instance ID and the DODAG ID fields identify the DODAG to which the sender node belongs to, in the case that multiple RPL instances or multiple DODAGs in the same RPL instance are present.

4 Our Implementation

Our standard-compliant Contiki implementation of the RPL security mechanisms is available from a public repository[1]. It extends the standard module ContikiRPL [13] and it has two possible configurations: a *light-security configuration* which implements only data confidentiality and data authenticity services; and a *full-security configuration* which also implements replay protection service.

Both configurations use a network-wide cryptographic key assumed to be already present in each node at boot time. This realizes the preinstalled security mode foreseen by the RPL specifications [16], which is of course vulnerable to key stealing through node compromise. However, the presented implementation is adaptable to the authenticated security mode, in which the nodes receive keys from some key authority. The authenticated mode will lose some performance with respect to the preinstalled one, since nodes have to undergo also

[1] https://github.com/arenantonio92/contiki.

Fig. 3. Join procedure in light-security configuration. The solid arrow represents a multicast message.

an authentication process when they join the network. Therefore, the performances presented in this paper represent also the best-case performances for the authenticated mode.

4.1 Threat Model

The light-security configuration defends against an adversary which tries to eavesdrop legitimate RPL messages to infer the topology, or forge malicious RPL messages to become part of the network and act as an internal adversary. This models a wide range of simple yet disruptive routing attacks [1,4,6,8,12,14].

The full-security configuration defends against an adversary which also tries to replay legitimate RPL messages to modify the topology. For example she can replay a legitimate DIO message originally sent by the root in a zone of the network where the root is not directly reachable. The victim nodes receiving such replayed message could think that they have a direct link with the root and could forward their upstream data along such link. Since the link does not actually exist, all the upstream communication of the victim nodes will be broken.

4.2 Light-Security Configuration

The light-security configuration simply includes a security section on each RPL message, which provides for integrity with a Message Authentication Code (MAC), and confidentiality with encryption. With such a configuration, the procedure by which a node acquires the first preferred parent (*join procedure*) is similar to that of the unsecured mode, except that DIO messages are secured. The join operation follows the sequence diagram shown in Fig. 3. The joined node sends a multicast secured DIO, which is received by the new node. The new node checks the validity of the MAC of the secured DIO, and possibly decrypts it. With the information carried by the DIO, the new node can choose the joined node as parent. This concludes the join procedure. The emission of secured DIO messages by the joined node is recurrent and regulated by the Trickle algorithm [9]. The asynchronous emission of secured DIOs can be requested through secured DIS messages.

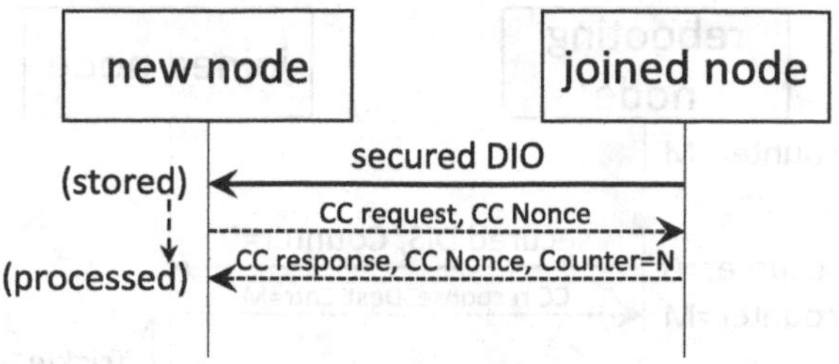

Fig. 4. Join procedure in full-security configuration. The dotted arrows represent unicast messages.

4.3 Full-Security Configuration

The full-security configuration provides for integrity and confidentiality with encryption, in the same way as the light-security configuration. In addition, it provides for replay protection by checking the Counter field of the Security Section of the incoming messages. Every node maintains a *counter watermark* for each neighbor, containing the highest Counter field received from that neighbor. Upon receiving a new message from that neighbor, if its Counter field is less than or equal to the counter watermark, then the message will be discarded. Otherwise, the message will be accepted and the counter watermark updated.

When a new node wants to join the DODAG, it does not have a counter watermark for the other nodes. Therefore, it cannot assess whether a DIO message received from them is a replay of an old message. In order to initialize the counter watermarks on the new node for the nodes already joined, we use CC requests and CC responses. The join operation follows the sequence diagram shown in Fig. 4. The joined node sends a multicast secured DIO, which is received by the new node. The new node checks the validity of the MAC of the secured DIO, and possibly decrypts it. Then, it stores the DIO into memory without processing it, waiting to assess whether it is a replay of an old message or not. The new node then initiates a CC challenge-response handshake with the joined node. In such a handshake, the new node sends a CC request carrying a fresh CC Nonce to the joined node, which answers with a CC response carrying back the same CC Nonce. The CC response also carries the current Counter N of the joined node. The fresh CC Nonce, together with the MAC protecting the whole CC response, assures the new node of the freshness of N and, retroactively, that the stored DIO message was not a replay. Upon having received the CC response, the new node checks the validity of its MAC, and possibly decrypts it. Then, the new node checks that the CC Nonce is the same of that transmitted in the CC request. Now the new node can initialize a counter watermark for the joined node to N. Finally, the new node checks if the stored DIO had a Counter field of $N - 1$, i.e., immediately precedent of the Counter field of the CC response. If it has, the DIO was not a replay, so it can be processed and the new node can choose the joined node as its parent. This concludes the join procedure. Note that, from now on, the new node has a counter watermark for its parent.

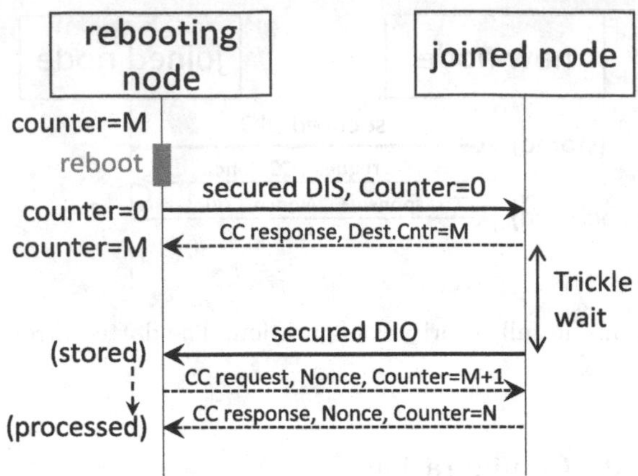

Fig. 5. Rejoin procedure in the full-security configuration.

This means that it can process future secured DIOs from such parent without repeating CC challenge-response handshakes.

In a low-power wireless network it can happen that a node reboots, as result of hardware failure, software bug or simply battery shortage. In this case, the node loses its current Counter value, and its transmitted messages will start again from a zero Counter field. As a consequence, the messages will be discarded by the neighbors as possible replays. To recover from this situation, we use a *rejoin procedure*, which follows the sequence diagram shown in Fig. 5. After rebooting, the node starts sending multicast secured DIS messages with a special Counter field of zero. If some neighbor is maintaining the value M of its past Counter and receives a secured DIS, then it sends a CC response to the rebooted node carrying a Destination Counter field of M. Then, the rest of the procedure is similar to the join procedure (see Fig. 4). After a wait imposed by the Trickle algorithm, the joined node sends a multicast secured DIO message. The rebooted node stores the DIO into memory without processing it, and initiates a CC challenge-response handshake with the joined node. After that, the rebooted node can initialize a counter watermark for the joined node to the Counter value N carried by the CC response. Finally, the new node checks if the stored DIO had a Counter field of $N - 1$. If it has, the DIO was not a replay, so it can be processed and the rebooted node can choose the joined node as its parent. This concludes the rejoin procedure.

5 Impact of Security on Performances

In order to evaluate the overhead introduced by the RPL security mechanisms, a performance evaluation based on simulations has been run. To this aim, three different RPL configurations are considered: the unsecured mode, which corresponds to the "vanilla" Contiki RPL implementation [13]; the preinstalled mode with light-security configuration; and the preinstalled mode with full-security configuration.

Simulations have been run exploiting COOJA [11], a network emulator which is available as part of the Contiki distribution. COOJA emulator provides a realistic simulation environment in which wireless nodes are emulated allowing to run the same binary image that would be executed on real nodes. In our simulations, COOJA has been configured to emulate the Zolertia-Z1 sensor mote[2], an MSP430-based board with an IEEE 802.15.4-compatible CC2420 radio chip. Wireless channel is simulated using the Unit Disk Graph Medium model, which implements a disk reception model. The reception/transmission range is set to 50 m and the interference range is set to 100 m.

In order to assess the performance of the RPL protocol in networks of different sizes, a regular grid topology with an increasing number of nodes is considered. Specifically, a 2×2 grid of 4 nodes, a 3×3 grid of 9 nodes, a 4×4 grid of 16 nodes, and a 5×5 grid of 25 nodes are considered. The distance between the nodes is fixed to 30 m in all the topologies. The node at the top-right corner is configured to behave as DODAG root. All the RPL settings are configured according to the Contiki default parameters. The radio duty cycling algorithm adopted by each node is the ContikiMAC one.

5.1 Impact on Network Formation

The additional complexity introduced by the security on the join operation influences mainly the initial network formation. The impact of the RPL security mechanisms on the network formation operation is assessed through the following metrics:

- *Network formation time*, defined as the time between the beginning of the simulation and the moment in which the last node joins the DODAG. This metric measures the time required by the network to become fully operational.
- *Power consumption*, defined as the average value of the power consumption of each node in the network. This metric includes both communication costs (for transmission and reception of messages) and computation costs (for cryptographic operations and message processing). The power consumption is evaluated using the Powertrace tool [2] and the nominal power consumption values reported in the Z1 datasheet.
- *RPL message overhead*, defined as the overall message size in bytes of all the RPL messages sent over the simulation by all the nodes in the network. This metric is adopted to assess the additional overhead in message size introduced by the security mechanisms.

Simulations are run for 60 min. In order to obtain statistically sound results, 32 independent replications with different seeds are run for each scenario. The average value of each metric with its 95%-confidence interval is reported.

[2] Zolertia Z1 website: http://zolertia.io/z1.

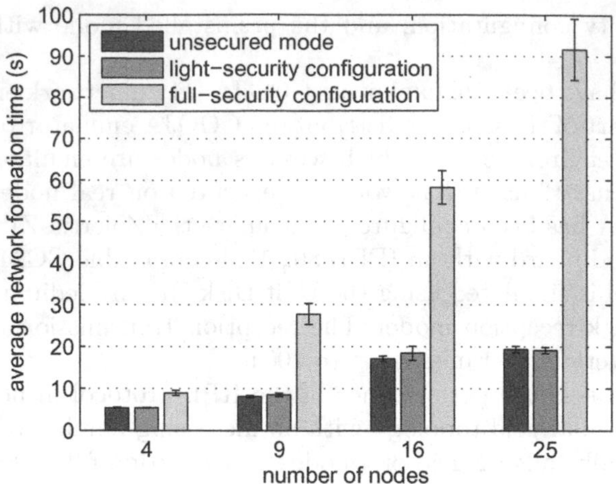

Fig. 6. Average network formation time. 95%-confidence intervals are displayed in error bars.

In Fig. 6 the average network formation time is reported. As expected, the network formation time increases with the network size. Regardless of the network size, just the usage of encrypted and authenticated RPL messages (light-security configuration) does not influence the formation time of the network. Instead, the introduction of the replay protection mechanism (full-security configuration) increases noticeably the network formation time. The overhead introduced by the replay protection increases with the network size. This can be explained considering the additional exchange of messages required before each node can join the DODAG. The network formation time with the full-security configuration can be probably reduced by employing algorithms to avoid collisions between CC challenge-response handshakes of joining nodes, e.g., with random waits. We plan to investigate this possibility in future work.

In Fig. 7 the average power consumption is reported. As expected the lowest power consumption is obtained with the unsecured mode, whereas the light-security configuration causes a slight increase in the average power consumption. This can be explained considering the computation overhead of the cryptographic operations and the increased size of the transmitted RPL messages due to the Security Section. With the full-security configuration, the power consumption increases noticeably, e.g., around 18% in the 5×5 scenario. This can be explained considering the additional number of messages exchanged during the join procedure executed by each node. As expected, results obtained with different network sizes show an increase of the power consumption when the network size grows. This is due to the larger number of messages exchanged among the nodes, which increases the overall time spent by each node in receiving and processing messages.

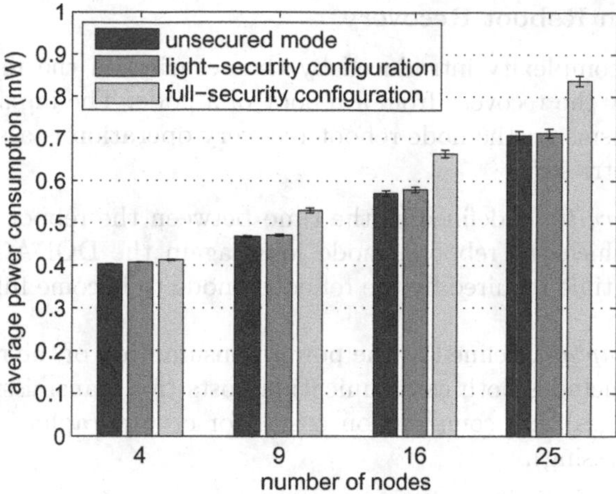

Fig. 7. Average power consumption.

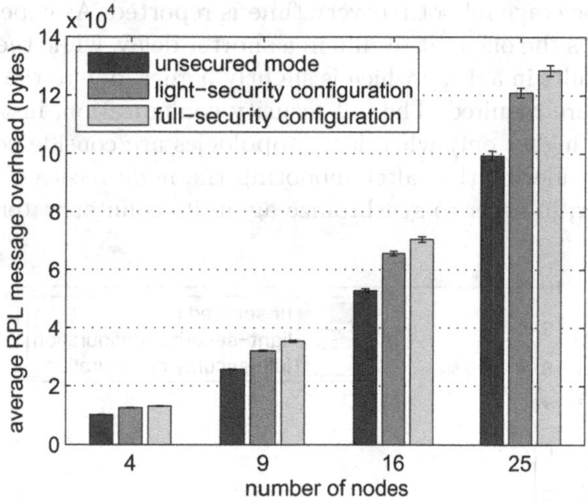

Fig. 8. Average RPL message overhead.

Figure 8 illustrates the average overhead due to RPL messages. The reported value includes the DIO and DIS messages, and also the CC messages exchanged in the full-security configuration. As expected, the lowest overhead is obtained with the unsecured mode. With the light-security configuration, instead, the overhead increases, as the presence of the Security Section increases the size of the RPL messages. With the full-security configuration, the overhead shows an additional slight increase, as additional messages are exchanged for the CC challenge-response handshakes. Coherently with the trend shown in Fig. 7, also in this case the overall overhead increases when larger networks are considered.

5.2 Impact on Reboot Recovery

The additional complexity introduced by the security on the rejoin operation influences mainly the recovery from a reboot of a node. The impact of the RPL security mechanisms on the node reboot recovery operation is assessed through the following metrics:

- *Reboot recovery time*, defined as the time between the reboot event and the moment in which the rebooted node joins again the DODAG. This metric measures the time required by the rebooted node to become fully operational again.
- *Power consumption*, defined as the power consumption of the rebooted node. This metric includes both communication costs (for transmission and reception of messages) and computation costs (for cryptographic operations and message processing).

For each simulation run, a non-root node is randomly selected to reboot after 10 min of simulation. In order to obtain statistically sound results, 32 independent replications with different seeds are run for each scenario.

In Fig. 9 the average reboot recovery time is reported. As expected, the RPL unsecured mode is the one that results in a shorter delay, while the light-security configuration results in a delay which is slightly higher, due to the cryptographic operations that are required. The full-security configuration, instead, results in a delay which is higher only when large topologies are considered. This can be explained by considering that after rebooting the node receives a CC response for each neighbor, in order to synchronize again its counter watermark.

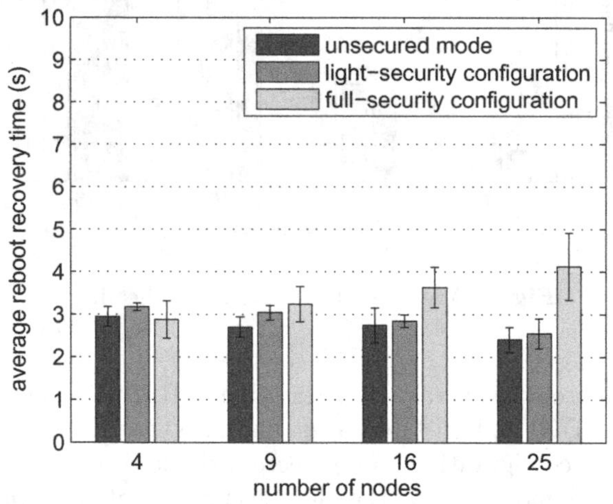

Fig. 9. Average reboot recovery time.

Finally, in Fig. 10 the average power consumption is reported. As expected, the full-security configuration increases significantly the energy consumption in

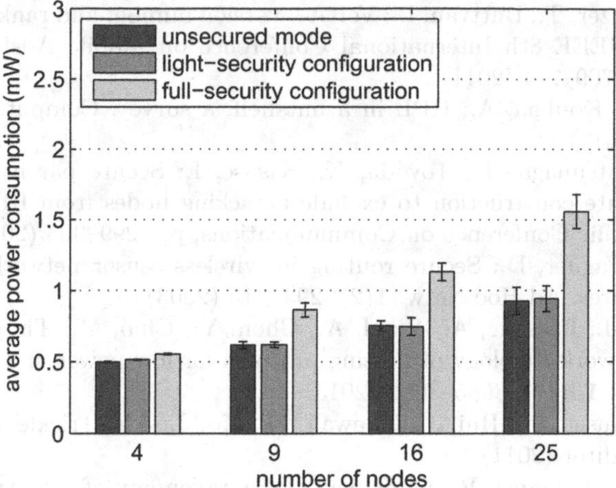

Fig. 10. Average power consumption.

large topologies, as it requires additional CC messages to be exchanged with each neighbor.

6 Conclusion

In this paper, we gave a first evaluation of the impact of the RPL security mechanisms on the WSAN performances. We developed a standard-compliant implementation of the RPL security mechanisms for the Contiki operating system [3], and we evaluated it by means of simulations. We showed that the RPL security mechanisms have a negligible impact on the performances in terms of network formation time and power consumption, if they do not have to defend against replay-based attacks. Otherwise, if also replay protection is needed, the impact on performances is more pronounced, especially on the network formation time which increases noticeably. To the best of our knowledge, this is the first implementation of the standard security mechanisms of RPL.

References

1. Airehrour, D., Gutierrez, J., Ray, S.K.: Securing RPL routing protocol from blackhole attacks using a trust-based mechanism. In: 26th International Telecommunication Networks and Applications Conference, pp. 115–120 (2016)
2. Dunkels, A., Eriksson, J., Finne, N., Tsiftes, N.: Powertrace: network-level power profiling for low-power wireless networks. Technical Report, Swedish Institute of Computer Science (2011)
3. Dunkels, A., Gronvall, B., Voigt, T.: Contiki—a lightweight and flexible operating system for tiny networked sensors. In: 29th Annual IEEE International Conference on Local Computer Networks, pp. 455–462 (2004)

4. Dvir, A., Holczer, T., Buttyan, L.: VeRA—version number and rank authentication in RPL. In: IEEE 8th International Conference on Mobile Ad-Hoc and Sensor Systems, pp. 709–714 (2011)
5. Gaddour, O., Koubâa, A.: RPL in a nutshell: a survey. Comput. Netw. **56**(14), 3163–3178 (2012)
6. Iuchi, K., Matsunaga, T., Toyoda, K., Sasase, I.: Secure parent node selection scheme in route construction to exclude attacking nodes from RPL network. In: 21st Asia-Pacific Conference on Communications, pp. 299–303 (2015)
7. Karlof, C., Wagner, D.: Secure routing in wireless sensor networks: attacks and countermeasures. Ad Hoc Netw. **1**(2), 293–315 (2003)
8. Le, A., Loo, J., Lasebae, A., Vinel, A., Chen, Y., Chai, M.: The impact of rank attack on network topology of routing protocol for low-power and lossy networks. IEEE Sens. J. **13**(10), 3685–3692 (2013)
9. Levis, P., Clausen, T., Hui, J., Gnawali, O., Ko, J.: The Trickle algorithm. RFC 6206, RFC Editor (2011)
10. Mayzaud, A., Badonnel, R., Chrisment, I.: A taxonomy of attacks in RPL-based Internet of Things. Int. J. Netw. Secur. **18**(3), 459–473 (2016)
11. Osterlind, F., Dunkels, A., Eriksson, J., Finne, N., Voigt, T.: Cross-level sensor network simulation with COOJA. In: 31st IEEE Conference on Local Computer Networks, pp. 641–648 (2006)
12. Perrey, H., Landsmann, M., Ugus, O., Wählisch, M., Schmidt, T.: TRAIL: topology authentication in RPL. In: International Conference on Embedded Wireless Systems and Networks, pp. 59–64 (2016)
13. Tsiftes, N., Eriksson, J., Dunkels, A.: Low-power wireless IPv6 routing with ContikiRPL. In: 9th ACM/IEEE International Conference on Information Processing in Sensor Networks, pp. 406–407 (2010)
14. Weekly, K., Pister, K.: Evaluating sinkhole defense techniques in RPL networks. In: IEEE 20th International Conference on Network Protocols, pp. 1–6 (2012)
15. Whiting, D., Housley, R., Ferguson, N.: Counter with CBC-MAC (CCM). RFC 3610, RFC Editor (2003)
16. Winter, T.: RPL: IPv6 routing protocol for low-power and Lossy Networks. RFC 6550, RFC Editor (2012)

A Ticket-Based Authentication Scheme for VANETs Preserving Privacy

Ons Chikhaoui, Aida Ben Chehida, Ryma Abassi(✉), and Sihem Guemara El Fatmi

SUP'Com, University of Carthage, Tunis, Tunisia
abassi.ryma@gmail.com

Abstract. In Vehicular Ad hoc NETworks (VANETs), vehicles exchange safety messages in order to enhance road safety. Because of their critical role, safety messages should be authenticated before being accepted while preserving the privacy of vehicles. However, misbehaving vehicles should be traced by legal authorities and evicted from the network. In this paper, we propose a new conditional privacy-preserving authentication scheme for VANETs. The proposed scheme is based on the use of temporary tickets to maintain the privacy of vehicles. An identity-based signature technique is employed for authentication. The trusted authority can trace misbehaving vehicles, given their tickets, and RSUs handle the task of evicting them from the network. An in-depth security analysis is performed to demonstrate the efficiency of our proposal.

Keywords: VANETs · Tickets · Conditional privacy preservation · Authentication · Security

1 Introduction

Vehicular Ad hoc NETworks (VANETs) are a sub-category of Mobile Ad hoc NETworks (MANETs) except that two types of nodes can interact: mobile nodes consisting of vehicles equipped with On-Board Units (OBUs) and fixed nodes consisting of infrastructure deployed alongside the roads called Road Side Units (RSUs).

Two types of communications can be used in VANETs: the Vehicular to Vehicular (V2V) communication in which vehicles communicate with each other directly, and the Vehicular to Infrastructure (V2I) communication in which vehicles communicate with RSUs. These two types of communications are conducted based on the Dedicated Short Range Communications (DSRC) [1] protocol.

The main purpose of VANETs is to enhance the safety in roads by exchanging safety related messages. These messages must be authenticated before being taken into consideration. In fact, as safety related messages have a direct influence on people's life, any successful attack can have devastating effects. Moreover, any messages authentication technique should also preserve the privacy of sending vehicles (i.e., their real identities should be protected from illegal disclosure). Privacy protects drivers' identities from multiple threats such as illegal tracking, related communications' eavesdropping, etc. However, the privacy preserving authentication technique should not prevent legal authorities from recovering the real identity of a misbehaving vehicle.

A. Puliafito et al. (Eds.): ADHOC-NOW 2017, LNCS 10517, pp. 77–91, 2017.
DOI: 10.1007/978-3-319-67910-5_7

The main contribution in this paper is then the proposition of a new conditional privacy preserving authentication scheme based on temporary tickets for VANETs. A ticket is formed through two stages. The first one is the offline stage in which vehicles obtain credentials and corresponding private keys from the Trusted Authority (TA). The second stage is the online stage in which vehicles request to sign their credentials and form their tickets from these latter. A vehicle should update its current ticket whenever it enters into a new domain (a domain contains few RSUs) and it should change its current ticket whenever the validity period of the ticket expires. Vehicles use the obtained tickets to anonymously communicate with each other. The identity-based signature is used for authentication between vehicles and RSUs and among vehicles. When a vehicle misbehaves, the TA recovers its real identity, decides its penalty according to the gravity of its misbehavior and includes its credential(s) into revocation list(s). In the beginning of each time slot, the related revocation list is sent to RSUs. These latter exclude misbehaving vehicles by not signing credentials belonging to received revocation lists.

An in-depth security analysis of our proposed scheme was also achieved proving the efficiency of our proposal in terms of authentication, non-repudiation, identity privacy preservation, short-term linkability, long-term unlinkability, traceability, identity revocation and defense against several types of attacks.

The remainder of this paper is organized as follows: in Sect. 2, some related works are reviewed. In Sect. 3, the scheme overview is presented. Section 4 describes the proposed scheme. In Sect. 5, a security analysis of this scheme is achieved. Finally, Sect. 6 concludes the paper.

2 Related Work

In this section, we present some existing schemes that were conducted to achieve anonymous authentication in VANETs. According to the method used in the work, we can categorize them into three classes: works based on asymmetric key cryptography, works based on symmetric key cryptography and works based on hybrid methods (i.e., combine asymmetric key cryptography with symmetric key cryptography).

Works based on asymmetric key cryptography: Fan et al. [2] used blind signature technique to deal with anonymous authentication in VANETs. RSUs blindly sign the safety messages of vehicles that exist in their coverage areas. However, a receiving vehicle must be in the coverage area of the signing RSU to be able to verify the validity of a received signature. In [3], vehicles use autonomously generated pseudonyms instead of their real identities to maintain their privacy. An identity-based signature method using bilinear pairing is utilized. A proxy vehicle verifies, in batch, the validity of signatures on messages of other vehicles. RSU checks, in batch, the verification results of proxy vehicles. A drawback of this scheme is its vulnerability to the Sybil attack as vehicles self-generate their own pseudonyms. In addition, a concern about batch verification technology is the fact that the batch verification fails if at least one invalid signature exists in the verified batch. In reality, it is infeasible to assume that all signatures are valid. Adversaries can attempt to negate the advantages of batch verification by polluting signatures within a batch [4]. In [5], vehicles also use self-generated pseudonyms to anonymously communicate with other vehicles and RSUs. The scheme uses an identity-based signature method without bilinear pairing to decrease the computational complexity of the bilinear pairing function. The batch verification of signatures is also possible. However, the scheme is vulnerable to the Sybil attack because of the self-generation of pseudonyms by vehicles. It is also vulnerable to the Global Positioning System (GPS) spoofing attack as no information is provided to prove the trustworthiness of a position supplied by a vehicle. The concern about bad batches exists as well. In [6], attribute based signature is used to achieve anonymous authentication in VANETs. A vehicle can generate a valid signature on a message if and only if its set of attributes satisfies the signing predicate of that message. The scheme is vulnerable to the Sybil attack as when a vehicle receives the same message multiple times, it cannot verify if the message was sent by different vehicles or repeatedly transmitted by the same vehicle. In [7], group signature is used for anonymous authentication among vehicles. Vehicles within the communication range of a same RSU form a group. Each group member receives a group certificate from its current RSU. The scheme presented a method to thwart the Sybil attack and hence to enable a receiving vehicle to accept a message only if it was confirmed by a certain threshold number of vehicles. The batch verification of group signatures is possible. However, a vehicle must provide its real identity to RSUs in order to get its group certificates. Besides, a misbehaving vehicle is only denied from obtaining new group certificates; however, its present group certificate could still be valid. The concern about bad batches is also existent.

Works based on symmetric key cryptography: In [8], vehicles of a same platoon generate a symmetric group key to anonymously authenticate each other. Although this scheme reduces the computation overhead by using symmetric key cryptography instead of asymmetric key cryptography, it is vulnerable to the repudiation attack as symmetric key cryptography does not satisfy the non-repudiation requirement.

Works based on hybrid methods: In [9], vehicles use ring signatures to anonymously establish symmetric keys between each other. This scheme is vulnerable to the repudiation attack as symmetric key cryptography does not provide non-repudiation.

To the best of our knowledge, there is no existing work satisfying several security requirements at the same time. As for our proposal, it satisfies message authentication, non-repudiation, identity privacy preservation, short-term linkability, long-term unlinkability, traceability, identity revocation and it resists the identity resolution attack, the impersonation attack, the Sybil attack, the modification attack, the GPS spoofing attack, the replay attack and the repudiation attack.

3 Scheme Overview

In this section, we present the network model, the used assumptions, the security goals and the preliminaries of the proposed scheme.

3.1 Network Modeling

As depicted in Fig. 1, the network modeling in our scheme is based on three entities:

The Trusted Authority. TA divides its territory into different domains containing a given number of RSUs and assigns to each domain a sequence number. Moreover, the TA divides the time domain into equal serial time slots, registers vehicles and provides them with credentials and their corresponding private keys. With each set of credentials and private keys, the TA supplies the related vehicle with a public key certificate containing a vehicle pseudonym that can be used latter to request, from the TA, a new set of credentials and the corresponding private keys. Furthermore, the TA registers RSUs and provides each one with an identity and a corresponding private key. Finally, the TA is in charge of resolving disputes, recovering the real identity of a misbehaving vehicle and deciding its penalty.

RSUs. They communicate with the TA via wired links and communicate with vehicles (OBUs) via wireless links by using the DSRC protocol. Moreover, they receive the revocation lists from the TA and verify the revocation status of vehicles as well as signing credentials of legitimate ones in order to relieve vehicles of downloading and checking the revocation lists.

Vehicles. Vehicles are mobile nodes equipped with OBUs enabling them to wirelessly communicate with other vehicles and RSUs based on the DSRC protocol and use Tamper Proof Devices (TPDs) to store their sensitive information. Moreover, they request, from the TA, for credentials and the corresponding private keys whereas they request, from RSUs, to sign their credentials. Finally, they form their tickets from their signed credentials and use them to anonymously communicate with each other. In fact, vehicles exchange safety messages, authenticate received ones and report, to the TA, any misbehaving behavior.

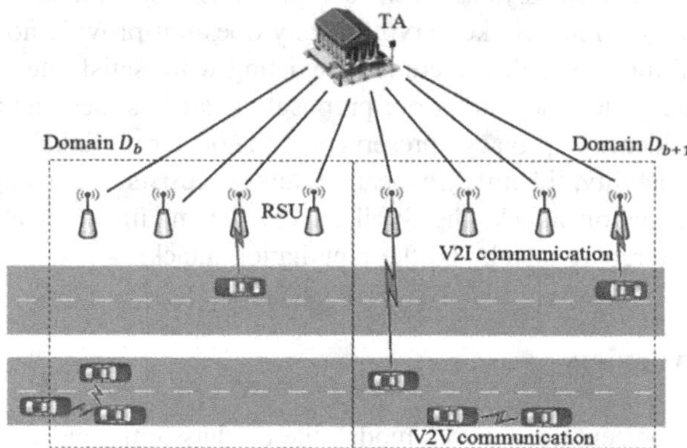

Fig. 1. Considered network modelling

3.2 Scheme Assumptions

Our scheme is based on the following assumptions:

(1) The TA is totally trusted by all the vehicles and the RSUs and cannot be compromised. The TA has high storage and computation capabilities.
(2) RSUs are densely deployed along the roads.
(3) Before the installation of each registered RSU, the TA assigns it with a public key certificate (that contains an RSU's unique identifier) and the corresponding private key. This certificate will be used for the secure communication between the TA and the related RSU.
(4) The division of the time domain is known by all RSUs in the network. Before the beginning of each new time slot, the TA broadcasts a reminder of the new time slot (i.e., the start and end time).
(5) Each vehicle is equipped with a clock used for time indicating and checking.
(6) The network can provide time synchronization and the Global Time (i.e., Greenwich Means Time: GMT) is used.

3.3 Security Goals

The following security goals are achieved:

- **Message authentication**: the receiver of a message should be able to verify the integrity of this message as well as the legitimacy of its origin.
- **Non-repudiation**: the sender of a message should not be able to deny having sent that message.
- **Identity privacy preservation**: the TA should be the only one able to disclose the real identity of a vehicle.
- **Short-term linkability**: when a same vehicle sends two or more safety messages in the same time slot, the receiver should be able to verify that these messages are generated by the same vehicle.
- **Long-term unlinkability**: apart from the TA, none should be able to link the relationship among two or more different tickets of the same vehicle.
- **Traceability**: the TA should be able to recover the real identity of a misbehaving vehicle from its tickets.
- **Identity revocation**: misbehaving vehicles should be evicted from the network.
- **Defense against several types of attacks**: the scheme should be able to resist multiple attacks such as the identity resolution attack, the impersonation attack, the Sybil attack, the modification attack, the GPS spoofing attack, the replay attack and the repudiation attack.

3.4 Preliminaries

We adopted the identity-based signature technique for its computation and communication efficiency. In identity-based signature schemes, the identity of an entity serves as the public key of that entity. The related private key is extracted from the identity by a trusted third party, called Private Key Generator (PKG). The entity uses the extracted private key to sign its outgoing messages. Receivers use the identity of the sender to verify the validity of received signatures.

In our scheme, we decided to use the identity-based signature scheme proposed in [10], called BNN-IBS scheme. This scheme possesses the following properties:

(1) It is based on elliptic curve cryptography.
(2) It does not use the time-consuming and expensive bilinear pairing and map-to-point hash functions.
(3) It has been proved to be existentially unforgeable against the chosen message and ID attacks (i.e., euf-cma-ida secure) in [10] under the discrete logarithm problem.

It is worth noting that an important advantage of our proposal is its re-usability, i.e., it can also be reutilized with other new identity-based signature schemes for security and performance improvements. For the description of the BNN-IBS scheme, we take the one provided in [11], (*Remark*: in the Key Extract algorithm, we have $s_u = r_u + c_u s$).

4 The Proposed Scheme

Our proposal is built upon five phases: network initialization phase, authentication phase, signature generation and verification phase, traceability phase and revocation phase. The notations used in our scheme are listed in Table 1.

4.1 Network Initialization Phase

TA plays the role of PKG and initializes the network by performing the following steps:

(1) The TA sets up the network parameters by performing the Setup algorithm of BNN-IBS.
(2) The TA publishes the network parameters $\{E/F_q, G, P, q, p, P_{TA}, H_1, H_2\}$ and keeps s secret.
(3) The TA divides its territory into different domains according to the number of RSUs and the direction of the road.
(4) The TA assigns to each domain a sequence number.
(5) The TA sets the identity of the RSU as its geographical coordinates concatenated with the sequence number of the domain to which it belongs. The identity of $RSU_{a,b}$ is: $ID_{RSU_{a,b}} = GC_{RSU_{a,b}} \| D_b$.
(6) The TA extracts the private key $s_{RSU_{a,b}}$ that corresponds to $ID_{RSU_{a,b}}$ by performing the Key Extract algorithm of BNN-IBS.
(7) The TA securely sends $<ID_{RSU_{a,b}}, R_{RSU_{a,b}}, s_{RSU_{a,b}}>$ to $RSU_{a,b}$. $RSU_{a,b}$ verifies the validity of its private key $s_{RSU_{a,b}}$ by checking whether $R_{RSU_{a,b}} + c_{RSU_{a,b}} P_{TA} = s_{RSU_{a,b}} P$ holds. The demonstration is provided in [12], (*Remark:* see [11] for the definition of $R_{RSU_{a,b}}$ and $c_{RSU_{a,b}}$).
(8) The TA divides the time domain into equal serial time slots. The length of each time slot is ΔT. Hence, one CP contains w time slots.

Table 1. Notations

Notation	Meaning
TA	The trusted authority
E/F_q	An elliptic curve E over a finite field F_q
q	The field size
p	A large prime number
P	A point of order p on the curve E
G	A cyclic group of order p under the point addition "+" generated by P
s/P_{TA}	The secret/public key of **TA**
H_i	Secure and collision resistance one-way hash function, i = **1, 2**
D_b	The bth domain
$RSU_{a,b}$	The ath RSU of D_b
$ID_{RSU_{a,b}}$	The identity of $RSU_{a,b}$
$GC_{RSU_{a,b}}$	The geographical coordinates of $RSU_{a,b}$
$s_{RSU_{a,b}}$	The private key of $RSU_{a,b}$ that corresponds to $ID_{RSU_{a,b}}$
ΔT	The length of a time slot
CP	A certain period chosen by **TA**
TS_x	The xth time slot
V_i	The ith vehicle
RID_{V_i}	The real identity of V_i
$Cred_{V_i,TS_x}$	The credential of V_i for TS_x
s_{V_i,TS_x}	The private key of V_i that corresponds to $Cred_{V_i,TS_x}$
TK_{V_i,TS_x}	The ticket of V_i in TS_x
PID_{V_i}	A pseudo-identity of V_i
$Cert_{V_i}$	A certificate of V_i that includes PID_{V_i}
$priv_{V_i}$	The private key of V_i that corresponds to $Cert_{V_i}$
$\sigma_{RSU_{a,b}}(\alpha)$	The signature on message a for $ID_{RSU_{a,b}}$
$\sigma_{V_i,TS_x}(\alpha)$	The signature on message a for $Cred_{V_i,TS_x}$
RL_{TS_x}	The revocation list that corresponds to TS_x
m	A safety message content
M_{V_i,TS_x}	A safety message of V_i in TS_x
$\|$	Message concatenation

(9) Each vehicle must register itself in the TA to request for its credentials required for tickets generation. The vehicle provides its real identity, which is its serial number, to the TA. The TA generates for the vehicle one credential for each time slot of one *CP*. Each credential contains two fields. The first field is the hash value, using a secure hash function such as SHA-2 [13], of the timestamp of the generation of the credential concatenated with a nonce. The nonce is added in order to more secure the scheme against the collision attack and the preimage attack on the hash function. The second field is the validity period of the credential which corresponds to the boundaries of the interval of the time slot to which the credential is dedicated. An example of a credential dedicated to the time slot TS_x is as follows:

H(timestamp ‖ nonce)	[time of start of TS_x, time of end of TS_x]

For a registered vehicle V_i with a set of credentials that starts from the time slot, TS_x, the set of credentials of V_i is $\{Cred_{V_i,TS_x}, Cred_{V_i,TS_{x+1}}, \ldots, Cred_{V_i,TS_{x+w-1}}\}$.

(10) The TA extracts the private key that corresponds to each one of V_i's credentials. For each credential $Cred_{V_i,TS_{x+k}}$ (where $0 \leq k \leq w-1$), the TA extracts the corresponding private key $s_{V_i,TS_{x+k}}$ by performing the Key Extract algorithm of BNN-IBS.

(11) The TA also generates, for V_i, a public key certificate $Cert_{V_i}$ and the corresponding private key $priv_{V_i}$. $Cert_{V_i}$ contains a pseudo-identity PID_{V_i} of V_i. PID_{V_i} is the hash value of the timestamp of the generation of $Cert_{V_i}$ concatenated with a nonce. $Cert_{V_i}$ also includes a validity period that corresponds to the boundaries of the interval of the *CP* to which the set of V_i's credentials is dedicated. V_i will use $Cert_{V_i}$ to securely request and obtain, from TA, a new set of credentials and the corresponding private keys before the expiration of the last credential of its current set.

(12) The TA retains the mapping between RID_{V_i}, $Cert_{V_i}$ and all the credentials $Cred_{V_i,TS_{x+k}}$, (*Remark*: The TA classifies the retained credentials, of the registered vehicles, according to their corresponding time slots in order to facilitate traceability of misbehaving vehicles later on).

(13) The TA securely sends, to V_i, <$Cred_{V_i,TS_{x+k}}$, $R_{V_i,TS_{x+k}}$, $s_{V_i,TS_{x+k}}$> and <$Cert_{V_i}$, $priv_{V_i}$>. V_i can verify the validity of its private key $s_{V_i,TS_{x+k}}$ by checking whether $R_{V_i,TS_{x+k}} + c_{V_i,TS_{x+k}} P_{TA} = s_{V_i,TS_{x+k}} P$ holds.

4.2 Authentication Phase

The mutual authentication between vehicles and RSUs should occur in these two cases:

Case 1: whenever a vehicle enters into a new domain, it should update its current ticket.
Case 2: whenever the current ticket of a vehicle expires (i.e., the ticket's corresponding time slot ends), so the vehicle should change its current ticket.

The mutual Authentication in Case 1.

(1) Each $RSU_{a,b}$ must periodically announce itself to the vehicles as follows:

- $RSU_{a,b}$ selects a timestamp t used for freshness and generates $\sigma_{RSU_{a,b}}(t)$ by performing the Sign algorithm of BNN-IBS.
- $RSU_{a,b}$ broadcasts <$ID_{RSU_{a,b}}$, t, $\sigma_{RSU_{a,b}}(t)$> within its coverage area.

(2) Once a vehicle V_i receives <$ID_{RSU_{a,b}}$, t, $\sigma_{RSU_{a,b}}(t)$>, it performs the following steps:

- V_i checks t.
- If t is fresh then V_i verifies $GC_{RSU_{a,b}}$ in $ID_{RSU_{a,b}}$ by using the GPS. Else V_i drops the message and exits.

- If $GC_{RSU_{a,b}}$ in $ID_{RSU_{a,b}}$ are correct then V_i verifies D_b in $ID_{RSU_{a,b}}$. Else V_i drops the message and exits.
- If D_b is a new domain for V_i then V_i verifies the validity of $\sigma_{RSU_{a,b}}(t)$ by performing the Verify algorithm of BNN-IBS. Else V_i drops the message and exits.
- If $\sigma_{RSU_{a,b}}(t)$ is valid then V_i uses its tuple $<Cred_{V_i,TS_{x+k}}, R_{V_i,TS_{x+k}}, s_{V_i,TS_{x+k}}>$ that corresponds to the current time slot TS_{x+k} in order to authenticate itself to $RSU_{a,b}$: V_i selects a timestamp t used for freshness and generates $\sigma_{V_i,TS_{x+k}}(t)$ by performing the Sign algorithm of BNN-IBS. Then, V_i sends $<Cred_{V_i,TS_{x+k}}, t, \sigma_{V_i,TS_{x+k}}(t)>$ to $RSU_{a,b}$. Else V_i drops the message and exits.

(3) Once $RSU_{a,b}$ receives $<Cred_{V_i,TS_{x+k}}, t, \sigma_{V_i,TS_{x+k}}(t)>$, it performs the following steps:

- $RSU_{a,b}$ checks t.
- If t is fresh then $RSU_{a,b}$ checks if the validity period indicated in $Cred_{V_i,TS_{x+k}}$ corresponds to the current time slot (i.e., TS_{x+k}). Else $RSU_{a,b}$ drops the message and exits.
- If the indicated validity period corresponds to the current time slot then $RSU_{a,b}$ checks $Cred_{V_i,TS_{x+k}}$ against $RL_{TS_{x+k}}$. Else $RSU_{a,b}$ drops the message and exits.
- If $Cred_{V_i,TS_{x+k}}$ is not in $RL_{TS_{x+k}}$ then $RSU_{a,b}$ verifies the validity of $\sigma_{V_i,TS_{x+k}}(t)$ by performing the Verify algorithm of BNN-IBS. Else $RSU_{a,b}$ drops the message and exits.
- If $\sigma_{V_i,TS_{x+k}}(t)$ is valid then $RSU_{a,b}$ generates $\sigma_{RSU_{a,b}}(Cred_{V_i,TS_{x+k}})$ by performing the Sign algorithm of BNN-IBS. Else $RSU_{a,b}$ drops the message and exits.
- $RSU_{a,b}$ sends $<ID_{RSU_{a,b}}, Cred_{V_i,TS_{x+k}}, \sigma_{RSU_{a,b}}(Cred_{V_i,TS_{x+k}})>$ to V_i.

(4) Once V_i receives $<ID_{RSU_{a,b}}, Cred_{V_i,TS_{x+k}}, \sigma_{RSU_{a,b}}(Cred_{V_i,TS_{x+k}})>$, it performs the following steps:

- V_i verifies the validity of $\sigma_{RSU_{a,b}}(Cred_{V_i,TS_{x+k}})$ by performing the Verify algorithm of BNN-IBS.
- If $\sigma_{RSU_{a,b}}(Cred_{V_i,TS_{x+k}})$ is valid then V_i sets its ticket $TK_{V_i,TS_{x+k}}$ as the concatenation of $ID_{RSU_{a,b}}$, $Cred_{V_i,TS_{x+k}}$ and $\sigma_{RSU_{a,b}}(Cred_{V_i,TS_{x+k}})$. Else V_i drops the message and exits.

Supposing that V_i was in D_{b-1}, its ticket $TK_{V_i,TS_{x+k}}$ will be updated from

$ID_{RSU_{f,b-1}}$	$Cred_{V_i,TS_{x+k}}$	$\sigma_{RSU_{f,b-1}}(Cred_{V_i,TS_{x+k}})$

To

$ID_{RSU_{a,b}}$	$Cred_{V_i,TS_{x+k}}$	$\sigma_{RSU_{a,b}}(Cred_{V_i,TS_{x+k}})$

The mutual Authentication in Case 2.
In this case, the same steps are performed as in **case 1** except that V_i omits the verification of D_b in $ID_{RSU_{a,b}}$, and it directly moves from verifying $GC_{RSU_{a,b}}$ in $ID_{RSU_{a,b}}$ to verifying the validity of $\sigma_{RSU_{a,b}}(t)$.

Supposing that V_i was using, in TS_{x+k-1}, the ticket $TK_{V_i,TS_{x+k-1}}$:

$ID_{RSU_{d,g}}$	$Cred_{V_i,TS_{x+k-1}}$	$\sigma_{RSU_{d,g}}(Cred_{V_i,TS_{x+k-1}})$

The new ticket $TK_{V_i,TS_{x+k}}$ of V_i in TS_{x+k} is:

$ID_{RSU_{a,b}}$	$Cred_{V_i,TS_{x+k}}$	$\sigma_{RSU_{a,b}}(Cred_{V_i,TS_{x+k}})$

4.3 Signature Generation and Verification Phase

In this phase, vehicles sign their outgoing safety messages and authenticate received ones (i.e., safety messages) as follows:

(1) After forming $TK_{V_i,TS_{x+k}}$, V_i uses <$Cred_{V_i,TS_{x+k}}$, $R_{V_i,TS_{x+k}}$, $s_{V_i,TS_{x+k}}$> to sign its safety messages, in TS_{x+k}, by performing the following steps:

- V_i selects a timestamp t used for freshness, calculates $M_{V_i,TS_{x+k}} = (m\|t)$ and generates $\sigma_{V_i,TS_{x+k}}(M_{V_i,TS_{x+k}})$ by performing the Sign algorithm of BNN-IBS.

(2) V_i sends <$TK_{V_i,TS_{x+k}}$, $M_{V_i,TS_{x+k}}$, $\sigma_{V_i,TS_{x+k}}(M_{V_i,TS_{x+k}})$> to other vehicles.

(3) Once a receiver V_j receives <$TK_{V_i,TS_{x+k}}$, $M_{V_i,TS_{x+k}}$, $\sigma_{V_i,TS_{x+k}}(M_{V_i,TS_{x+k}})$>, it should perform the following steps in order to authenticate $M_{V_i,TS_{x+k}}$:

- V_j checks t.
- If t is fresh then V_j checks the validity period indicated in $TK_{V_i,TS_{x+k}}$. Else V_j drops the message and exits.
- If $TK_{V_i,TS_{x+k}}$ is still valid then V_j verifies the validity of $\sigma_{RSU_{a,b}}(Cred_{V_i,TS_{x+k}})$ by performing the Verify algorithm of BNN-IBS, (*Remark*: V_j gets $ID_{RSU_{a,b}}$ from $TK_{V_i,TS_{x+k}}$). Else V_j drops the message and exits.
- If $\sigma_{RSU_{a,b}}(Cred_{V_i,TS_{x+k}})$ is valid then V_j verifies the validity of $\sigma_{V_i,TS_{x+k}}(M_{V_i,TS_{x+k}})$ by performing the Verify algorithm of BNN-IBS, (*Remark*: V_j gets $Cred_{V_i,TS_{x+k}}$ from $TK_{V_i,TS_{x+k}}$). Else V_j drops the message and exits.
- If $\sigma_{V_i,TS_{x+k}}(M_{V_i,TS_{x+k}})$ is valid then V_j accepts the message. Else V_j drops the message and exits.

4.4 Traceability Phase

In cases of misbehaviors (i.e., a misbehaving vehicle V_i sends a bogus and misleading message $M_{V_i,TS_{x+k}}$ to other vehicles), the TA recovers the real identity of V_i as follows:

(1) The TA receives from vehicles reports that contain <$TK_{V_i,TS_{x+k}}$, $M_{V_i,TS_{x+k}}$, $\sigma_{V_i,TS_{x+k}}(M_{V_i,TS_{x+k}})$>.

(2) The TA investigates the event.
(3) The TA retrieves $Cred_{V_i,TS_{x+k}}$ from $TK_{V_i,TS_{x+k}}$.
(4) The TA determines TS_{x+k} from $Cred_{V_i,TS_{x+k}}$.
(5) The TA scans the retained credentials, that correspond to TS_{x+k}, of all the registered vehicles until finding $Cred_{V_i,TS_{x+k}}$.
(6) The TA recovers RID_{V_i} from the retained mapping between the real identity of each registered vehicle and all its credentials.

4.5 Revocation Phase

According to the level of gravity of the misbehavior, the TA decides whether to temporarily or permanently revoke V_i.

In the temporary Revocation

(1) The TA chooses the number of time slots of the revocation.
(2) The TA includes the credential(s) of V_i that correspond(s) to the time slot(s) of the revocation in the related revocation list(s).

In the permanent Revocation

(1) The TA includes all the remaining credentials of V_i in the revocation lists that correspond to their time slots.
(2) The TA will also deny V_i from getting new credentials in the next credentials' refill.

The revocation process is as follows:

(1) In the beginning of each time slot, the TA broadcasts to the RSUs in the network, the revocation list that corresponds to that time slot.
(2) When V_i requests for a signed credential while this credential is included in the revocation list, its request will be denied by all RSUs.

5 Security Analysis

In this section, a security analysis of our scheme is provided.

5.1 Message Authentication

Each node (an RSU/a vehicle) has to sign its outgoing messages. The validity of the identity-based signature of an RSU $RSU_{a,b}/a$ vehicle V_i on a message ensures to a receiver that the signing RSU $RSU_{a,b}$/vehicle V_i, using an identity ($ID_{RSU_{a,b}}$)/a credential ($Cred_{V_i,TS_{x+k}}$), has the private key that corresponds to $ID_{RSU_{a,b}}/Cred_{V_i,TS_{x+k}}$. When dealing with safety messages, a receiving vehicle V_j has also to verify the validity of the identity-based signature of an RSU on the credential of the sending vehicle in order to check the legitimacy of this latter since RSUs sign only credentials that are not included in revocation lists. For the integrity requirement, the validity of an identity-based signature on a message guarantees to a receiver the integrity of the signed message.

5.2 Non-repudiation

In our scheme, identity-based signature is used in order to fulfil non-repudiation.

5.3 Identity Privacy Preservation

Vehicles use credentials to authenticate themselves to RSUs. A credential is stripped of any identifying information and it is regularly changed with each new time slot (i.e., whenever its validity period expires). Vehicles also use tickets to send their safety messages. A ticket is formed by appending, to a verified credential, the identity of the verifying RSU and its identity-based signature on the credential. Hence, the obtained ticket does not include the real identity of its owner vehicle. Besides, a ticket is regularly changed with the expiration of its related credential.

5.4 Short-Term Linkability

A required property in some applications of VANETs is that in the short-term, a receiver be capable to link messages generated by the same vehicle. In our scheme, a vehicle uses the same credential over one time slot. Hence, its messages signed by the private key that corresponds to that credential can be linked to each other.

5.5 Long-Term Unlinkability

The credentials of a vehicle do not include any connecting information to each other or to the vehicle's real identity. Besides, each credential of a vehicle expires with the end of the corresponding time slot. Hence, in each slot, the related credentials of all the vehicles expire simultaneously. Thus, all the vehicles change their tickets in the beginning of each new time slot. In view of this reasoning, except for the TA, no other third party can reveal the relation among two or more different tickets of the same vehicle.

5.6 Traceability

The TA retains the mapping between the real identity of each registered vehicle, its public key certificate and the set of all its credentials. Hence, the TA can recover the real identity of a misbehaving vehicle given its ticket.

5.7 Identity Revocation

The TA includes credentials of misbehaving vehicles in revocation lists. RSUs do not sign credentials included in revocation lists. Hence misbehaving vehicles are evicted.

5.8 Defense Against Several Types of Attacks

Our scheme can resist the following types of attacks:

- **Identity resolution attack**: our scheme can resist the identity resolution attack, according to the abovementioned analysis about identity privacy preservation and long-term unlinkability.
- **Impersonation attack**: each vehicle and RSU in the network has to sign its outgoing messages. Hence, an adversary cannot assume the identity of another vehicle or RSU as it will not be able to use the convenient private key.
- **Sybil attack**: a malicious vehicle may intend to appear as many vehicles by simultaneously using different tickets. However, in our scheme each vehicle has in its set of credentials only one credential (hence only one ticket) for each time slot. In this situation, a malicious vehicle might attempt to get multiple sets of credentials from the TA in only one ***CP*** in order to have multiple credentials related to the same time slot. However, the TA maintains the mapping between the real identity of each registered vehicle, its public key certificate and the set of all its credentials. Whenever a vehicle contacts the TA for a new set of credentials, the vehicle must provide its real identity. The TA will check the identity of the requesting vehicle against the table of real identity, public key certificate and credentials mappings. If the requesting vehicle has a set of credentials that still contain non-expired credentials, the TA will revoke them before giving a new set of credentials to the requesting vehicle. Thus, our scheme can thwart the Sybil attack.
- **Modification attack**: all vehicles and RSUs in the network sign their data before sending them. Receivers can detect the modification of received data by verifying whether the corresponding identity based signature is valid.
- **GPS spoofing attack**: a malicious vehicle may try to misguide other nodes in the network by faking its actual location. In our scheme, each ticket of a vehicle includes the identity of the RSU that signed the related credential. A credential should be signed whenever its corresponding time slot starts or the vehicle enters into a new domain. The validity period of a credential (hence the related ticket) is limited (only one time slot). Hence, a vehicle cannot escape the step of signing credentials by RSUs. Otherwise, receivers will not accept the vehicle's safety messages. The identity of an RSU is composed of its geographical coordinates concatenated with the sequence number of the domain to which it belongs. Thus, the receiver can have an idea about the vicinity of the sending vehicle.
- **Replay attack**: each vehicle and RSU in the network includes a timestamp in each message it sends. Receivers can detect the replay of a message by verifying the freshness of the incorporated timestamp.
- **Repudiation attack**: our scheme can resist the repudiation attack, according to the abovementioned analysis about non-repudiation.

6 Conclusion

In VANETs, the privacy of vehicles should be preserved when authenticating their safety messages. However, misbehaving vehicles should be traced by legal authorities and discarded from the network. Meeting these requirements raises a serious challenge.

In this paper, we proposed a new conditional privacy preserving authentication scheme that relies on temporary tickets, in order to deal with this challenge. A detailed security analysis was performed proving the efficiency of our scheme. As a future work, we intend to implement the proposed scheme.

References

1. Dedicated Short Range Communications (DSRC). http://grouper.ieee.org/groups/scc32/dsrc/index.html
2. Fan, C.I., Sun, W.Z., Huang, S.W., Juang, W.S., Huang, J.J.: Strongly privacy-preserving communication protocol for VANETs. In: 2014 Ninth Asia Joint Conference on Information Security, pp. 119–126. IEEE Press, Wuhan (2014). doi:10.1109/AsiaJCIS.2014.24
3. Liu, Y., Wang, L., Chen, H.H.: Message authentication using proxy vehicles in vehicular Ad Hoc Networks. IEEE Trans. Veh. Technol. **64**(8), 3697–3710 (2015). doi:10.1109/TVT.2014.2358633
4. Chen, J., Yuan, Q., Xue, G., Du, R.: Game-theory-based batch identification of invalid signatures in wireless mobile networks. In: 2015 IEEE Conference on Computer Communications, pp. 262–270. IEEE Press, Hong Kong (2015). doi:10.1109/INFOCOM.2015.7218390
5. He, D., Zeadally, S., Xu, B., Huang, X.: An efficient identity-based conditional privacy-preserving authentication scheme for vehicular Ad-hoc Networks. IEEE Trans. Inf. Forensics Secur. **10**(12), 2681–2691 (2015). doi:10.1109/TIFS.2015.2473820
6. Mrabet, K., El Bouanani, F., Ben-Azza, H.: A secure multi-hops routing for VANETs. In: 2015 International Conference on Wireless Networks and Mobile Communications, pp. 1–5. IEEE Press, Marrakech (2015). doi:10.1109/WINCOM.2015.7381299
7. Shao, J., Lin, X., Lu, R., Zuo, C.: A threshold anonymous authentication protocol for VANETs. IEEE Trans. Veh. Technol. **65**(3), 1711–1720 (2016). doi:10.1109/TVT.2015.2405853
8. Mejri, M.N., Achir, N., Hamdi, M.: A new group Diffie-Hellman key generation proposal for secure VANET communications. In: 2016 13th IEEE Annual Consumer Communications and Networking Conference, pp. 992–995. IEEE Press, Las Vegas (2016). doi:10.1109/CCNC.2016.7444925
9. Büttner, C., Bartels, F., Huss, S.A.: Real-world evaluation of an anonymous authenticated key agreement protocol for vehicular Ad-Hoc Networks. In: 2015 IEEE 11th International Conference on Wireless and Mobile Computing, Networking and Communications, pp. 651–658. IEEE Press, Abu Dhabi (2015). doi:10.1109/WiMOB.2015.7348024
10. Bellare, M., Namprempre, C., Neven, G.: Security proofs for identity-based identification and signature schemes. In: Advances in Cryptology-EUROCRYPT 2004. LNCS, vol. 3027, pp. 268–286. Springer, Heidelberg (2004)

11. Yasmin, R., Ritter, E., Wang, G.: Provable security of a pairing-free one-pass authenticated key establishment protocol for wireless sensor networks. Int. J. Inf. Secur. **13**(5), 453–465 (2014). doi:10.1007/s10207-013-0224-7
12. Islam, S.K., Khan, M.K.: Provably secure and pairing-free identity-based handover authentication protocol for wireless mobile networks. Int. J. Commun Syst (2014). doi:10.1002/dac.2847
13. SHA-2. http://csrc.nist.gov/publications/fips/fips180-2/fips180-2withchangenotice.pdf

A Trust Based Communication Scheme for Safety Messages Exchange in VANETs

Ryma Abassi(✉) and Sihem Guemara El Fatmi

Higher School of Communication, SUP'Com, University of Carthage, Tunis, Tunisia
abassi.ryma@gmail.com

Abstract. VANET (Vehicular Ad hoc NETwork) is a self-organized network formed by connecting vehicles. It can be used in several cases such as internet connection to obtain real time news, traffic, and weather reports, etc. However, the main applications of VANET are oriented to safety issues (e.g., traffic services, alarm and warning messaging). In such context, a security problem can have disastrous consequences. In fact, an attacker can be tempted to redirect all vehicles to a given location for malicious reasons. Hence, it is necessary to make sure that exchanged packets are not inserted or modified by the attacker. The main contribution of this paper is then, the proposition of a secure communication scheme for safety message exchange in VANET.

1 Introduction

Nearly 3.400 people die on the world's roads every day, tens of millions of people are injured or disabled every year, these are some statistics given by the WHO [1]. Hence, improving road safety and guaranteeing the efficiency of transport are the key challenges posed to road traffic today. Vehicular Ad hoc NETwork (VANET) emerged in this context.

A VANET is a self-organized network that can be formed by connecting vehicles equipped with on-board units (OBUs) [2]. It is utilized for a broad range of safety applications such as collision warnings and non safety applications such as road navigation. Two types of communication are provided in VANET: Vehicular to Vehicular (V2V) and Vehicular to Infrastructure (V2I). In the first communication type, vehicles communicate directly whereas in V2I, vehicles communicate through routers called Road Side Unit (RSU). Trusted authorities (TA) control the network.

Due to the criticality of communication in such network, security is vital. In fact, an attacker could be tempted to create congestion on a given road or to liberate another one for malicious reasons. Moreover, security requirements in VANET are also different from other networks [8]. They have to deal with mobility, scalability, privacy, heterogeneity, volatility, etc.

Several aspects have been investigated for securing VANET communications mainly: cryptography based security solutions [10], Vehicular Public Key Infrastructure (VPKI) [7], trust models [3–5]. Although cryptography solutions

A. Puliafito et al. (Eds.): ADHOC-NOW 2017, LNCS 10517, pp. 92–103, 2017.
DOI: 10.1007/978-3-319-67910-5_8

allow vehicles authentication, they do not provide any reliability of the data. VPKI constitute a promising solution by securing communications however their deployment is arduous and costly. Trust models are used in order to detect malicious vehicles through the maintaining of reputations [5]. However, in VANET maintaining reputations is very difficult even unfeasible since vehicles are moving quickly preventing them from establishing trust relations or storing reputations. In fact, VANET environment is quite different from other networks due to its high speed mobility nodes and distributed nature; security requirements in VANET are also different from other networks [8]. Hence, an ideal security scheme should secure exchanged data while preserving privacy and limiting the infrastructure involvement.

In this paper, we propose a secure communication scheme for safety messages exchange in VANET. Our scheme was designed in order to ensure reliability in received messages based on three modules: a recommendation module, an opinion module and an alert module.

The rest of this paper is organized as follows. In Sect. 2 some existing works are recalled. In Sect. 3 the main proposition is presented, a VANET authentication scheme preserving privacy. In Sect. 4, a formal validation of our proposal is presented mainly through two inference systems handling the proposed modules and some verification procedures proving the soundness and correctness properties. Finally, Sect. 5 concludes this paper.

2 Related Work

Few works dealt with trust in VANET. Most of them were interested by trust establishment in VANET that relies on a security infrastructure and most often makes use of certificates [5] However, in this work, we focus on trust models that do not fully rely on the static infrastructure and thus can be more easily deployed.

In [6], TEAM a Trust-Extended Authentication Mechanism for Vehicular Ad hoc Networks was proposed. It involves eight procedures: initial registration, login, general authentication, password change, trust-extended authentication, key update, key revocation and secure communication. However, the main drawback of this proposition is that if an adversary node is authenticated as trustful, it may authenticate other misbehaving nodes.

In [3], authors proposed TRIP (a trust and reputation infrastructure-based proposal for vehicular ad hoc networks), aimed to quickly and accurately distinguish malicious or selfish nodes spreading false or bogus messages throughout the network. Hence, a reputation score is computed for each node taking into account three different sources of information, namely: direct previous experiences with the targeting node, recommendations from other surrounding vehicles and, when available, the recommendation provided by a central authority.

Recently, [9] proposed an efficient pseudonym generation technique. The vehicles receive a small number of long-term secrets to compute pseudonyms/keys to be used in reporting the events without leaking private information about the drivers. Moreover, they proposed a scheme to identify the vehicles that use their pool of pseudonyms to launch Sybil attacks without leaking private information to road side units.

Some other works dealt with other aspects of secure vehicular communications [11–13]. Ben Jaballah et al. proposed in [11], a secure solution is effective in mitigating the position cheating attack. In fact, they analyzed the vulnerabilities of a representative approach named Fast Multi-hop Algorithm (FMBA) to the position cheating attack. Then, they devised a fast and secure inter-vehicular accident warning protocol which is resilient against the position cheating attack. In [12], authors employed each RSU to maintain and manage an on-the-fly group within its communication range. More precisely, vehicles entering the group can anonymously broadcast V2V messages, which can be instantly verified by the vehicles in the same group as well as from neighboring groups. Later, if the message is found to be false, a third party can be invoked to disclose the identity of the message originator. In [13], authors dealt with broadcasting techniques which are used for sending safety messages, traffic information, or comfort messages. When a packet is broadcast, it is received by all nodes within the sender's coverage area (provided that no interference or radio channel trouble occurs). Every receiver will decide to relay or not the packet depending on its own broadcasting strategy. This hop-to-hop communication would lead to a full coverage of the network.

Our proposition is intended to minimize the infrastructure involvement for safety message exchange in VANET. It is close to the proposition of [4] where authors proposed a trust management scheme for the vehicular networks built upon three phases. In the first one, receiver nodes calculate confidence value based on location closeness, time closeness and location verification. In the second phase, a trust value is calculated for each message related to the same event. In the last phase, receiver takes decision of acceptance of the message. Our proposition is similar this work in that sense that it uses the same criteria e.g. location closeness and time closeness but is different since trust evaluation is made using direct observations, neighbors recommendation as well as location, and time closeness.

3 Proposition

The main contribution in this paper is a communication scheme designed in order to exchange safety messages securely. This scheme is built upon three modules: a recommendation module, an opinion module and an alert module as depicted by Fig. 1.

The opinion module is used in order to assess the veracity of a received message based on the estimation of location and time closeness as well as their verification.

The recommendation module is in charge of exchanging appreciations between vehicles as well as the calculation of similarity degree between these latter.

The alert module is intended to inform RSU and TA about a potential misbehaving vehicle.

As detailed by Algorithm 1, each vehicle detecting a particular event calculates a veracity score based on the event information. A safety message is then sent including this score. Each vehicle receiving this message sends a recommendation request asking for other vehicles' appreciations. These latter are sent back using recommendation responses. A similarity degree is then calculated. If this degree is below a given threshold, then the message is rejected and an alert message is sent to the trusted authority through RSUs otherwise it is forwarded with the addition of the new veracity score calculated by this vehicle. Each vehicle receiving a safety message from a forwarder (not a witness), verifies its trustworthiness through the set of veracity scores included in this latter. This is made by calculating the veracity scores average while excluding the max and the min values.

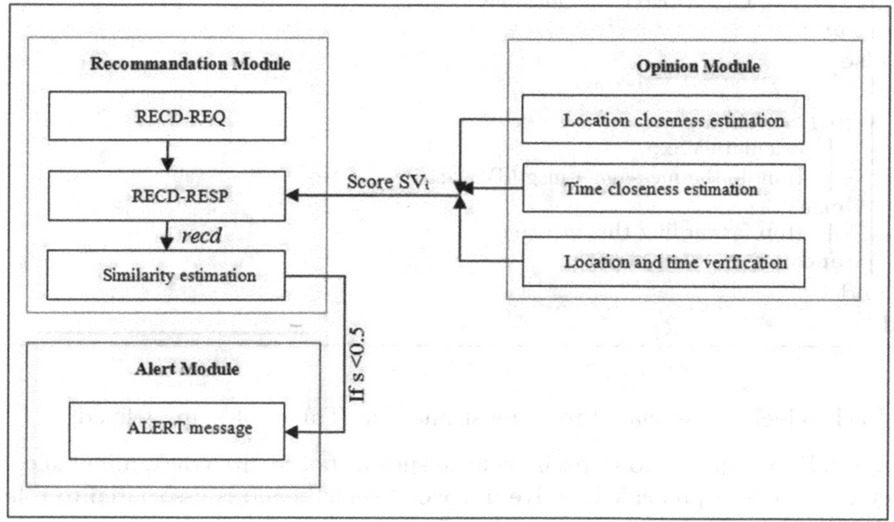

Fig. 1. Proposed modules

3.1 Basic Assumptions

An event is defined as follows:

$$Event : l||t||tp||timestamp$$

where l is the event location; t is the event time, tp is the event type $\in$ *accident*, *road liberation*, *traffic information* and *timestamp* is used to keep track of time and location validity.

Algorithm 1. Safety Message reception

```
Data: r: a vehicle role; svi: the veracity score calculated by vi; n: the number of for-
      warders
Result: message handled
begin
    r := verify-role (Vc);
    if r == special then
        Forward safety message;
    else if r == witness then
        Vi:= neighbors: RECD-REQ;
        Neighbors:= Vi: RECD-REQ = <recd_n>;
        similar:= verify similarity(∑ recd_n);
        if similar == 1 then
            calculate VSi;
            Forward <msg||VSi4||cred>;
        else
            stop forwarding this message;
            V = CA: ALERT = <message, Vc>;
        end
    else
        t = ( ∑VSi−minVSi−maxVSi) / n ;
        if t > δ then
            calculate VSx;
            Handle the message <msg||SVx>;
        else
            stop forwarding this message;
        end
    end
end
```

Each vehicle is associated to at most one role. Three roles are defined:

- Special: corresponds to a special vehicle such as police, fire truck, ambulance, etc. This role is preloaded, active in special vehicles and is associated to role flag $r - flag$ '0'.
- Observer: corresponds to the event witness. It is preloaded in vehicles but is triggered when needed. It is associated to role flag $r - flag$ '1'.
- Forwarder: corresponds to a vehicle forwarding a received message. It is preloaded in vehicles but is triggered when needed. It is associated to role flag $r - flag$ '2'.

Each vehicle receiving a message verifies its role flag $r - flag$ as depicted by Algorithm 2.

3.2 Opinion Module

A vehicle receiving a message verifies its veracity by calculating a score VS. According to this latter, the message can be forwarded or stopped. The veracity score VS is defined as follows:

Algorithm 2. Role verification

```
Data: r-flag: role flag of vehicle x
Result: vehicle role
begin
    if r-flag(X)==0 then
        return special;
    else if r-flag(X)==1 then
        if timestamp(X) is correct then
            return observer;
        end
    else
        return forwarder;
    end
end
```

$$VS = (Lc + Tc) * Fc \tag{1}$$

where Lc is the location closeness, Tc the time closeness and Fc is the number of forwarders.

More precisely, this score satisfies the following hypotheses:

- More the event sender is close to the event location, more the score increase.
- More the time closeness decreases, more the score decreases.
- More the number of senders increases, more the Fc decreases and consequently, the score decreases.
- if s ≥ 0 then, the message is considered trustworthy, then the score is added to the message and is forwarded otherwise it is untrustworthy and is simply stopped.

Location Closeness. Location closeness (L_c) estimates the closeness of the sender to the reported event. It is formalized as follows:

$$L_c = \begin{cases} \frac{1}{x_s - x_e} \; if \; (x_s - x_e)^2 + (y_s - y_e)^2 < \Delta^2 \\ 0 \quad otherwise \end{cases} \tag{2}$$

The event location is used as the origin ($x_e = 0$, $y_e = 0$). Any vehicle located at (x, y) around the event position within a radius of Δcan be trusted with a confidence decreasing with the increase of $(x - x_e)$. Figure 2 shows the implementation of Eq. (1) with $\Delta = 10$.

Time Closeness. Time closeness (Tc) estimates the freshness of the reported event. It is formalized as follows:

$$T_c = \begin{cases} 1 - \frac{1}{|t_r - t_e|} \; if \; |t_r - t_e| < \delta_i \\ 0 \quad otherwise \end{cases} \tag{3}$$

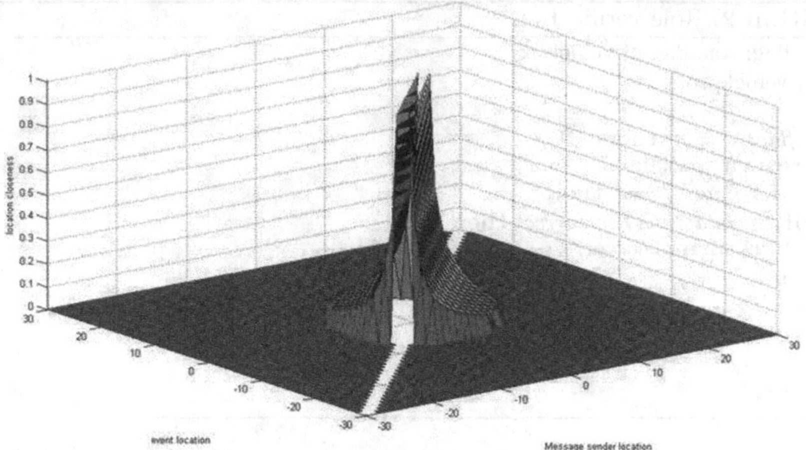

Fig. 2. Location closeness ($\triangle = 10$)

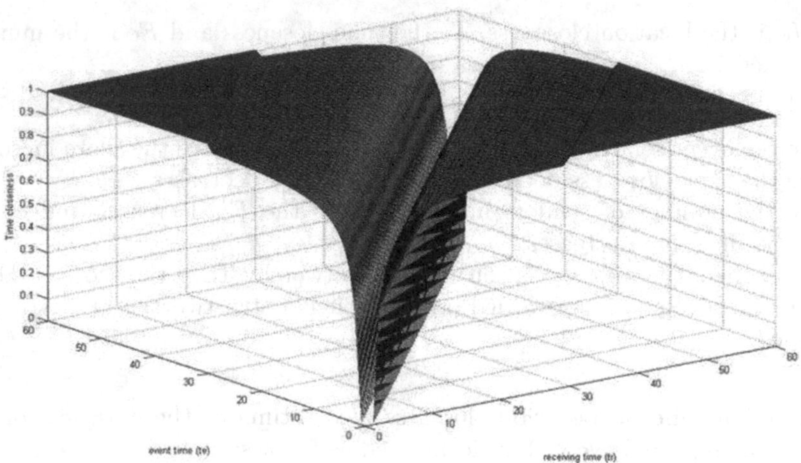

Fig. 3. Time closeness

Figure 3 shows the implementation of Eq. (2).

One can observes that if the time difference between the event occurrence and its reception increases, then the time closeness decreases.

Forwarding Chain Length. Forwarding chain closeness (F_c) estimates the number of vehicles having forwarded the reported event. It is formalized as follows:

$$F_c = \begin{cases} \frac{1}{n} \; if \, n < \delta_n \\ 0 \; otherwise \end{cases} \tag{4}$$

3.3 Recommendation Module

Each vehicle V_i sends a recommendation request $RECD-REQ$ to its neighbors asking for their appreciations regarding a given event.

$$V_i \rightarrow Multicast\, RECD - REQ : location,\, time,\, type$$

Each vehicle V_c receiving a $RECD - REQ$ sends back a $RECD - RESP$

$$V_c \rightarrow Unicast\, V_i\, RECD - RESPQ : recd_c$$

Based on received recommendations, a similarity s is calculated as follows:

$$s = \frac{min\,(recd_1, \ldots, recd_n)}{max\,(recd_1, \ldots, recd_n)} \tag{5}$$

where $s \geq 0,5$ similar; $s = 1$ totally similar; $s < 0,5$ not similar.

3.4 Alert Module

Only the CA can match the pseudonym with the original identity of the node. The CA maintains a reputation table for vehicles as follows (Table 1):

Table 1. Reputation table

TID	ID	rep
5	65	0
6	43	−2

The alert is sent by a vehicle to the CA and contains its temporary ID, the temporary ID of the denounced vehicle,

$$V_i \rightarrow Unicast\, CA\, ALERT(TID_s, TID_d)$$

Such as depicted by Algorithm 3, after receiving this alert, the CA associates the TID_s and TID_d with the real ID and verifies the reputation of the sender rep_s. Two cases are conceivable: (1) if the reputation $rep_s \leq 0$ meaning that the vehicle is not trustworthy, then the alert is ignored. (2) if the reputation $rep_s > 0$, then rep_d is decreased until reaching 0. In such case, the vehicle is blacklisted and the RSUs are informed.

4 Formal Specification and Validation

In the following, we propose a formal and automated expression of the proposed communication scheme using an inference system. This system is based on the use of logical rules consisting of a function which takes premises, analyzes their applicability and returns a conclusion. The second part of this section concerns the validation task proving the correctness and the soundness of the various propositions.

Algorithm 3. Alert handling

```
Data: an Alert message sent by node TID_s concerning a node TID_d
Result: updated reputation
begin
    associates (TID_s) with its real idendity;
    associates (TID_d) with its real idendity;
    verifies (rep_s);
    if rep_s < 0 then
        ignores alert;
    else
        rep_d − −;
        if rep_d == 0 then
            blacklists v_d;
            informs RSU;
        end
    end
end
```

4.1 Formal Specification

The proposed inference system describing our routing protocol is presented in Table 3 . It is based on a set of rules called inference rules. Used notations in the proposed inference systems are defined in Table 2.

Table 2. Used notations

Symbol	Meaning
E	The set of detected event by a given vehicle
R	Vehicles roles: forwarder, witness, special
VS	Veracity score
RECD-REQ	A recommendation request message
RECD-RESP(recd)	A recommendation response message containing the recommendation *recd*
S	A similarity score
s_{avg}	The average similarity score
M	Exchanged messages
δ	Similarity threshold
cred	Presented credential

Let us note consider that the following assumptions:

- The authentication were already made.
- The sending vehicle was successfully authenticated.

Such as depicted in Table 3, inference rules apply to triples $(VS, M, \emptyset)$ whose first component VS is a set of veracity scores. The second component, M represents the set of exchanged messages. The third component $RECD - REQ$ is the set of sent recommendation requests. Initially $RECD - REQ$ is empty.

Four inference rules are proposed. $Event_{wdet}$ handling an event detection by w witness vehicle, $Opinion_{req}$ addressing research of other vehicles opinions, $Success$ addressing the case where the alert is accepted and $Failure$ concerned with alert rejection.

In the following, each of the proposed inference rule is detailed.

- $Event_{wdet}$ inference rule deals with the detection of an event by a witness vehicle. In such case, this latter calculates a veracity score $VS \sqcup \{vs\}$ corresponding to the detected event $\{e\} \sqcup E$, builds and sends a safety message $M \sqcup \{m\}$such that $m =< m \parallel sv_i \parallel cred >$.
- $Opinion$ inference rule deals with the sending of recommendation requests $RECD - REQ \sqcup \{recd - req\}$to other vehicles when a message $\{m\} \sqcup M$ is received.
- $Success$ inference rule deals with he case where calculated score is greater than δ. In such case, the message is handled i.e. forwarded or executed.
- $Failure$ inference rule deals with the case where calculated score is smaller than δ. In such case, the message is simply ignored and the system stops until the next safety message reception.

Table 3. Proposed inference system: witness vehicle

init	$\dfrac{}{VS, M, \emptyset}$	
$Event_{wdet}$	$\dfrac{\{e\}\sqcup E, \{r\}\sqcup R, M, S, \emptyset}{VS\sqcup\{vs\}, M\sqcup\{m\}, \emptyset}$	$if\ r = \text{“}witness\text{”}\ where \begin{cases} VS = [(Lc + Tc) \times Fc] \times \beta \\ m \equiv < m \parallel vs_i \parallel cred > \end{cases}$
$Opinion_{req}$	$\dfrac{E, \{r\}\sqcup R, \{m\}\sqcup M, S, \emptyset}{VS, M\backslash\{m\}, RECD-REQ\sqcup\{recd-req\}}$	
Success	$\dfrac{E, R, M, \{s\}\sqcup S, recd-resp(recd)\sqcup RECD-RESP}{VS\sqcup\{vs\}, M\sqcup\{m\}, \emptyset}$	$if\ s > \delta\ where\ s = \dfrac{min\,(recd_1,\ldots,recd_n)}{max\,(recd_1,\ldots,recd_n)}$
Failure	$\dfrac{E, R, M, \{s\}\sqcup S, recd-resp(recd)\sqcup RECD-RESP}{fails}$	$if\ s < \delta\ where\ s = \dfrac{min\,(recd_1,\ldots,recd_n)}{max\,(recd_1,\ldots,recd_n)}$

4.2 Validation

In this subsection, the verification of the soundness, correctness and completeness of the proposed inference system is achieved. Soundness is proved by showing that the failure of the system implies that received messages are not trustworthy. Correctness is proved by showing that reported events are correct.

We denote by $\vdash^*$ the reflexive application of inference rules of Table 3.

(witness correctness). For $vs \sqcup VS$, $m \sqcup M$, $recd{-}resp(recd) \sqcup RECD{-}RESP$, if V, M, $\emptyset \vdash^*$ *success* then the reported event is sound and the received message is trustworthy.

By applying the recurscall $event_{wdet}$, each vehicle calculates a veracity score vs and appends it to the whole message including the veracity scores of other vehicles already appended. This score is timestamped in order to ensure its time and location validity and a credential is also added in order to prove the good will of the witness vehicle.

Each vehicle receiving such message, verifies its similarity and thus by calculating a similarity score.

Since the application of the inference always terminates, and the outcome can only be success or failure, it follows immediately from the witness correctness that if the received message is trustworthy, then V, M, $\emptyset \vdash^* success$.

(soundness of failure). if V, M, $\emptyset \vdash^* fails$ then the received message is not trustworthy.

By applying the recurscall $opinion_{req}$, each vehicle sends $recd-req$ messages asking for other vehicles appreciation and a score s is then calculated. Two cases are conceivable: if s is greater than the threshold δ, then the message is accepted and handled otherwise, the message is rejected.

Since the application of the inference always terminates, and the outcome can only be success or failure, it follows immediately from the soundness of failure that if the received message is not trustworthy, then V, M, $\emptyset \vdash^* fails$.

5 Conclusion

VANET are used for a broad range of safety applications such as collision avoidance, road liberation, etc. In such context, information liability is vital in order to be sure that exchanged messages were not modified or inserted for example.

In this paper, we proposed a secure communication scheme for safety message exchange in VANET. This scheme is based on three modules: a recommendation module, an opinion module and an alert module. More precisely, a vehicle detecting a given event calculates its appreciation using its opinion module and includes it in the safety message. Each vehicle receiving such message, asks for other vehicles appreciations using recommendation request/response. A similarity degree is then calculated. Two cases are conceivable: the degree is below a threshold or greater than this latter. In the first case, the message is stopped and the RSU is informed whereas in the second case a new veracity score is calculated and is added to the message before forwarding the message. Finally, any vehicle receiving a forwarded safety message, estimates its trustworthiness using the included veracity scores and the same previously introduced reaction procedure is made i.e. alert or forward. Our second contribution was validating our scheme formally. Hence, we proposed adequate inference system. Next, we built a validation process using the proposed inference systems and proving soundness and correctness of our proposal.

In the future works, this proposition will be implemented and evaluated against attacks such as collusion attack, privacy violation, etc.

References

1. The website of the World Health Organization. http://www.who.int
2. Raw, R.S., Kumar, M., Singh, N.: Security challenges, issues and their solutions for VANET. Int. J. Netw. Secur. Appl. **5**(5), 95–105 (2013)
3. Mármol, F.G., Pérez, G.M.: TRIP, a trust and reputation infrastructure-based proposal for vehicular Ad Hoc networks. J. Netw. Comput. Appl. **35**(3), 934–941 (2012). ISSN 1084-8045
4. Shaikh, R.A., Alzahrani, A.S.: Trust management method for vehicular Ad Hoc networks. In: 9th International Conference on Heterogeneous Networking for Quality, Reliability, Security and Robustness, Greater Noida, India, pp. 801–815 (2013)
5. Wex, P., Breuer, J., Held, A., Leinmuller, T., Delgrossi, L.: Trust issues for vehicular Ad Hoc networks. In: Proceedings of the 67th IEEE Vehicular Technology Conference (2008)
6. Chuang, M.-C., Lee, J.-F.: TEAM: trust-extended authentication mechanism for vehicular Ad Hoc networks. Syst. J. IEEE **8**(3), 749–758 (2014)
7. Maxim, R., Papadimitratos, P., Hubaux, J.-P.: Securing vehicular communications. In: IEEE Wireless Communications Magazine, Special Issue on Inter-Vehicular Communications 13.LCA-ARTICLE-2006-015, pp. 8–15 (2006)
8. Gillani, S., Shahzad, F., Qayyum, A., Mehmood, R.: A survey on security in vehicular ad hoc networks. In: Berbineau, M., Jonsson, M., Bonnin, J.-M., Cherkaoui, S., Aguado, M., Rico-Garcia, C., Ghannoum, H., Mehmood, R., Vinel, A. (eds.) Nets4Cars/Nets4Trains 2013. LNCS, vol. 7865, pp. 59–74. Springer, Heidelberg (2013). doi:10.1007/978-3-642-37974-1_5
9. Rabieh, K., Mahmoud, M.M.E.A., Azer, M., Allam, M.: A secure and privacy-preserving event reporting scheme for vehicular Ad Hoc networks. In: Security and Communication Networks (2015)
10. Boneh, D., Boyen, X., Shacham, H.: Short group signatures. In: Proceedings of Advances in Cryptology, CRYPTO 2004, Santa Barbara, pp. 41–55 (2004)
11. Ben Jaballah, W., Conti, M., Mosbah, M., Palazzi, C.: A secure alert messaging system for safe driving. Comput. Commun. **46**, 29–42 (2014)
12. Zhang, L., Wu, Q., Solanas, A., Domingo-Ferrer, J.: A scalable robust authentication protocol for secure vehicular communications. IEEE Trans. Veh. Technol. **59**(4), 1606–1617 (2010)
13. Abdou, W., Darties, B., Mbarek, N.: Priority levels based multi-hop broadcasting method for vehicular Ad Hoc networks. Ann. Telecommun. Ann. des télécommun. **70**(7–8), 359–368 (2015)

Smart City

Mobility as the Main Enabler of Opportunistic Data Dissemination in Urban Scenarios

Jorge Herrera-Tapia[1,3(✉)], Anna Förster[2], Enrique Hernández-Orallo[1], Asanga Udugama[2], Andrés Tomas[1], and Pietro Manzoni[1(✉)]

[1] Universitat Politècnica de València, Valencia, Spain
jorherta@doctor.upv.es, {ehernandez,pmanzoni}@disca.upv.es, antodo@upv.es
[2] University of Bremen, Bremen, Germany
{anna.foerster,udugama}@uni-bremen.de
[3] Universidad Laica Eloy Alfaro de Manabí, Manta, Ecuador
jorge.herrera@live.uleam.edu.ec

Abstract. The use of opportunistic communications to disseminate common interest messages in an urban scenario have various applications, like sharing traffic status, advertising shop offers, spread alarms, and so on. In this paper, we evaluate the combined use of fixed and mobile nodes to establish an optimal urban opportunistic network aimed at the distribution of general interest data.

Our results not only contradict current assumptions about the combination of fixed and mobile nodes, but also provide interesting general-purpose observations about the dynamics of opportunistic networks. First of all, we found that mobility is not the hindering and challenging property of these networks, but probably their main enabler. Moreover, we determined that if we want to increase the performance of opportunistic networks by increasing the node density, we should increase the number of mobile and not the fixed nodes since adding fixed nodes only increases the overhead.

Finally, our evaluation approach goes beyond the state of the art and is based on using two different simulators, the ONE and OMNeT++, and two different mobility traces from cities with different structure.

Keywords: Opportunistic networks · Vehicular networks · Vanets · Epidemic protocol · Delay tolerant networking · Wireless Ad Hoc networks

1 Introduction

Opportunistic communications initially enforced the fact that a stable or fixed infrastructure is not required. First attempts relied on solutions like the Mobile Ad-Hoc Networks (MANETs) [1,2] or Vehicular Ad-hoc Networks (VANETs) [3–5], both self-forming and self-healing types of networks that provided communication links without the support of fixed infrastructure. However, to handle long periods of connectivity lack, Delay/disruption Tolerant Networks (DTNs) [6]

A. Puliafito et al. (Eds.): ADHOC-NOW 2017, LNCS 10517, pp. 107–120, 2017.
DOI: 10.1007/978-3-319-67910-5_9

and Vehicular Delay Tolerant Networks (VDTNs) [7] were later proposed as an alternative to disseminate and share information among mobile users. By “long periods” we refer to a duration of at least various minutes up to hours. Some authors, such as [8], propose their utilization in rural areas or catastrophe zones.

In this paper, we explore the interplay between static and mobile nodes in an opportunistic communication scenario and show which are the properties driving better performance. This is crucial for better understanding of real-life opportunistic services and deployments. We consider the combined use of fixed and mobile nodes to establish an urban opportunistic network aimed to push common interest messages in this scenario. Fixed nodes were used either as relays and as message producers/consumers. We used two very large realistic GPS vehicular traces belonging to cities with different street structures, namely Rome [9] and San Francisco (SFO) [10]. Through simulations, we compare the message delivery probability, latency and the network overhead. These experiments were carried out using two different simulators, namely ONE [11] and OMNeT++. The ONE simulator was designed and built to specifically evaluate DTN (Delay Tolerant Networks) protocols and applications, and it is focused on the network layer without considering the particularities of lower layers such as physical and Media Access Control (MAC). OMNeT++ is a modular and general purpose discrete event simulator. The simulator provides the building blocks to develop a number of network nodes that could be deployed with different protocols layers (e.g., HTTP, UDP, TCP, 802.11, etc.).

We believe that this paper contributes significantly to provide a better understanding of the dynamics of mobile opportunistic networks. The results show that mobility and density are the critical factors that shape the performance of opportunistic networks. These results are very important for the development of future services, as they contradict the trivial and widely spread assumption that using fixed nodes always “helps”. Moreover, we also consider that the approach we used, based on the combined evaluations through two different simulators and using real traces, helps in advancing in the definition of a more exact methodology for evaluating opportunistic networks.

The outline of the paper is as follows: an overview of related works about opportunistic vehicular networks and message diffusion is presented in Sect. 2, our experimental scenarios are described in Sect. 3, and the experiments evaluation and our main findings are presented in Sect. 4. Finally, Sect. 5 contains some conclusions and future work.

2 Related Work

Opportunistic networking is mainly characterized by the ephemeral and variable duration of contacts between pairs of mobile devices. Thus, various solutions to increase the message delivery probability were proposed in order to take the maximum advantage of the contact duration time and the mobility of the users. Proposals such as [12,13] evaluated the message dissemination behavior of the

Epidemic protocol by focusing on the mobility patterns of the nodes. In these works, the authors determined the relationship between factors such as speed, mobility model, and nodes density. The analytic modeling of the diffusion of messages in social groups taking into account the transmission time of the messages was proposed by the authors of [14–17].

The authors of [18–20] exploited the social interaction as an indicator to improve the message ratio delivery and to optimize the energy resources. In this same direction, the authors of [21] after analyzing various device buffer management strategies, propose a practical scheme called Friendly Sharing, where the users interested in sharing information are suggested to stop for a bounded time period to increase the delivery rate. The previous studies emphasized on data traffic and cooperation in pedestrian contexts. Their results cannot be easily extended to vehicular contexts where the mobility scenario changes drastically due to the higher relative speed of nodes. But authors like [22] showed the importance of vehicles participating actively in the distribution of messages, since they can provide: a relative abundance of memory, constant energy availability, computational resources and the availability of geographical information through GPS devices.

Related with this area, in [23], the authors characterize a total of three vehicular traces in China, two from Shanghai (bus and taxis), and one from Shenzhen. In [24] the authors offer a wide application of vehicular networks detailing where and how to employ certain communication approaches. They clearly establish the differences between MANETs, VANETs and VDTNs (Vehicular Delay Tolerant Network), concluding that the high mobility of vehicles leads to short contact durations limiting the amount of data transferred. They explore new routing protocols and new interaction mechanisms to improve the collaboration and data transmission in VANETs and VDTNs.

Improving data sharing in this context requires understanding the mobility of vehicles in terms of speed and their relative distance, to address realistically the analysis of message diffusion in order to clearly identify the challenges when the objective is pushing common interest messages in an urban scenario.

The authors of [25] used another trace set of 4,000 taxis, validated the collected data, and created their own mobility model called Shanghai Urban Vehicular Network (SUVnet). They evaluated message diffusion performance by comparing through simulation two different DTN routing protocols: the non-geographic pure Epidemic routing and their new proposed geographic DTN routing, the Distance-Aware Epidemic Routing (DAER). They also compared, as mobility models, the Random Way Point model, a SUMO-generated model, and their own SUVnet. Results show that the use of location information can improve DTN routing performance significantly. Their evaluation also demonstrate that the performance of DTN routing in VANET depends mainly on the underlying mobility models.

Other important works in the area, that analyze the message delivery probability based on real traces, are [26–28]. Here, the authors examine the performance of protocols in opportunistic networks considering GPS information of large cities, like Rome, Berlin, Beijing, among others. In [29,30] the authors propose improvements to diffusion protocols using analytical models tested by simulations.

Differently from the above cited works, where the authors have improved the message diffusion ratio by different approaches, we offer a different perspective on how these types of systems should be designed and deployed and give indications on an evaluation methodology aimed to obtain better results. In this paper we evaluate the combined use of fixed and mobile nodes to establish an urban opportunistic network aimed at the distribution of general interest data. We consider two very large realistic vehicular traces belonging to cities with different street organizations, namely Rome and San Francisco. Through simulations, we compare the message delivery probability and the network overhead. These experiments were carried out using two different simulators, ONE and OMNeT++.

3 Use Case Scenario

The first goal of this section is to describe the different opportunistic communication strategies that were considered to exchange the messages in our scenarios detailing its parameters. Then, we evaluate the combined use of fixed and mobile nodes to establish an urban opportunistic network aimed at the distribution of general interest data.

3.1 General Scenario

The main goal of this paper is to experimentally validate the assumption that fixed nodes can potentially improve the performance of mobile opportunistic networks. Moreover, we also investigate the opposite scenario: can mobile nodes assist non-connected fixed nodes? In the first case, we speak about mobile nodes serving as sources and destinations, while in the second case the fixed nodes are used either as sources or destinations.

For all the experiments we assume that data messages are created at all nodes according to a Poisson traffic generator with an average value of one message every two hours. The size of each message is 1 kB and has a single random destination (a node ID). We suppose that, when two nodes meet, they interchange the data messages they do not have in their respective buffers. Once a message gets delivered to its final destination, it is not propagated further and it is marked as received. These events are used to calculate the **delivery ratio** and the **delivery latency**. The parameters we consider are the following (see also Table 1).

In the following paragraphs we detail the structure of the different source/destination scenarios and the procedures used for the placement of the fixed nodes.

Table 1. Simulation parameters varied to evaluate message diffusion.

Parameter	Values
Buffer size	1 GB
Routing	Epidemic
Mobile nodes	316 (Rome), 536 (San Francisco)
Fixed nodes	0, 27, 55, 108, 216, 337
Interval	1 msg per node/7200 s
Time to live	12, 48 h
Bandwidth	2 Mb/s (WiFi-Direct)
Tx-Range	50 m (WiFi-Direct)

3.2 Fixed Nodes Assisting Mobile Nodes

The first approach is based on using the vehicles (in this scenario the mobile nodes are concretely taxi cabs) as "Producers", "Consumers" and "Relays" of messages **(P/C/R)**, and the fixed nodes only act as relays **(R)** of the messages. In other words, the fixed nodes support the mobile nodes.

The main motivation behind this scenario is the widely spread assumption in opportunistic networks research (for example in [31]) that fixed nodes at some strategic places, such as tourism offices, police or train stations, would significantly improve the delivery delay and delivery ratio by relaying messages also when none of the mobile nodes are around. In other words, if you arrive late at the train station, your messages will be waiting for you there, even if none of your friends is around. As we will show later in the paper, this assumption does not hold in general.

3.3 Mobile Nodes Assisting Fixed Nodes

In this second approach, the fixed nodes are the "Producers", "Consumers" and "Relays" of messages **(P/C/R)**, and the vehicles act as mules or relay nodes **(R)**, bringing messages from one fixed node (*source*) to another fixed node, until the final fixed node destination is reached (*destination*).

The motivation here is derived from the world of wireless sensor networks and their use of "data mules" for sensed data delivery. However, the main difference is that our data mules are not controlled and travel on their own without being aware that they are serving the purposes of our application. However, many scenarios are possible where the mules can be very helpful: for example to send messages from traffic lights to the police station in a Smart City context.

3.4 Node Placement Strategy

A good question is where to place the fixed nodes. This problem was faced by taking as reference a grid, in order to avoid concentration of fixed nodes, and by looking

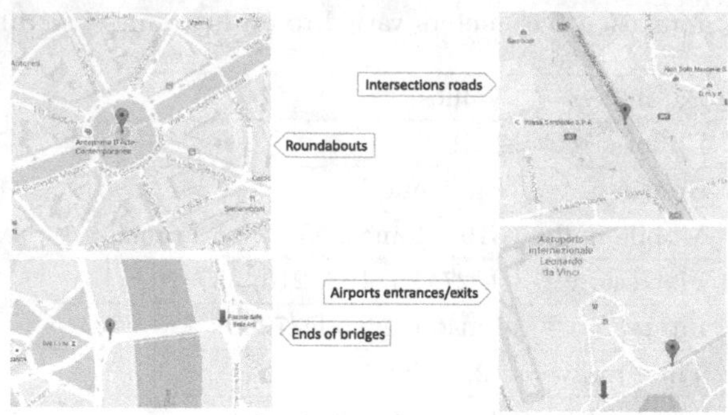

Fig. 1. Criteria to place the fixed nodes.

for specific strategic geographical locations in the target scenario. We considered areas with high user concentration probability, like touristic places (squares, main buildings, museums, etc.), and areas with high traffic probability, such as roundabouts, road intersections, ends of bridges, among others; Fig. 1 illustrates the criteria. Considering the transmission range, in our case 50 m, when the selected places were too large we put more than one fixed node in order to cover the whole area.

4 Performance Evaluation

In this section we explore the dynamics of opportunistic networks using the ONE simulator [11] with two large real vehicular movement traces, San Francisco (SFO) and Rome, and the OMNeT++ simulator [32] with one of these traces (SFO), and generating a network load based on a Poisson traffic generator with message size 1 kB. We decided to explore the behavior of the network with two different simulators to exclude the possibility that the results are tool-biased.

In the next paragraphs, we first present the properties of the GPS traces themselves. Next, we describe our main results by presenting the experimental data and their evaluation.

4.1 The Vehicular Traces

We use two sets of vehicular traces in our experiments. The first trace (Rome) has about 21 million points and belongs to a network formed by 316 taxi cabs in the area of Rome during a complete month (30 days) [9]. We employ a traverse Mercator projection [33] to convert the GPS data-set to Cartesian coordinates, centered near the Coliseum covering an area of 100 km $\times$ 100 km.

The second GPS trace (SFO) has about 11 million points and comes from 536 cabs in San Francisco during 24 days [10]. The covered area is, however, smaller than the Rome trace with only 40 km $\times$ 40 km. Thus, the average density of this trace is approximately 10 times larger than the Rome trace. We again converted

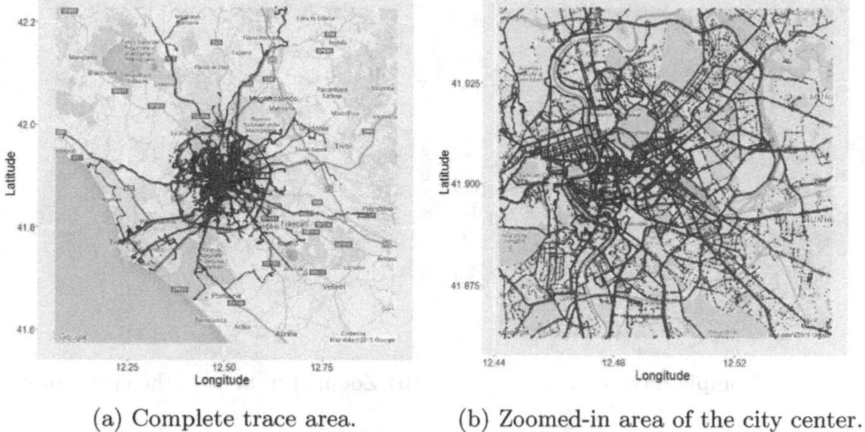

(a) Complete trace area. (b) Zoomed-in area of the city center.

Fig. 2. Rome trace sample of 6 h of GPS traces. Total length of trace is 30 days, 316 taxis in an area of 10,000 km^2 (0.0316 nodes per km^2).

the GPS records to Cartesian coordinates, centered at San Francisco's business center.

Figures 2a and 3a show how the vehicular traces are distributed around the metropolitan areas in Rome and San Francisco cities respectively. Figures 2b and 3b are zoomed views of the previous ones, showing how the main urban area of the cities are almost fully covered by the traces.

The parameters we use for our simulations (both OMNeT++ and ONE simulators) are summarized in Table 1 and correspond to typical scenarios for opportunistic networks.

4.2 Fixed Nodes Do Not Help

The first question we asked is whether the trivial assumption that fixed nodes potentially support mobile nodes in the dissemination of data is true or not. Our results show that this assumption is not true. Figure 4 shows the experimental results which lead us to this conclusion. The scenario is as follows: all mobile nodes of the traces serve as sources/destinations/relays, while the fixed nodes "support" the mobile ones with relaying messages. The more fixed nodes we add, the larger the overhead (c), but neither the latency decreases (b) nor the delivery ratio increases (a) significantly. At the same time, we see some slight impact on the Rome traces, which are less dense. What we also see is our next observation: the denser the network, the better the performance.

4.3 The Denser the Network, the Better

We can see in Fig. 5 that higher nodes densities leads to a dramatic increase in the number of contacts, from around 150–200 contacts per hour for the Rome

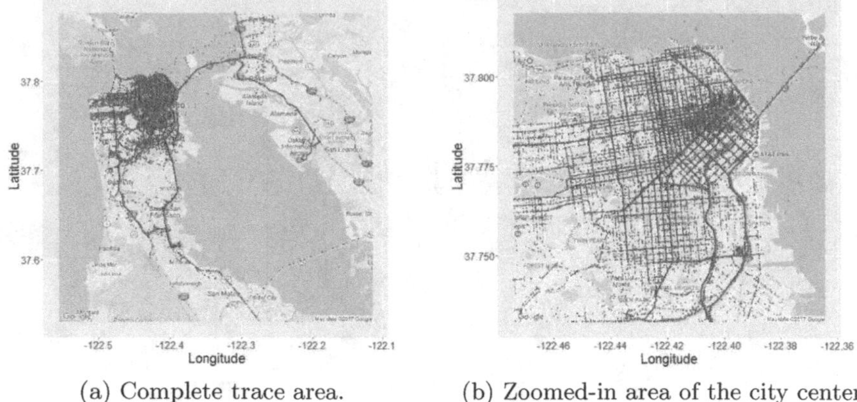

(a) Complete trace area. (b) Zoomed-in area of the city center.

Fig. 3. San Francisco trace sample of 6 h of GPS traces. Total length of trace is 24 days, 536 taxis in an area of 1,600 km^2 (0.335 nodes per km^2)

scenario to over 12,000 contacts for SFO. This is a very important observation, as the density is still very low compared to the fixed network nodes, but already delivers quite a good performance, as we will also see later. This observation is not surprising by itself, but the level of performance increase was unexpected. The Rome scenario has a mean density of 0.0316 nodes per square kilometer, while the SFO scenario has a density of 0.355 nodes per square kilometer.

4.4 Mobility Helps

Now, let us revert the scenario as explained in Sect. 3 and explore what happens with a fixed network, supported by more and more mobile nodes. The results are again as expected, but not at that level of significance. The experimental results are shown in Fig. 6. While the fixed network is completely disconnected by itself, as can be seen by the first point in the graph with no mobile nodes at all, even a small number of mobile nodes improve dramatically the delivery rate and the latency. For example, the delivery ratio increases from 8% for the Rome traces with a TTL of 48 h and only 10 mobile nodes to 50% for the same experiment but with 100 mobile nodes. This is an increase by a factor of approx. 6. The overhead increases from 30 to 100, a factor of only 3.3 - half of the increase in the delivery rate. **Thus, we can conclude that adding mobile nodes increases significantly the performance of the network at a relatively low cost.**

4.5 Mobile Nodes Are More Important Than Fixed Nodes

In order to further explore and understand the dynamics of mobile against fixed nodes, we performed the following experiment with an opportunistic network that uses the Randomized Rumor Spreading (RRS) [34] as its data propagation

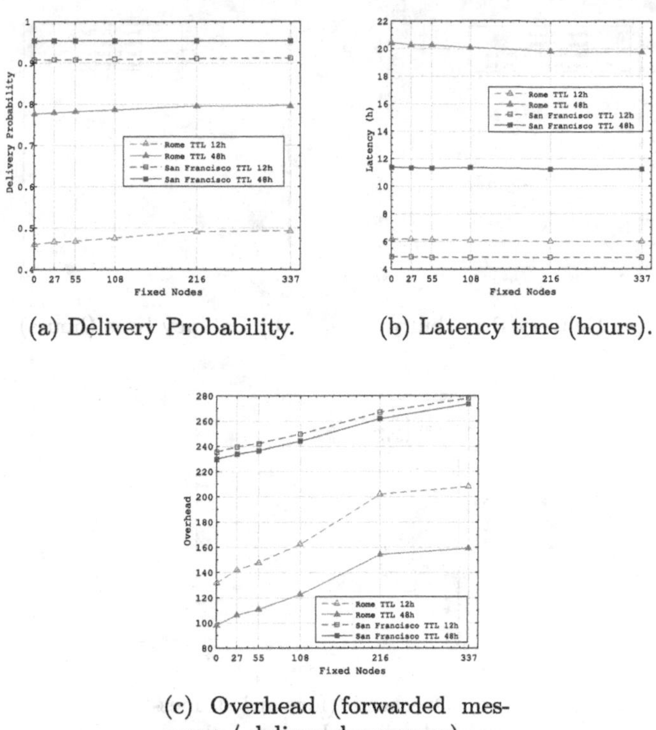

(a) Delivery Probability.

(b) Latency time (hours).

(c) Overhead (forwarded messages / delivered messages).

Fig. 4. Average delivery success ratio, latency and overhead for mobile nodes as sources and destinations and fixed nodes as relays.

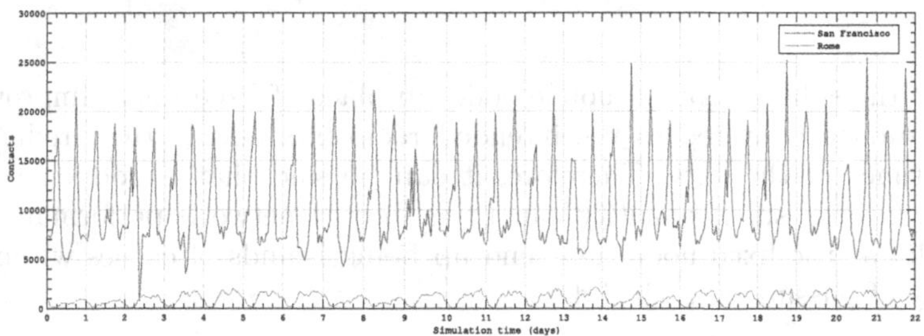

Fig. 5. Contacts per hour generated by the simulation for each taxi trace.

strategy. Concretely, we simulated a number of network scenarios where the number of nodes in the network was kept constant (100) while changing the ratio of fixed nodes to mobile nodes from 100% to 0% in each of the scenarios. The data was generated always by 30% of the nodes to keep the number of generated data constant for all the scenarios. The mobile nodes use the SFO traces to obtain their movement pattern while the fixed nodes are positioned in a grid across the mobility area. Figure 7 shows the obtained results. The results

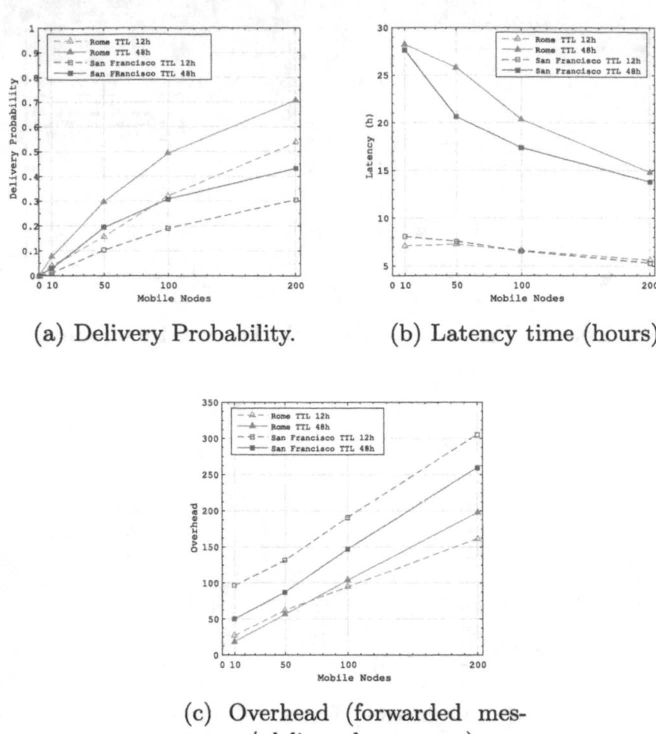

(a) Delivery Probability.

(b) Latency time (hours).

(c) Overhead (forwarded messages / delivered messages).

Fig. 6. Average delivery success ratio, latency and overhead. In these experiments mobile nodes act as relays, and fixed nodes are the sources and destinations of messages.

show that the substitution of mobile nodes in place of fixed nodes improves the amount of data delivered to the intended recipients and further, reduces the delivery times of the data. Therefore, it can be said that introducing mobile nodes, even in a small quantity can bring better network performance than fixed nodes as the fixed nodes may end up being islands of caches without the possibility of passing on cached data.

4.6 Delay Is Always Significant

The community has been discussing for a long time now what are the advantages and disadvantages of opportunistic networks. In the last paragraphs, we have shown how much potential mobility offers by easily connecting very sparse networks. However, we confirmed also one large disadvantage: the delay is significant in all cases. This is an intrinsic property of opportunistic networks and probably will remain the main challenge for future research. However, it has to be mentioned here also, that the delay is probably due to the destination-oriented way of diffusing messages. If you consider an alert application, where everybody

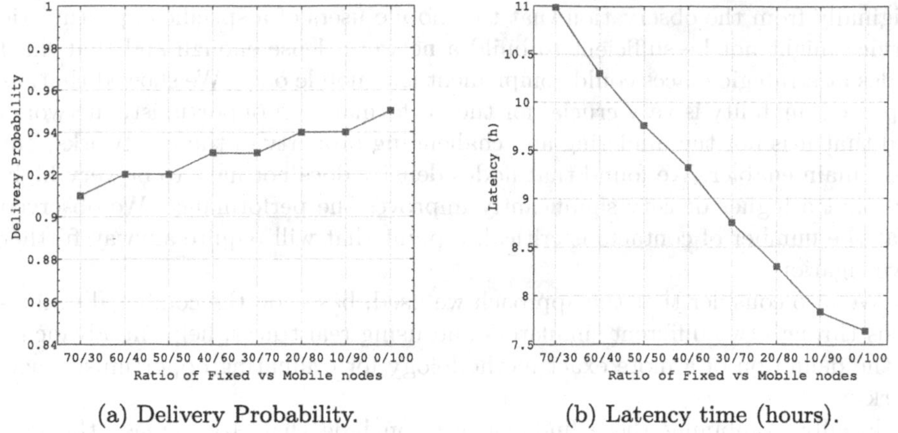

(a) Delivery Probability. (b) Latency time (hours).

Fig. 7. A network with 100 nodes with different percentages of mobile against fixed nodes. There are always exactly 30 nodes producing data in the network and random destinations. The scenario is San Francisco.

in a certain radius around has to be notified quickly, we expect the delay to be really low. We will focus on this in our future work.

4.7 OMNeT++ Versus ONE

As mentioned in the beginning of this section, we did the experiments described here both in the OMNeT++ and the ONE simulators. Even if the used models (mobility traces, Epidemic routing, transmission radius of 50 m, etc.) were the same, the implementations are very different due to different programming languages and different simulation kernels. It is good to report that the results can be considered identical, with differences below 0.1%. However, it needs to be also mentioned that both simulators have advantages and disadvantages. While ONE has already many opportunistic protocols ready to use, for OMNeT++ we needed to implement those protocols. On the contrary, OMNeT++ runs faster: for a simulation of 50 nodes (among the smallest we had here) ONE needs more than 11 h, while OMNeT++ finishes in 2.5 h on the same hardware configurations. This is probably due to the fact that OMNeT++'s kernel is written in C++.

5 Conclusions and Future Work

In this paper we evaluated the combined use of fixed and mobile nodes to establish an urban opportunistic network aimed at the distribution of general interest data. We used simulations based on two real-world movement traces from taxicabs of Rome and San Francisco, with a workload characterizing a large text messages applications. We used WiFi-Direct as the data transmission technology.

We determined that fixed nodes only increase the overhead while mobile nodes boost the performance significantly. The idea of adding fixed nodes came originally from the observation that the mobile users of a specific opportunistic service might not be sufficient to build a network dense enough and that fixed nodes at strategic places could complement the mobile ones. We showed that, as expected, mobility is very crucial for the performance of opportunistic networks and that it is not the hindering and challenging property of these networks, but their main enabler. We found that nodes density does not have to be very high, but that a higher density significantly improves the performance. We observed that the number of contacts is critical, a point that will require anyway further investigation.

We also consider that the approach we used, based on the combined evaluations through two different simulators and using real traces, helps in advancing in the definition of a more exact methodology for evaluating opportunistic networks.

Finally, combining these findings, we conclude that, to increase the performance in opportunistic networks by increasing the node density, should be obtained by adding mobile nodes - not fixed ones. Mobile nodes are able to boost the performance, while fixed nodes only increase the overhead. Thus, when needed, adding mobile nodes like for example on buses, trains or taxis, is the best strategy to efficiently establish an urban opportunistic network.

Acknowledgments. This work was partially supported by the *Ministerio de Economía y Competitividad, Programa Estatal de Investigación, Desarrollo e Innovación Orientada a los Retos de la Sociedad, Proyectos I+D+I 2014*, Spain, under Grant TEC2014-52690-R, the *Generalitat Valenciana*, Spain, under Grant AICO/2015/108, the Secretaría Nacional de Educación Superior, Ciencia, Tecnología e Innovación del Ecuador (SENESCYT), the Universidad Laica Eloy Alfaro de Manabí, Ecuador, and the University of Bremen, Germany.

References

1. Conti, M., Giordano, S.: Mobile ad hoc networking: milestones, challenges, and new research directions. IEEE Commun. Mag. **52**(1), 85–96 (2014)
2. Poonguzharselvi, B., Vetriselvi, V.: Survey on routing algorithms in opportunistic networks. In: 2013 International Conference on Computer Communication and Informatics, ICCCI 2013 (2013)
3. Khilar, P.M., Bhoi, S.K.: Vehicular communication: a survey. IET Netw. **3**(3), 204–217 (2014)
4. Al-Sultan, S., Al-Doori, M.M., Al-Bayatti, A.H., Zedan, H.: A comprehensive survey on vehicular ad hoc network. J. Netw. Comput. Appl. **37**(1), 380–392 (2014)
5. Vegni, A.M., Campolo, C., Molinaro, A., Little, T.D.C.: Modeling of intermittent connectivity in opportunistic networks: the case of vehicular ad hoc networks. In: Woungang, I., Dhurandher, S., Anpalagan, A., Vasilakos, A. (eds.) Routing in Opportunistic Networks. Springer, New York (2013)
6. Thakur, G.S., Kumar, U., Helmy, A., Hsu, W.-J.: On the efficacy of mobility modeling for DTN evaluation: analysis of encounter statistics and spatio-temporal preferences. In: 2011 7th International Wireless Communications and Mobile Computing Conference (IWCMC), Istanbul, Turkey, pp. 510–515 (2011)

7. Dias, J.A., Rodrigues, J.J., Zhou, L.: Cooperation advances on vehicular communications: a survey. Veh. Commun. **1**(1), 22–32 (2014)
8. Martín-Campillo, A., Crowcroft, J., Yoneki, E., Martí, R.: Evaluating opportunistic networks in disaster scenarios. J. Netw. Comput. Appl. **36**, 870–880 (2013)
9. Bracciale, L., Bonola, M., Loreti, P., Bianchi, G., Amici, R., Rabuffi, A.: CRAWDAD dataset roma/taxi (2014). Accessed 17 July 2014
10. Piorkowski, M., Sarafijanovic-Djukic, N., Grossglauser, M.: CRAWDAD dataset epfl/mobility (2009). Accessed 24 Feb 2009
11. Keränen, A., Ott, J., Kärkkäinen, T.: The ONE simulator for DTN protocol evaluation. In: Proceedings of the Second International ICST Conference on Simulation Tools and Techniques, Rome, Italy (2009)
12. Natalizio, E., Loscrí, V.: Controlled mobility in mobile sensor networks: advantages, issues and challenges. Telecommun. Syst. **52**(4), 2411–2418 (2013)
13. Neena, V.V., Rajam, V.M.A.: Performance analysis of epidemic routing protocol for opportunistic networks in different mobility patterns. In: 2013 International Conference on Computer Communication and Informatics, Coimbatore, India, pp. 1–5 (2013)
14. Hernández-Orallo, E., Herrera-Tapia, J., Cano, J.-C., Calafate, C.T., Manzoni, P.: Evaluating the impact of data transfer time in contact-based messaging applications. IEEE Commun. Lett. **19**, 1814–1817 (2015)
15. de Abreu, C.S., Salles, R.M.: Modeling message diffusion in epidemical DTN. Ad Hoc Netw. **16**, 197–209 (2014). Benidorm, Spain
16. Zhang, Y., Zhao, J.: Social network analysis on data diffusion in delay tolerant networks. In: Proceedings of the Tenth ACM International Symposium on Mobile Ad Hoc Networking and Computing - MobiHoc 2009, pp. 345–346 (2009)
17. Herrera-Tapia, J., Hernández-Orallo, E., Manzoni, P., Tomas, A., Calafate, C.T., Cano, J.-C.: Evaluating the impact of data transfer time and mobility patterns in opportunistic networks. In: 2016 International IEEE Conferences on Ubiquitous Intelligence & Computing, Advanced and Trusted Computing, Scalable Computing and Communications, Cloud and Big Data Computing, Internet of People, and Smart World Congress (UIC/ATC/ScalCom/CBDCom/IoP/SmartWorld), pp. 25–32 (2016)
18. Herrera-Tapia, J., Hernández-Orallo, E., Tomas, A., Manzoni, P., Calafate, C.T., Cano, J.-C.: Improving Message Delivery Performance in Opportunistic Networks Using a Forced-Stop Diffusion Scheme, pp. 156–168. Springer, Cham (2016)
19. Förster, A., Garg, K., Nguyen, H.A., Giordano, S.: On context awareness and social distance in human mobility traces. In: Third ACM International Workshop on Mobile Opportunistic Networks, Zürich, Switzerland, pp. 5–12 (2012)
20. Boldrini, C., Conti, M., Passarella, A.: Modelling data dissemination in opportunistic networks. In: Proceedings of the third ACM Workshop on Challenged Networks - CHANTS 2008, San Francisco, USA, pp. 89–96 (2008)
21. Herrera-Tapia, J., Hernández-Orallo, E., Tomás, A., Manzoni, P., Calafate, C.T., Cano, J.-C.: Friendly-sharing: improving the performance of city sensoring through contact-based messaging applications. Sensors **16**(9), 1523 (2016)
22. Costa, P., Gavidia, D., Koldehofe, B., Miranda, H., Musolesi, M., Riva, O.: When cars start gossiping. In: Proceedings of the 6th Workshop on Middleware for Network Eccentric and Mobile Applications - MiNEMA 2008, pp. 1–4 (2008)
23. Zhu, H., Li, M.: Dealing with vehicular traces. In: Studies on Urban Vehicular Ad-hoc Networks, pp. 15–21. Springer New York (2013)

24. Sanguesa, J.A., Fogue, M., Garrido, P., Martinez, F.J., Cano, J.C., Calafate, C.T.: A survey and comparative study of broadcast warning message dissemination schemes for VANETs. Mob. Inf. Syst. **2016**, 18 (2016)
25. Luo, P., Huang, H., Shu, W., Li, M., Wu, M.-Y.: NET 07–2 - performance evaluation of vehicular DTN routing under realistic mobility models. In: 2008 IEEE Wireless Communications and Networking Conference, pp. 2206–2211 (2008)
26. Amici, R., Bonola, M., Bracciale, L., Rabuffi, A., Loreti, P., Bianchi, G.: Performance assessment of an epidemic protocol in VANET using real traces. Procedia Comput. Sci. **40**, 92–99 (2014)
27. Bischoff, J., Maciejewski, M., Sohr, A.: Analysis of Berlin's taxi services by exploring GPS traces. In: 2015 International Conference on Models and Technologies for Intelligent Transportation Systems, MT-ITS 2015, December 2012, pp. 209–215 (2015)
28. Fu, Q., Zhang, L., Feng, W., Zheng, Y.: DAWN: a density adaptive routing algorithm for vehicular delay tolerant sensor networks. In: 2011 49th Annual Allerton Conference on Communication, Control, and Computing, Allerton 2011, pp. 1250–1257 (2011)
29. Marquez-Barja, J.M., Ahmadi, H., Tornell, S.M., Calafate, C.T., Cano, J.C., Manzoni, P., DaSilva, L.A.: Breaking the vehicular wireless communications barriers: vertical handover techniques for heterogeneous networks. IEEE Trans. Veh. Technol. **64**(12), 5878–5890 (2015)
30. Chen, Q.: Multi-metric opportunistic routing for VANETs in urban scenario. In: 2011 International Conference on Cyber-Enabled Distributed Computing and Knowledge Discovery, pp. 118–122 (2011)
31. Förster, A., Udugama, A., Görg, C., Kuladinithi, K., Timm-Giel, A., Cama-Pinto, A.: A novel data dissemination model for organic data flows. In: 7th EAI International Conference on Mobile Networks and Management (MONAMI), Santander, Spain (2015)
32. Mallanda, C., Else, S., Suri, A., Kunchkarra, V., Iyengar, S., Kannan, R., Durresi, A.: Simulating wireless sensor networks with OMNeT++. IEEE Computers (2005)
33. Karney, C.F.F.: Transverse Mercator with an accuracy of a few nanometers. J. Geodesy **85**(8), 475–485 (2011)
34. Karp, R., Schindelhauer, C., Shenker, S., Vocking, B.: Randomized rumor spreading. In: Proceedings of 41st Annual Symposium on Foundations of Computer Science, pp. 565–574. IEEE (2000)

Analysis and Classification of the Vehicular Traffic Distribution in an Urban Area

Jorge Luis Zambrano-Martinez[1](✉), Carlos T. Calafate[1](✉), David Soler[2], Juan-Carlos Cano[1], and Pietro Manzoni[1]

[1] Department of Computer Engineering (DISCA), Universitat Politècnica de València, Valencia, Spain
jorzamma@doctor.upv.es, {calafate,jucano,pmanzoni}@disca.upv.es

[2] Institute of Multidisciplinary Mathematics (IMM), Universitat Politècnica de València, Valencia, Spain
dsoler@mat.upv.es

Abstract. Nowadays, one of the main challenges faced in large metropolitan areas is traffic congestion. To address this problem, an adequate traffic control could produce many benefits, including reduced pollutant emissions and reduced travel times. If it were possible to characterize the state of traffic by predicting traffic conditions, measures could be taken to preventively mitigate the effects of congestion and related problems. This paper performs an experimental study of the traffic distribution in the city of Valencia, characterizing the different streets of the city in terms of vehicle load with respect to the travel time during rush hour traffic conditions. Experimental results based on realistic vehicular traffic traces show that most of the street segments under analysis present a good fit under quadratic regression, although a large number of street segments fall under other categories mainly due to lack of traffic. Based on this study, a clustering analysis study associated to the different streets shows how these streets can be classified into four independent categories, evidencing an uneven traffic distribution throughout the city.

Keywords: Traffic prediction · Traffic behavior · Clustering · Urban traffic · Valencia

1 Introduction

Traffic congestion can significantly increase the travel time of vehicles, being directly associated to increased delays, an inefficient use of fuel, and an increase of CO_2 emissions, being all of them critical issues for city authorities [1]. As we gradually move towards a new paradigm centered on automated vehicles, we are able to empower traffic administrators with more sophisticated ways of regulating traffic compared to the usual strategies based on semaphore timing regulations, or the deployment of traffic agents on site. Among these novel techniques of handling traffic, centralized route management emerges as a solution

A. Puliafito et al. (Eds.): ADHOC-NOW 2017, LNCS 10517, pp. 121–134, 2017.
DOI: 10.1007/978-3-319-67910-5_10

offering authorities full control of the traffic flow [2], allowing traffic optimization to reach maximum levels of effectiveness by deciding the route to be followed by each individual vehicle. Another approach would be to rely on vehicles themselves to decide on their preferred route, while endowing them with knowledge of the traffic congestion status beforehand. This approach, although not so efficient as the centralized approach, would still allow exploring alternative routes, which would alleviate congestion in the city and so improve the overall traffic flow.

When progressing towards these new traffic management paradigms, it becomes mandatory to gain full awareness of the behavior of the different street segments of a city in terms of how the travel time can vary depending on the number of vehicles simultaneously traveling on a particular segment. Such knowledge is critical in order to properly model a city, being such modeling effort a prerequisite to enable the envisioned traffic management systems.

In this work we rely on realistic traffic models describing the traffic behavior in the city of Valencia during rush hours, as detailed in our previous work [3]. In particular, we start from induction loops measurements made available by the City Hall of Valencia [4], and by using the DFROUTER tool, along with an heuristic that iteratively refines the output produced by this tool, we generate an O-D traffic matrix that resembles the real traffic distribution. In this paper we use this reference traffic as input load to the SUMO mobility simulation tool, to analyze how traffic becomes distributed along the city, gathering details about the number of vehicles traveling along the different street segments, as well as their travel times. Post-processing of the gathered data allows merging segments when excessive fragmentation is detected, characterizing the different streets in terms of travel time behavior under variable traffic loads. In particular, we sample the travel times of vehicles entering a segment along with the number of vehicles already in that segment, allowing to extract a relation between street occupation and delay. Through regression we show that a quadratic fit is adequate for the most representative cases. Additionally, a clustering analysis of the load with respect to the travel-time associated to the different streets allows detecting four independent categories, whose characteristics are then properly discussed.

The paper is organized as follows: Sect. 2 presents some related work regarding studies that predict traffic behavior using different approaches. Section 3 provides information on the SUMO and DFROUTER tools, along with our iterative heuristic. Section 4 describes the methodology that has been used to achieve the desired per-segment traffic modeling, starting from induction loop data and ending in a classification of the different street segments. The results obtained in terms of traffic characterization through clustering are then shown in Sect. 5. Section 6 discusses the relevance of the results obtained, and their potential regarding future traffic management systems. Finally, Sect. 7 concludes the paper.

2 Related Work

Some works dealing with traffic prediction involve learning algorithms such as fuzzy logic [5], although facing some problems such as low accuracy and efficiency.

For instance, Onieva et al. [6] present a case study in which an automated vehicle must cooperate with a driver to achieve cross-road maneuvers without risk, and a three layer hierarchical fuzzy rule-based system is developed. The first layer detects the type of maneuver that is needed, the second layer is the appropriate speed to cross an intersection, and the third layer is the actual speed of the vehicle. Hodge et al. [7] present a binary neural network algorithm for short-term traffic flow prediction using univariate and multivariate data from a single traffic sensor with temporal delays, and combining information from multiple traffic sensors with time series prediction or spatial-temporal lags. Porikli et al. [8] train a set of Hidden Markov Model chains corresponding to five traffic patterns (stop, heavy congestion, open flow, moderate, and empty congestion), and then use a Maximum Likelihood criterion to determine the state of the separated Hidden Markov Models. Differently from previous works, Kunt et al. [9] focus on predicting the severity of motorway traffic accidents by employing twelve accident-related parameters in an artificial neural network, genetic algorithm, and pattern search methods.

Sananmongkhonchai et al. [10] propose an algorithm based on cells to predict the travel time, estimating the traffic conditions when having multiple GPS receivers integrated in taxis. However, GPS accuracy depends on additional factors such as satellite geometry, signal blocking, atmospheric conditions, and receiver design features. In addition, other studies such as [11] and [12] involve vehicle probes for the prediction and detection of incidents in an automated manner. The problem of using raw probe data is that the estimation accuracy is primarily based on driver behavior.

In this paper, a method that accounts for the number of existing vehicles to determine the expected travel time when entering a street is developed. This method allows achieving a detailed characterization for each specific street segment, thereby accounting for its singularities, and allowing us to predict traffic flow levels at any specific traffic load on a microscopic basis. This way our technique enables improving travel time estimations, thereby reducing traffic congestion.

3 Overview of the Simulation Tools Used

In this section we provide some details about the SUMO traffic simulator [13]. We will also introduce DFROUTER [14], and briefly explain how it allows generating a traffic matrix detailing origins and destinations (typically known as O-D matrix) for SUMO based on induction loop data.

3.1 SUMO Synopsis

Usually, the traffic model consists of obtaining some variables, such as the departure and arrival times, and the streets that vehicles pass through.

SUMO [13] performs the simulation of vehicular mobility through a detailed microscopic modeling of cities and vehicles. In fact, being an open source simulator, it is constantly improved, being widely accepted by the scientific community.

Its features include support for different map formats including OpenStreetMap, importing road networks in multiple formats, and generating routes with multiple sources. Also, it offers high-performance simulation capabilities through the TraCI interface, allowing it to perform interactive simulations, and enabling many more features when coupled with another simulator like OMNeT++ [15].

The flow of traffic is simulated microscopically, meaning that each vehicle movement within the road network is individually modeled, which allows us to know its location, speed, acceleration, time of departure, and time of arrival. By default, each time step has a duration of one second, which allows a discrete simulation of continuous mobility in space.

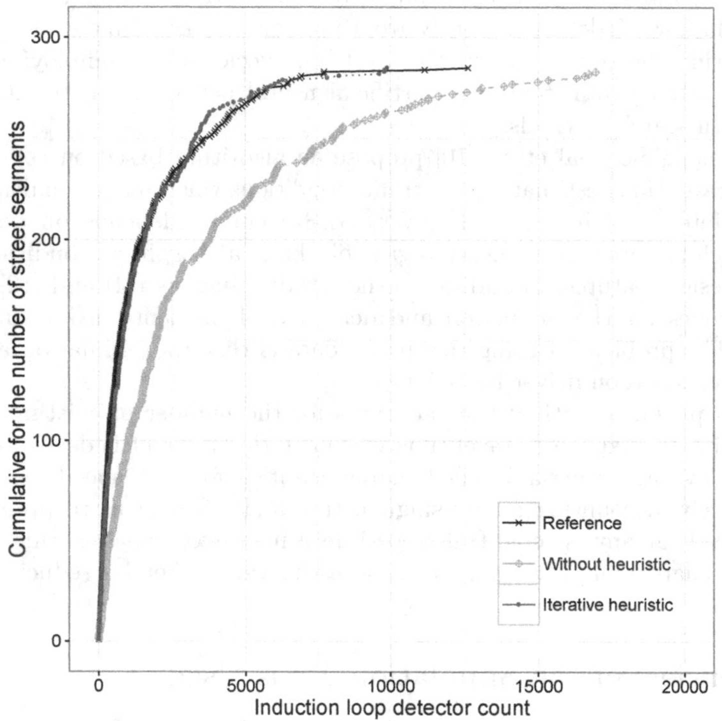

Fig. 1. Traffic flow modeling for Valencia. Results with and without our iterative heuristic [3].

3.2 O-D Matrix Generation with DFROUTER

One of the packages included by the SUMO simulator is the DFROUTER tool. This tool has been designed for road scenarios based on the main idea that roads are equipped with induction loops that allow measuring the inflow and outflow of the roads. DFROUTER can reconstruct the number of vehicles and routes to be injected into the simulator of the road network, based on the data obtained from induction loops such as number of vehicles, flows, and speeds, to achieve the desired O-D traffic matrix. In other words, this tool allows, starting induction loop counts for the different roads of a city, is able to estimate the possible vehicle routes that match such input.

In a previous work [3] we have used induction loop data provided by the City Hall of Valencia, Spain, and corresponding to a period between 8 and 9 a.m. for a typical Monday, as input to DFROUTER. In that work we noticed that there was a significant mismatch between the traffic generated and the original data, requiring the introduction of an iterative heuristic that compensates for this error by refining the output provided by this tool in order to achieve an O-D matrix that resembles the real traffic distribution. Figure 1 shows that, as a result, the output of the iterative process achieves a high level of matching with the reference data, resulting in an error lower than 0.0001, being significantly better than the initial DFROUTER output.

4 Methodology

In this section we describe the procedure followed to characterize the traffic in the city of Valencia, Spain, from a microscopic perspective, and starting from OpenStreetMap road layouts. In particular, our goal is to characterize individual street segments in terms of average travel times experienced by vehicles for different degrees of congestion, being the latter estimated based on the number of vehicles found ahead by a vehicle just entering a segment. To achieve this goal, we found necessary to first perform some preprocessing, as in many cases the presence of micro-segments (streets unnecessarily partitioned in many short segments) impeded an adequate analysis of the behavior of vehicles traversing particular streets, as we can see in Fig. 2. Then, we used the SUMO tool coupled

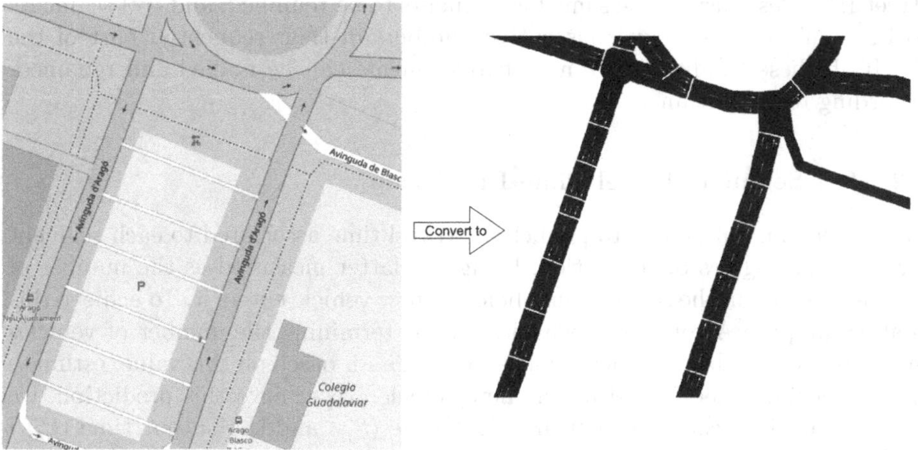

Fig. 2. Example of unnecessary street partitioning.

with the OMNeT++ simulator to study the traffic flow for the entire city of Valencia based on a realistic traffic trace, as described in the previous section.

Below we describe the methodology followed to characterize and predict traffic for the different street segments. Our proposed methodology first unifies segments whenever required, then it allows predicting the number of vehicles in each segment, and finally, it characterizes the different street segments through regression. Afterward, we detail the algorithms we proposed to unify segments, predict traffic time, and finally characterize the different segments according to the traffic travel time.

4.1 Unifying Segments

Usually, when the city map is converted to a format accepted by SUMO for simulation, certain characteristics of the map must be eliminated, such as bicycle paths, pedestrian paths, train tracks, etc. This conversion has a drawback because it causes the streets to be intercepted by other ways, different from those used by vehicles, and the SUMO simulator acts by partitioning those streets. This inconvenience causes, in many cases, that (i) streets are partitioned into tiny segment sizes, often measuring less than 7.5 m (size of a vehicle plus inter-vehicular security gap), (ii) such small sizes do not allow to characterize the segment profile correctly, and (iii) inconsistent graphs are obtained when applying the polynomial regression to predict traffic behavior.

To understand the solution adopted to address this issue, we should mention that the ID that represents a street segment is composed of two parts, where the first part is the code identifying the street, and the second part is the sequential code assigned by SUMO to each street partition. The proposed procedure tries to unify those street segments whenever possible, if certain conditions are met. These conditions are: (i) the street to be reunified must be partitioned, (ii) the adjacent segment should not have another segment that intersects it, (iii) the street ID codes must be the same for segments to be reunified, and (iv) segments to be reunified must have consecutive numbers in their sequential part of the ID. If all these conditions are met, two segments can be unified and renamed, according to Algorithm 1.

4.2 Per-Segment Travel Time Prediction

In this section, our goal is to predict the travel time associated to each segment for different degrees of congestion, being the latter measured as the number of vehicles located in the segment just before a new vehicle enters it. To achieve this goal we propose Algorithm 2, which allows determining the number of vehicles in a segment (ν_n) before a new vehicle joins it, and, based on this value, estimate the travel time experienced by the new vehicle. To achieve this prediction, we have to take into consideration the input time (t_{in}^{ν}) and the output time (t_{out}^{ν}) of the vehicle in the segment, as well as the number of lanes of the segment (l_n) where the vehicle is traveling. Input times (t_{in}) for vehicles entering a segment

Algorithm 1. Reunification of segments.

```
Require: Road Network file, edges files
Ensure: Reunified segment file
edgeConnectedNoIntersection[] ← dictionary that stores all edges without intersections
for all edge in Road Network file do
    edge_id ← store the edge id of the road network file
    connections[] ← dictionary that stores all connections for that edge id
    for all connection in Road Network file do
        connection_from ← store the edge id(from) of the road network file
        connection_to ← store the edge id(to) of the road network file
        if (connection_from = edge_id) and (connection_to not in connections) then
            connections[edge_id] ← connection_to
        end if
    end for
    if edge_id partition = TRUE then
        lenEdgeConnect ← length of dictionary in a specific edge id
        street_id ← code of the street
        for i = 0 to length(connections[edge_id]) do
            if (lenEdgeConnect = 1) and (connections[edge_id][i] partition = TRUE) and (street_id in edge_id) and (street_id in connections[edge_id][i]) then
                edgeConnectedNoIntersection[edge_id] ← connections[edge_id][i]
            else
                edge_id has some intersection
            end if
        end for
    else
        edge_id is not split
    end if
end for
```

on lane l_n are registered in matrix (l_n, t_{in}), while output times (t_{out}) for vehicles leaving the segment at lane l_n are then registered in matrix (l_n, t_{out}).

The number of vehicles in the segment before a vehicle joins it will increase as long as the t_{in}^{ν} is less than t_{out}^{ν}, and both refer to the same lane. Then, the travel time of each vehicle in the segment will be obtained (Δt).

As a final step, according to Algorithm 2, an average of the travel times $(\bar{\Delta} t)$ associated to different degrees of congestion (number of vehicles in the segment before a new vehicle enters that segment) will be included in a file, along with the number of vehicles in that segment (ν).

4.3 Segment Behavior Characterization

Once the process described above to estimate travel times in a segment for different degrees of congestion is completed, the next step is to characterize and

Algorithm 2. Extraction of travel times vs. load samples.

Require: Reunified segment file, Segment-info files
Ensure: Statistical learning by segment files
for segment **in** Reunified segment file **do**
  segmentConnected[] ← vector that store all edges id connected
  for s=0 **to** length(*segmentConnected*) **do**
    segment_info ← Read lines *segmentConnected*[s] in Segment-info files
    segmentSorted[][] ← sort_by_t_{in}(*segment_info*)
    for t_{in}=0 **to** length(*segmentSorted*) **do**
      ν_n ← number vehicles per segment in each lane
      for $t_{out} = t_{in}$ **to** t_{out} >= 0 **step -1 do**
        if (*segmentSorted*[t_{in}][l_n] = *segmentSorted*[t_{out}][l_n]) **and** (*segmentSorted*[t_{in}][t^{ν}_{in}] <= *segmentSorted*[t_{out}][t^{ν}_{out}]) **then**
          $\nu_n = \nu_n + 1$
        end if
      end for
    end for
    $\bar{\Delta t}$ ← average of the travel times
    ν ← number of vehicles in the segment before a new vehicle enters the segment
  end for
end for

classify the behavior of the different segments. In particular, we seek to determine the relationship between the number of vehicles in a segment (x values), and the average travel time of vehicles (y values) for each particular segment. To achieve this goal we perform regression to obtain the best curve fit describing the nonlinear relationship between segment congestion and travel time. To perform the fitting we used function $f(x) = ax^2 + c$, which belongs to the second-order polynomial family, as traffic theory in general considers that the relationship between traffic load and travel time tend to vary quadratically. Thus, the chosen expression is able to adequately represent this parabolic behavior starting from the free flow travel time (no congestion), represented by constant c in this function, and then increasing as the number of vehicles ahead in a segment (represented by x) increases.

Once the regression results for all the segments tested where obtained, we observed that the expected quadratic behavior was indeed taking place in many of the segments, although other special cases were also detected. Figure 3 presents different representative cases corresponding to the patterns we observed. The first class of regression curves, which illustrates the expected pattern according to traffic engineering theory, is shown in Fig. 3a. As can be seen, when a vehicle entering a segment finds many vehicles ahead, it will on average experience much higher travel times, with differences up to 1000% being possible and expected. However, other patterns were also obtained, as traffic flow properties also cause other types of behavior to take place, especially when modeling a very large city like Valencia. For instance, the second class of curve represents a behavior that

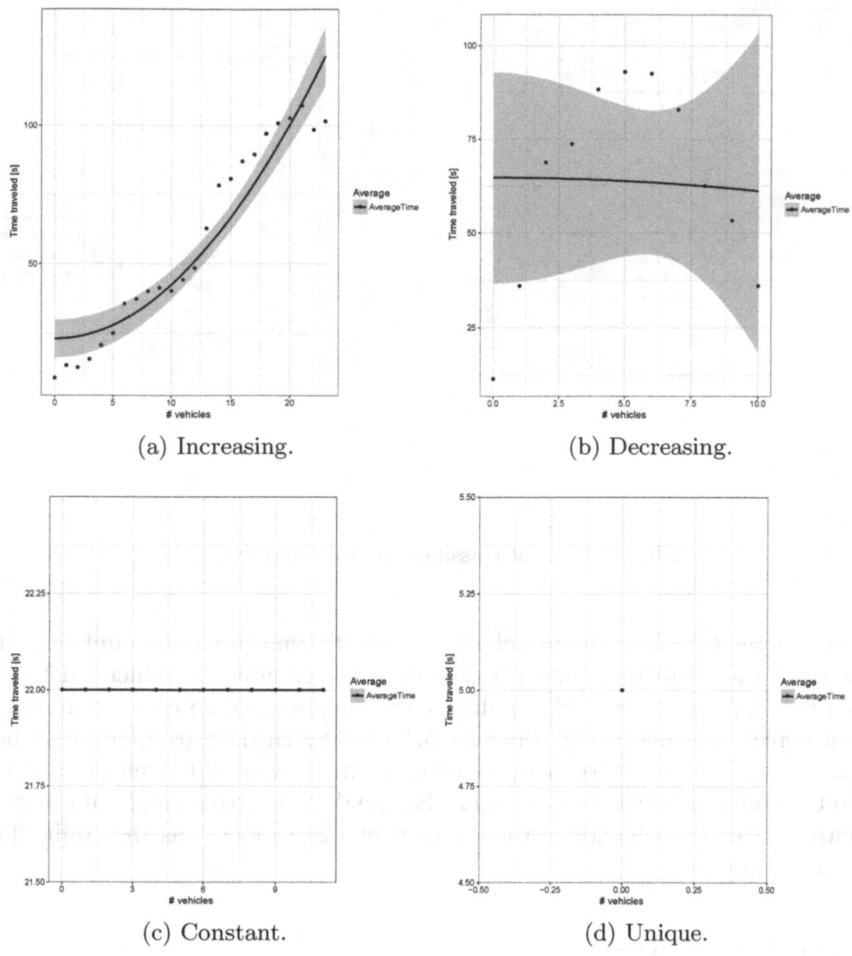

(a) Increasing. (b) Decreasing.

(c) Constant. (d) Unique.

Fig. 3. Segment classification.

is just the opposite compared to the previous one. As shown in Fig. 3b, this kind of curve shows an increase followed by a decrease in the time traveled as we increase the number of vehicles ahead. Such behavior is explained by different factors, including the departure of vehicles from the segment, as they turn to join other segments, and, more important, the presence of traffic lights that tend to accumulate vehicles on the segment, being that vehicles finding many vehicles ahead usually means that the accumulation period was long, and the semaphore is about to turn green.

In addition to the two types of behavior described above, there are also other cases taking place, as exemplified in Figs. 3c and d. Regarding the behavior observed in Fig. 3c, we can see that the travel time remains constant regardless of the number of vehicles in the segment, typically meaning that there are

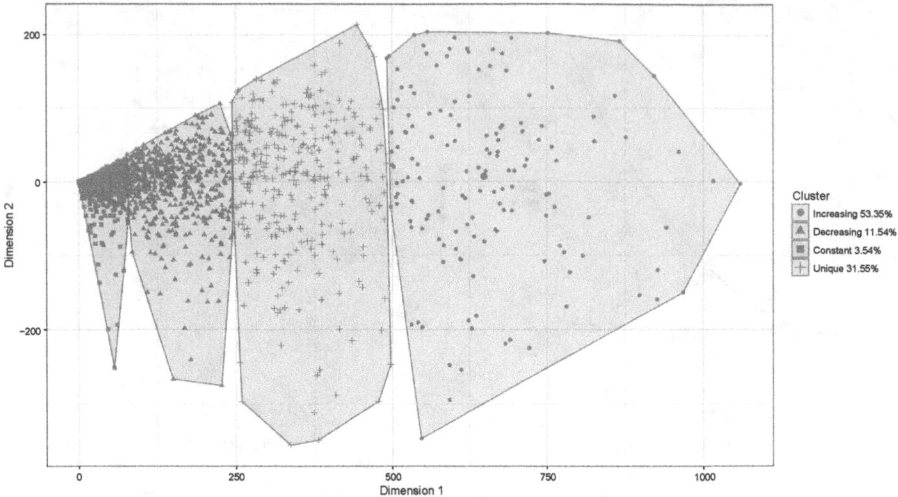

Fig. 4. Segment classification by clustering.

no semaphores (areas in the periphery of the city), no junctions, and that the capacity of the segment is much higher than the number of vehicles detected during the simulation (typically multi-lane segments), and so congestion effects are not perceived. Finally, regarding the behavior for the last group, as described in Fig. 3d, it corresponds to one-lane segments rarely visited by vehicles according to the traffic patterns used as input. So, we do not have enough information to characterize the behavior of the segment at higher loads, as the traffic flow levels are minimal.

5 Clustering Results

In Sect. 4, we empirically detected some patterns based on a quick overview of the regression outcomes. However, to properly classify those patterns, systematic and automated approach is required. To this end, in this section we apply clustering techniques to formally group the different street segments according to their behavior. This approach meets our requirements, as it is able to provide an automatic classification of the segments belonging to the city of Valencia according to their profile in terms of congestion and traffic load.

The number of segments in this scenario is 8126. To properly categorize these segments in their respective groups, we will use a popular machine learning technique for classification called K-means [16]. In particular, and reminding that equation $f(x) = ax^2 + c$ was used for regression, we will use the regression variables to automatically categorize the segments, and thereby group them into different categories according to their behavior. Regarding variable a, it will allow discriminating between increasing, constant, and decreasing trends. Similarly, variable c can help at distinguishing between segments according to their free-flow speeds. Finally, we also used $f(x)_{max}$ as an input to the classification procedure, since it represents the highest travel time associated to a particular segment.

Once the input variables for the exploratory data technique are defined, we obtained which groups can be automatically generated. As a result, the K-means algorithm clearly identified four clusters. In addition, to make the result easier to represent graphically, we adopted a statistical procedure called Principal Component Analysis (PCA) to transform the initial input variables into a reduced set of variables that are still able to explain the target phenomenon, being this reduced variable set a linear combination of the original data known as principal components [17]. Thus, since each input variable is considered as different dimension, we used PCA to reduce the dimensionality of the multivariate data to two dimensions, thereby allowing to visualize the resulting partitioning graphically, and with a minimum information loss. In fact, the reduction to two dimensions is still able to explain 98% of the segment classifications obtained. The resulting chart can be seen in Fig. 4.

Overall we can see that, as a result of applying the clustering to the city of Valencia, we have a majority of segments (53.35%) exhibiting the expected congestion behavior, with increasing travel times when facing a higher number of vehicles ahead. Likewise, it can be observed that the percentage of segments that tend to be decongested is 11.54%, which would indicate that the city remains nevertheless congested in most parts. In addition, we find that only 3.54% of the segments are characterized by a constant travel time under the simulated conditions, which corresponds to segments without intersections. Finally, we have situations where only travel times without the presence of other vehicles could be retrieved; this situation corresponds to 31.55% of segments of the city, and is mostly associated to very small segments that persist despite the segment reunification procedure proposed in the paper.

6 Discussion

The characterization of real streets in terms of free flow speeds and expected travel times under different loads is a very complex issue since it accounts for all the details concerning that particular street, from lane widths to the presence of semaphores, among many other factors. Thus, this is a very costly analysis, usually only considered for the main streets of a city during the planning phase (previous to actual construction). The presence of accurate street maps, along with vehicular mobility simulators, provides another approach to this street characterization problem that enables a more global and faster analysis, as well as a characterization of the travel delay behavior expected by vehicles under real traffic loads. The main drawback of such approaches is that the accuracy of street characterization is limited to the actual accuracy of the maps adopted for the studies. Fortunately, map accuracy and details have been significantly improved in the last years, allowing to perform studies with a high degree of representativity.

From the results obtained in this paper, a complete characterization of all streets/segments in the city of Valencia is provided. Gaining access to such detailed characterization enables to predict travel times in real situations much more accurately by accounting for the expected load, which, as seen in the paper, can cause very substantial variations on the travel times compared to the free flow speeds typically used by navigation tools. In addition, it also enables traffic authorities to have a better view of traffic conditions, allowing to predict upcoming congestion problems. In a near future, it will also enable accounting for traffic loads by performing centralized route planning, providing advanced management features including traffic balancing throughout a city.

7 Conclusions and Future Work

Having a realistic traffic model for a specific target city is a key requirement to obtain meaningful simulation results when issues like traffic density and traffic patterns can have an impact on the conclusions derived from experiments. Achieving such realistic models typically requires describing traffic in terms of Origin-Destination (O-D) matrices. In addition, if aimed at developing advanced traffic management solutions, it becomes further necessary to have a more in-depth understanding of how traffic is distributed in a particular city, which basically requires performing a correct analysis and classification of such traffic.

Valencia is one of the many cities where the analysis of traffic is of primary importance. However, traffic authorities do not provide an O-D matrix describing traffic to enable further studies, nor do authorities provide a classification and characterization of vehicular travel times for the different streets of the city, and at different degrees of congestion.

The starting point of this paper is a realistic traffic model for Valencia derived in a previous work. Then, the contribution of this paper is the characterization of all the street segments in Valencia in terms of travel times when vehicles face different degrees of congestion. To achieve this characterization, we start by processing the map of the target area in order to merge segments of a same street whenever unnecessary fragmentation is detected and can be reversed. Then, we perform simulation experiments using SUMO to retrieve the travel times of vehicles when facing different degrees of saturation on the travelled segment. Finally, using polynomial regression, we perform curve adjustment to obtain an expression that allows characterizing these travel times.

Once all segments where characterized through a polynomial, our next contribution was to apply clustering in order to automatically classify segments according to their travel delay behavior. In particular, we applied the K-means technique to generate the clusters, followed by a Principal Component Analysis in order to extract the main clustering features that enable visual representation. The results of the clustering process clearly define four independent categories: segments with incremental traffic delays (the majority), segments with decremental traffic delays, segments with constant delays (typical loads do not cause congestion), and single value results corresponding to small segments rarely visited by vehicles.

As future work, we plan to develop a centralized traffic management platform that, based on the per-segment travel delay characterization provided in this paper, is able to globally minimize vehicle travel times by accounting for congestion and by performing load balancing.

Acknowledgments. This work was partially supported by Valencia's Traffic Management Department, by the "Ministerio de Economía y Competitividad, Programa Estatal de Investigación, Desarrollo e Innovación Orientada a los Retos de la Sociedad, Proyectos I+D+I 2014", Spain, under Grant TEC2014-52690-R, and the "Programa de Becas SENESCYT" de la República del Ecuador.

References

1. Jabali, O., Woensel, T., de Kok, A.G.: Analysis of travel times and CO_2 emissions in time-dependent vehicle routing. Prod. Oper. Manag. **21**(6), 1060–1074 (2012)
2. Djahel, S., Doolan, R., Muntean, G.M., Murphy, J.: A communications-oriented perspective on traffic management systems for smart cities: challenges and innovative approaches. IEEE Commun. Surv. Tutorials **17**(1), 125–151 (2015)
3. Zambrano, J.L., Calafate, C.T., Soler, D., Cano, J.C., Manzoni, P.: Using real traffic data for ITS simulation: procedure and validation. In: 2016 International IEEE Conferences on Ubiquitous Intelligence and Computing, Advanced and Trusted Computing, Scalable Computing and Communications, Cloud and Big Data Computing, Internet of People, and Smart World Congress (UIC/ATC/ScalCom/CBDCom/IoP/SmartWorld), pp. 161–170. IEEE, July 2016. doi:10.1109/UIC-ATC-ScalCom-CBDCom-IoP-SmartWorld.2016.0045
4. Calafate, C.T., Soler, D., Cano, J.C., Manzoni, P.: Traffic management as a service: the traffic flow pattern classification problem. Math. Prob. Eng. (2015). Article ID 716598. doi:10.1155/2015/716598
5. Zhang, X., Onieva, E., Perallos, A., Osaba, E., Lee, V.: Hierarchical fuzzy rule-based system optimized with genetic algorithms for short term traffic congestion prediction. Transp. Res. Part C Emerg. Technol. **43**, 127–142 (2014)
6. Onieva, E., Milanés, V., Villagra, J., Pérez, J., Godoy, J.: Genetic optimization of a vehicle fuzzy decision system for intersections. Expert Syst. Appl. **39**(18), 13148–13157 (2012)
7. Hodge, V.J., Krishnan, R., Jackson, T., Austin, J., Polak, J.: Short-term traffic prediction using a binary neural network. In: 43rd Annual UTSG Conference, York, January 2011
8. Porikli, F., Li, X.: Traffic congestion estimation using HMM models without vehicle tracking. In: 2004 IEEE Intelligent Vehicles Symposium, pp. 188–193. IEEE, June 2004
9. Kunt, M.M., Aghayan, I., Noii, N.: Prediction for traffic accident severity: comparing the artificial neural network, genetic algorithm, combined genetic algorithm and pattern search methods. Transport **26**(4), 353–366 (2011)
10. Sananmongkhonchai, S., Tangamchit, P., Pongpaibool, P.: Cell-based traffic estimation from multiple GPS-equipped cars. In: 2009 IEEE Region 10 Conference on TENCON 2009, pp. 1–6. IEEE, January 2009
11. Kerner, B.S., Rehborn, H., Aleksic, M., Haug, A.: Traffic prediction systems in vehicles. In: Proceedings of Intelligent Transportation Systems, 2005, pp. 72–77. IEEE, September 2005

12. Basnayake, C.: Automated traffic incident detection with GPS equipped probe vehicles. In: ION GNSS Proceedings of the 17th International Technical Meeting of the Satellite Division of the Institute of Navigation, pp. 1–10, September 2004
13. Behrisch, M., Bieker, L., Erdmann, J., Krajzewicz, D.: SUMO—Simulation of Urban Mobility: an overview. In: Proceedings of the Third International Conference on Advances in System Simulation, SIMUL 2011. ThinkMind (2011)
14. Nguyen, T.R.V., Krajzewicz, D., Fullerton, M., Nicolay, E.: DFROUTER—Estimation of vehicle routes from cross-section measurements. In: Behrisch, M., Weber, M. (eds.) Modeling Mobility with Open Data. LNM, pp. 3–23. Springer, Cham (2015). doi:10.1007/978-3-319-15024-6_1
15. Varga, A., Hornig, R.: An overview of the OMNeT++ simulation environment. In: Proceedings of the 1st International Conference on Simulation Tools and Techniques for Communications, Networks and Systems and Workshops, p. 60. ICST (Institute for Computer Sciences, Social-Informatics and Telecommunications Engineering), March 2008
16. Jain, A.K.: Data clustering: 50 years beyond K-means. Pattern Recogn. Lett. **31**(8), 651–666 (2010)
17. Jolliffe, I.: Principal Component Analysis. Wiley, New York (2014). doi:10.1002/9781118445112.stat06472

User-Space Network Tunneling Under a Mobile Platform: A Case Study for Android Environments

Dario Bruneo[1], Salvatore Distefano[1,2], Kostya Esmukov[2], Francesco Longo[1], Giovanni Merlino[1](✉), and Antonio Puliafito[1]

[1] Dipartimento di Ingegneria, Università degli Studi di Messina, Messina, Italy
{dbruneo,sdistefano,flongo,gmerlino,apuliafito}@unime.it

[2] Social and Urban Computing Group, Kazan Federal University, Kazan, Russia
s_distefano@it.kfu.ru, kostya@esmukov.ru

Abstract. The IoT ecosystem is taking the whole ICT world by storm and, in particular for currently hot topics such as Smart Cities, it is becoming one of the key enablers for innovative applications and services. When talking about end users, or even citizens, mobiles enter the picture as the ultimate personal gadget, as well as relevant outlets for most of the duties (sensing, networking, edge computing) IoT devices are typically envisioned in the first place. Smartphones, tablets and similar accessories are even more powerful in terms of hardware capabilities (and function diversity) than typical embedded systems for IoT, but it is typically the software platform (e.g., the OS and SDK) which limits choices for the sake of security and control on the user experience. Even a relatively open environment, such as Android, exhibits these limits, in stark contrast to the otherwise very powerful and feature-complete functionalities the underlying system (i.e., Linux) natively supports. In this work the authors describe a fully user-friendly and platform-compliant approach to let users break free from some of these limitations, in particular with regard to network virtualisation, for the purpose of extending an IoT-ready Smart City use case to mobiles.

Keywords: Stack4Things · OpenStack · Fog computing · IoT · Cloud · Network virtualization · VPN · Reverse tunneling

1 Introduction

Information and communication technologies (ICT) and solutions are progressing very quickly, radically changing the landscape. Recent trends, on the one hand, pushed towards more and more powerful computing infrastructure such as the Cloud ones, providing customizable computational resources as services through the Internet, elastically, on demand. This allowed to think about new offloading patterns where business logic processing is outsourced, ubiquitously offloaded to remote server while lightweight thin client are running on local

A. Puliafito et al. (Eds.): ADHOC-NOW 2017, LNCS 10517, pp. 135–143, 2017.
DOI: 10.1007/978-3-319-67910-5_11

nodes. This, on the other hand, favoured the widespread deployment of devices, initially mainly conceived for pervasively probing real-word/physical phenomena, gradually becoming more and more powerful and 'smart'. Network, Internet and mobile computing solutions then allowed to consider this ensemble as an ecosystem of smart objects or things, giving rise to the Internet of Things (IoT). Tens of billions of things already populate the IoT ecosystem, allowing to think about novel application domains such as Smart Cities, Industry 4.0, intelligent transportation systems, e-government, to name a few, while posing challenges related to heterogeneity, networking, scalability, security, high level management, among others.

In particular, after its first inception, mobiles and personal devices triggered a new wave for IoT. Indeed, rapid advances in embedded systems (especially mobile devices), wireless sensors, and mobile communications have led to the development of more and more powerful personal devices. Modern tablets and even smartphones, in terms of processing capabilities, can be compared to 4–5 years old computers, so they can be also used as effective computing systems. This introduces further complexity and uncertainty in the widely heterogeneous IoT world, but mainly enables novel unexplored opportunities. Albeit resource-constrained, a challenge is to think on how personal device processing, storage, sensing and networking capabilities could be exploited in IoT applications. In this direction, fog-edge-mist computing [17], aiming at exploiting onboard processing capabilities to compute sensed data locally, as well as mobile crowdsensing [12], implementing participatory and opportunistic contribution patterns for crowdsourced applications through mobile devices, are good examples. But several other opportunities can arise from mobiles in IoT, by properly exploiting their resources, ubiquity and unique combination of (built-in) technologies such as the wide range of communication subsystems. To this extent, the good news is that mobile hardware is ready for lots of scenarios. The bad one is that mobile OS/-platforms are naturally constrained due to security/privacy concerns and tight control and restrictions from manufacturers and telcos.

In this paper, we mainly focus on networking aspects, aiming at unlocking and properly exploiting communication capabilities natively provided by personal, mobile devices in IoT contexts. This is not trivial since in geographical contexts such as the IoT one several network issues can arise when interconnecting things from different administrative domains, subject to NAT, firewalls and similar restrictions. Advanced network features, widely available in opensource environments such as Linux, are definitely required by several IoT scenarios (e.g. smart city), but severely impaired under mobile OS such as Android. This is the case of network virtualization, among others. IoT network virtualization implies reconfiguration capabilities on mobile devices [10, 11], since the physical environment (channels, topologies, routers) in IoT scenarios is not always under control of the designer of the infrastructure, and may be opportunistically established, e.g., volunteer-contributed. Furthermore, the configuration of most deployments (or their extent, ownership, etc.) is usually not completely known in advance, in contrast to Wireless Sensor Networks (WSN) where some network virtualization solutions have been proposed [14].

Focusing on such specific aspects, the main contribution of this paper lies in relaxing these limitations by proposing a level-2/level-3 tunneling-based solution for IoT virtual networking adopting standard UNIX tools and mobile-native services under Android environments.

To this purpose, the remainder of the paper is organised as follows: Sect. 2 lays out the problem and related work in literature, in Sect. 3 the approach to the solution is described, Sect. 4 outlines a usage scenario together with a brief evaluation of the core mechanisms here outlined, and Sect. 5 closes the work with a summary of the results and hints to future work.

2 Background and Related Work

Network virtualization is a quite hot topic in IoT. Several works address related issues from different perspectives, including software defined networking (SDN), network functions virtualization and service function chains. A survey on the topic is available in [8], mainly focusing on the application of SDN paradigm to IoT contexts. An interesting solution is proposed in [14], integrating nodes (resource-constrained or otherwise) into a secured virtual network or IoTVN, enabling end-to-end communication among IoT nodes. The main focus is on providing a lightweight solution able to run on resource constrained device, while security and other issues are left to future work. A slightly different approach that could be somehow related to the networking topic is the opportunistic IoT one. In [13] an IoT framework is proposed extending opportunistic networking towards participatory sensing with the main aim of enabling information sharing among things to support mobile social networking. Similarly, opportunistic mobile networking is addressed in [19] dealing with low level data forwarding issues through a framework able to support and optimize opportunistic sensing. The underlying technologies are mainly based on ad-hoc network solutions.

Although promising, all these approaches partially solve the problem of connecting remote nodes in a geographical IoT context with network barriers. Ad-hoc networks and similar approaches have limited scope, while, on the other hand, the SDN approach is not lightweight and could be hardly adapted to resource constrained devices. Furthermore, restrictions from operating systems and software platform running on mobile devices require to attack the problem case by case, dealing with specific platform-dependent issues. The target of this paper is Android, aiming to implement virtual network/VPN mechanisms for IoT.

To this purpose, there are various solutions for VPN functionality under UNIX-derived systems, and even OS-native (i.e., kernel-space) VPN-enabling interfaces are available under ubiquitous Linux-based systems specifically. Indeed, under Linux tunnels may be natively instantiated using `/dev/tun` interface [2,6]. Such a mechanism has the following modes of operation:

- `tun` (OSI level 3 frames should be written to the device and read from it)
- `tap` (OSI level 2 frames)

On Android accessing `/dev/tun` requires root permissions, which cannot be granted to an application without "rooting" the device. Moreover, any functionality implemented in kernel-space requires the ability to install loadable kernel modules (LKM), another option which is not available without administrative privileges.

3 The Proposed Solution

In order to implement virtual networking/VPN mechanisms for mobiles in the IoT context, the authors base on existing solution coming from IoT and Cloud, more specifically on Stack4Things (S4T) [15]. Stack4Things is an OpenStack-based IoT framework that helps in managing IoT device fleets without caring about their physical location, their network configuration, their underlying technology. It is a Cloud-oriented horizontal solution providing IoT object virtualization, customization, and orchestration.

Among S4Tcloud-enabled services are node remoting and network virtualization. The basic remoting mechanisms are based on the creation of generic TCP tunnels over WebSocket (WS), a way to get client-initiated connectivity to any server-side local (or remote) service. In this sense, the authors devised the design and implementation of an incremental enhancement to standard WS-based facilities, i.e., a *reverse* tunneling technique, as a way to provide server-initiated, e.g., Cloud-triggered, connectivity to any board/mobile-hosted service. Beyond mere remoting, level-agnostic network virtualization needs mechanisms to overlay network- and datalink-level addressing and traffic forwarding on top of such a facility. In [16] the authors describe a model of tunnel-based layering employed for Cloud-enabled setup of virtualized bridged networks among nodes across the Internet, and Fig. 1 shows a conceptual depiction of the aforementioned model.

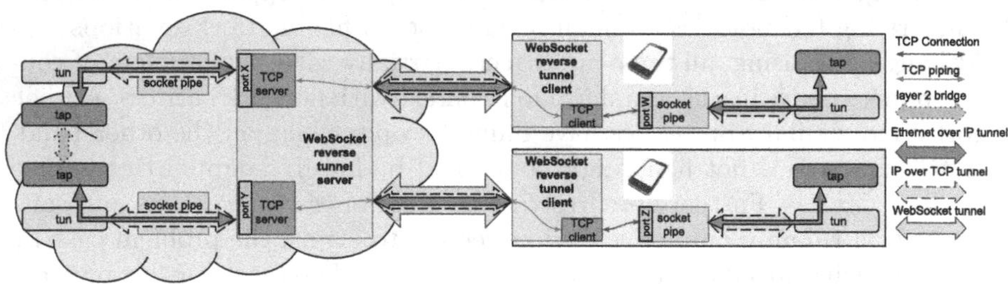

Fig. 1. Stack4Things tunnel-based layering: model.

Thus the ability to establish virtual networks among mobiles (as well as between a mobile and a server) requires at the very least core mechanisms to instantiate VPN-like tunnels between the corresponding endpoints. One way to actually create a tunnel between two Linux-like machines is using SOCAT [5].

This is a very simple program which can pipe two open file descriptors, like sockets, back-and-forth.

The most suitable way of creating tunnels on Android is to extend the `VpnService` class from Android API [7]. This class provides a simple way to create a L3 `tun` device. Actually VpnService is a thin wrapper around `/dev/tun`, which is opened in `tun` mode [4]. By default, a `/dev/tun`-based tunnel prepends (and expects to be prepended) each packet with the Packet Info (PI) structure: that is a 4 byte [3,6] struct with 2 fields: flags (2 bytes) and ether type (2 bytes). It doesn't seem very useful though, especially in L2 mode, but in L3 mode it allows to send IPv6 packets along with IPv4. Without PI only IPv4 packets might be sent. PI can be disabled by opening a tunnel with `iff-no-pi` option. Thus, in order to emulate `/dev/tun` on Android fully, `tap` mode along with PI support have been implemented leveraging the Java-based Android SDK.

Getting back to standard UNIX tooling, as leveraged in our solution, an essential requirement is fulfilled by Termux[1]. Termux is an open source app providing an Android terminal emulator and, more importantly for our purposes, a minimal Android-hosted Linux environment, and a fully Linux-compatible userland, unlike the Android-native environment. A unique feature of Termux is that it works directly with no rooting, or other complex procedure required (e.g., no user input). A base system is installed automatically by the app itself, and additional packages are then available by means of the Debian-compatible APT package manager.

A software running in Termux may then expect availability of standard, widespread userland UNIX tools, such as SOCAT, to be already installed in the environment, or at least readily provisionable by just invoking the package manager to download and install the corresponding software package. Plain SOCAT cannot be used on Android for piping to `/dev/tun` for an already stated reason: root permissions are required for that. Instead, we have created a script, which emulates `socat` binary, but actually creates a tunnel using Java application. For that, we used Android Broadcasts [1] to pass tunnel creation intentions to the APK which would actually create them.

On a first tunnel creation intent, Android forces the application to ask for consent from a user to open a tunnel. This is a security measure to prevent easy sniffing of the user traffic by an application. The consent dialog is created by Android OS and it explains to a user possible consequences of giving to an application access to tunneling the traffic through it. Consequently, to use a SOCAT-emulation script, the APK GUI should be opened first, to show the consent confirmation dialog to the user. Once consent is given, tunnels may be emulated behind scenes, thus not bothering further the user.

4 A Smart City Use Case

As described in [9], a Smart City developer may rely on an effective approach for interaction among Cloud-controlled smart objects, e.g., to let a mobile entity

[1] https://termux.com/.

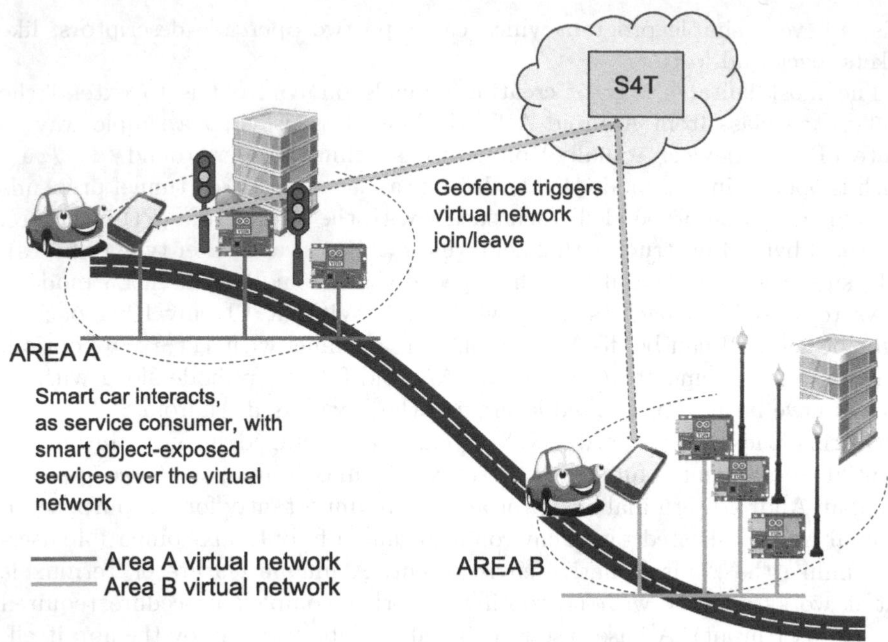

Fig. 2. VPN join, followed by service discovery

dynamically discover services and directly consume them, without the mediation of the Cloud at the application level, in a totally distributed fashion. However, for some popular service discovery frameworks, e.g., AllJoyn[2], the underlying technologies may require the discovery phase to be carried out within a single broadcast domain. Packing all services for the Smart City within such a global scope may incur in scalability issues.

Leaving aside other functionalities the Cloud, as described in [9], may provide (e.g., plugin injection, complex event processing), the virtual network instantiation and deallocation functionalities provided by Stack4Things can be very helpful in this regard. In fact, an, e.g., smart car, featuring an in-dash infotainment system based on a mass-market mobile OS (i.e., Android), may be dynamically added to virtual networks of much smaller scope, depending on the city area (e.g., a geofence-delimited one) it is traversing, discovering and consuming only the services that are listening within the broadcast domain associated with that specific area.

A service discovery-enabled client needs to be deployed dash-side for the set of services that the smart car is supposed to discover and consume, as well as discovery-enabled services on the relevant Smart City objects, such as, e.g., smart traffic lights and smart streetlights.

[2] http://allseenalliance.org.

Fully client-side geofences takes care of detecting the car entering specific areas of interest. Accordingly, addition or removal of the smart car to/from specific virtual networks is thus triggered by the smart car itself, the latter invoking the Cloud and passing the predefined (geofenced) area as parameter. As soon as the smart car gets added to a new virtual network, it discovers the services in the area and consumes them (see Fig. 2). At the application layer, the interactions between the smart car and a specific City-provided smart object are direct, without any mediation by the Stack4Things Cloud, which only provides communication services at the network layer.

A preliminary evaluation in this scenario is provided in the following, focusing on specific key performance indices, namely latency and throughput, related to the aforementioned core (enabling) mechanisms: wrapped SOCAT-based tunneling. This is tasked at quantifying the performance of the proposed tunneling mechanism, although it is a technology enabling a new feature in the IoT, thus without terms of comparison in this context so far.

Table 1. Throughput and latency measurements over WiFi.

Topology/Technology	iperf: *throughput [**Mbps**]*	ping: *latency (RTT) [**ms**]*
direct	92.5	0.23
vpn	81.75	1.501
s4t	79.7	0.625

Table 1 reports on the set of experiments that have been conducted. The experiments are based on the *iperf3* [18] tool for measuring throughput and the ubiquitous *ping* tool to gauge latency, the latter by means of ICMP echo requests to obtain Round-Trip Time values as estimation of delay. iPerf3 works by repeatedly sending an array of *len* bytes for *time* seconds, where *len* by default is 128 KB for TCP, and the default for *time* is 10 s. In both cases, the test setup consists in having a server ready, and generating traffic over TCP (iPerf3) or ICMP requests (ping) from the client (an Android mobile, or its emulated instance), which collects partial and final statistics. Values in the table are averages computed over a number of 1000 samples, where each chunk of 10 samples represent a single run for the tool. Variance values have not been included because negligible.

The first column indicates the kind of technology employed and under which topology: in particular, *direct* refers to tests between two hosts directly connected, over (WiFi) LAN. The *vpn* abbreviation refers to an OpenVPN server to which an OpenVPN client is connected, under the same roles as the two aforementioned hosts. Same happens for *s4t*, in this case with two hosts set up as if controlled by the S4T Cloud thus connected directly over a SOCAT-based tunnel.

For the sake of comparing under the most relevant conditions, OpenVPN has been tested in TCP mode, and various parameters (e.g., MTU of the TUN interfaces, MSS for both iperf and the TCP tunnels) had already been tuned for S4T in the implementation phases, as also discussed below. As latencies are quite aligned in the various scenarios, the discussion has been focused on the more interesting and insightful values obtained for the throughput metric. Albeit actually latency may be considered the most relevant metric for the use case under consideration, as it is key for near real-time (e.g., multimedia) applications, and the most reliable metric in general for embedded systems, considering that throughput is naturally more susceptible to other factors, e.g., high CPU load or RAM usage, differences in the media interface, etc., the latter has been chosen.

It may be noticed that throughput for the S4T-based setups degrades only slightly, as can be seen in Table 1, most likely due to the overhead of inter-process piping. This as a result of trading off raw performance, at the price of high application-level complexity and an (internal) ad-hoc architecture, as is the case for OpenVPN, with the simplicity and flexibility of off-the-shelf tools acting as separate subsystems and taking care of different facets of the communication model, in line with the UNIX philosophy of using one (good) tool for each job.

5 Conclusions

In summary, in this work the authors have pursued the extension of an embedded system-powered IoT use case, developed to enable geo-localized Smart City services, to Android mobiles.

The involvement of personal devices featuring a tightly controlled environment, such as Android, called for the implementation of adaptations to the core Android-supported network virtualization mechanisms, lacking, e.g., L2 interface emulation. Other implementation choices, such as wrapping userspace Linux networking tools, derived from the requirement to preserve the same behavior, interfaces and layering model already employed by Stack4Things for embedded systems.

In terms of future work, it should be noted, that, unlike `/dev/tun` on Linux, `VpnService` cannot open multiple tunnels at once. However, this limitation might be overcome by multiplexing connections to several endpoints onto the single `VpnService` tunnel, in terms of Java-developed logic.

References

1. Broadcasts—Android Developers. https://developer.android.com/guide/components/broadcasts.html. Accessed 23 June 2017
2. linux/drivers/net/tun.c - Elixir - Free Electrons. http://elixir.free-electrons.com/linux/v4.11.5/source/drivers/net/tun.c. Accessed 23 June 2017
3. linux/include/uapi/linux/if_tun.h - Elixir - Free Electrons. http://elixir.free-electrons.com/linux/v4.11.5/source/include/uapi/linux/if_tun.h#L87. Accessed 23 June 2017

4. platform_frameworks_base/com_android_server_connectivity_Vpn.cpp at 52eb4e01a 49fe2e94555c000de38bbcbbb13401b android/platform_frameworks_base - GitHub. https://github.com/android/platform_frameworks_base/blob/52eb4e01a49fe2e945 55c000de38bbcbbb13401b/services/core/jni/com_android_server_connectivity_Vpn.cpp#L66. Accessed 23 June 2017
5. socat - multipurpose relay. http://www.dest-unreach.org/socat/. Accessed 23 June 2017
6. Universal TUN/TAP device driver. https://www.kernel.org/doc/Documentation/networking/tuntap.txt. Accessed 23 June 2017
7. VpnService—Android Developers. https://developer.android.com/reference/android/net/VpnService.html. Accessed 23 June 2017
8. Bizanis, N., Kuipers, F.A.: SDN and virtualization solutions for the Internet of Things: a survey. IEEE Access **4**, 5591–5606 (2016)
9. Bruneo, D., Distefano, S., Longo, F., Merlino, G., Puliafito, A., DAmico, V., Sapienza, M., Torrisi, G.: Stack4things as a fog computing platform for smart city applications. In: 2016 IEEE Conference on Computer Communications Workshops (INFOCOM WKSHPS), April 2016
10. Chowdhury, N.M.K., Boutaba, R.: A survey of network virtualization. Comput. Netw. **54**(5), 862–876 (2010). http://www.sciencedirect.com/science/article/pii/S1389128609003387
11. Fischer, A., Botero, J., Till Beck, M., de Meer, H., Hesselbach, X.: Virtual network embedding: a survey. Commun. Surv. Tutor. IEEE **15**(4), 1888–1906 (2013)
12. Ganti, R.K., Ye, F., Lei, H.: Mobile crowdsensing: current state and future challenges. IEEE Commun. Mag. **49**(11), 32–39 (2011)
13. Guo, B., Zhang, D., Wang, Z., Yu, Z., Zhou, X.: Opportunistic IoT: exploring the harmonious interaction between human and the Internet of Things. J. Netw. Comput. Appl. **36**(6), 1531–1539 (2013)
14. Ishaq, I., Hoebeke, J., Moerman, I., Demeester, P.: Internet of Things virtual networks: bringing network virtualization to resource-constrained devices. In: 2012 IEEE International Conference on Green Computing and Communications, pp. 293–300, November 2012
15. Longo, F., Bruneo, D., Distefano, S., Merlino, G., Puliafito, A.: Stack4things: a sensing-and-actuation-as-a-service framework for IoT and cloud integration. Annales des Telecommun./Ann. Telecommun., pp. 1–18 (2016). https://www.scopus.com/inward/record.uri?eid=2-s2.0-84976292948&partnerID=40&md5=4f9e9d4a88b9d4e6b020331c7f92689c, cited by 0, Article in Press
16. Merlino, G., Bruneo, D., Longo, F., Distefano, S., Puliafito, A.: Cloud-based network virtualization: an IoT use case. In: Mitton, N., Kantarci, M.E., Gallais, A., Papavassiliou, S. (eds.) ADHOCNETS 2015. LNICSSITE, vol. 155, pp. 199–210. Springer, Cham (2015). doi:10.1007/978-3-319-25067-0_16
17. Shi, W., Cao, J., Zhang, Q., Li, Y., Xu, L.: Edge computing: vision and challenges. IEEE Internet Things J. **3**(5), 637–646 (2016)
18. Tirumala, A., Qin, F., Dugan, J., Ferguson, J., Gibbs, K.: iPerf: the TCP/UDP bandwidth measurement tool. http://software.es.net/iperf/ (2005)
19. Zhao, D., Ma, H., Tang, S., Li, X.Y.: COUPON: a cooperative framework for building sensing maps in mobile opportunistic networks. IEEE Trans. Parallel Distrib. Sys. **26**(2), 392–402 (2015)

Mobile Crowd Sensing as an Enabler for People as a Service Mobile Computing

Paolo Bellavista[1], Javier Berrocal[2(✉)], Antonio Corradi[1], and Luca Foschini[1]

[1] Department of Computer Science and Engineering, Scuola di Ingegneria, Università di Bologna, 40135 Bologna, Italy
{paolo.bellavista, antonio.corradi.it, luca.foschini}@unibo.it

[2] Department of Computer and Telematic Systems Engineering, Escuela Politécnica, University of Extremadura, 10003 Cáceres, Spain
jberolm@unex.es

Abstract. Mobile Crowd Sensing (MCS) is a new sensing paradigm exploiting the capabilities of smart devices (smartphones, wearables, etc.) to gather large volume of data. Gathering contextual information is a very expensive activity in terms of mobile device resource consumption, so limiting this consumption is essential for user satisfaction. The architectural style applied to the MCS platform largely affects the consumption of these resources. A server-centric MCS is more efficient when there are many entities interested on the gathered information, whilst a mobile-centric architecture has lower consumption when real-time information is required. In this paper, we propose a platform combining both architectural styles. This allows us to reduce the resource consumption of mobile devices, since it is easier to take advantage of the benefits of each style, and to better facilitate user aggregation, being able to group users both at the server and at the client-side depending on the freshness of the required information and the sensing task to be assigned. Finally, we have evaluated this platform for two different case studies, obtaining very promising results.

Keywords: Mobile crowd sensing · Server-centric · Mobile-centric

1 Introduction

The increasing adoption and sensing capabilities of smartphones has fostered the rise of the Mobile Crowd Sensing (MCS) paradigm. MCS is a paradigm for gathering information on people and their surrounding [7]. This information can be obtained passively, using the sensing capabilities of the owners' smartphones, or actively, asking owners for that information. Indeed, this is a new way of creating a collective intelligence that can be exploited in different scenarios such as smart cities (in order to know how people interact with public services, for instance), well-being (e.g., for helping cognitive diseased people to move around the city) or industry 4.0 (to better coordinate the people working in a factory).

A. Puliafito et al. (Eds.): ADHOC-NOW 2017, LNCS 10517, pp. 144–157, 2017.
DOI: 10.1007/978-3-319-67910-5_12

During the last few years, different platforms with different architectural styles have been developed to support this paradigm. There are some platforms, such as PartcipAct [6] and Vita [9], following a server-centric architectural style. These platforms gather the users' contextual information using their smartphones (or any other Internet-connected sensor-equipped device). All the captured information is then uploaded and stored into a server so that complex algorithms can be executed to obtain high level information of the users, even combining their data. Therefore, all the detected information is used by the platform's backend to control the segmentation of the users (i.e., to group users in smaller subsets according to specific characteristics) and to coordinate different social activities.

Instead, People as a Service (PeaaS) [8] is a mobile-centric computing paradigm focused on harnessing the processing and storage capabilities of smartphones for gathering, storing, computing and providing the user's contextual information. NimBees is its commercial implementation. This platform gathers the users' context using their smartphones and keep that data on them. In addition, it exploits the smartphones' computing capabilities to process it in order to obtain high-level information to segment and coordinate users from the client-side, rather than from the server-side.

Finally, there are other approaches, such as [16, 17], designing community-centric frameworks to detect users' activity and needs based on big data analysis. These proposals allow developers to reduce the burden on smartphones and, even, to obtain a highly valuable information by consulting other sources.

One of the great challenges of these platforms is to be accepted by users, so that they would be willing to share their personal information in exchange for the value these platforms provide them. One of the initial requirements for getting user acceptance is the sustainability of their functionalities in terms of the devices' resources. It is well known that resource consumption, in particular battery use [14] and network traffic [11], is a factor determining the success of mobile applications [1].

Some previous work allowed us to demonstrate that the architectural style determines the consumption patterns of these systems [4]. The mobile-centric approach normally entails lower consumption when the system requires real-time information (since the information is obtained by the sensors but does not have to be uploaded to the server). Instead, the server-centric approach consumes fewer resources when the gathered information does not have to be fully up-to-date and when there are many entities interested on this information.

In this paper, we propose the combination of two of these platforms with different architectural styles: nimBees and ParticipAct. In the combined platform, nimBees is responsible for gathering the real time information (whenever necessary), computing it (to obtain high level information) and segmenting users based on that information at the client-side. Periodically, the gathered information is uploaded to the ParticipAct's backend. In the backend, the information is further processed to obtain more complex data and to facilitate the management and coordination of the sensing task. This combination reduces the resource consumption, improving the user acceptance. On the other hand, it allows us to have different levels of inferred information (both at the server and at the client-side), improving the user segmentation process.

This platform has been evaluated in two case studies one related to the care of cognitive disable people and another one improving the efficiency of the task assignment in smart factories, allowing us to demonstrate that the platform can be applied in different environments and to validate the benefits it provides.

The rest of the paper is structured as follows. Section 2 further details the ParticipAct and nimBees platforms. Section 3 describes the platform combining the server-centric and mobile-centric approaches. In Sect. 4, we show how the combined platform can be applied to different case studies. Section 5 reports some of the results collected from the application of the platform. In Sect. 6, we detail some related works, and Sect. 7 contains the conclusions and future works.

2 Mobile Crowd Sensing and People as a Service

Crowd sensing is a paradigm to coordinate group of people to gather different kind of data. This data can be either some information passively collected by different sensors, or data actively provided by users. MCS is based on the power of user-companioned Internet-connected devices (such as smartphones, wearables devices, etc.) to gather these data. This paradigm is especially important when the system has some mobility requirements, since the mobility of these devices increases the system versatility. Currently, there are different platforms supporting the MCS paradigm. Most of them are based on a server-centric architecture, storing and computing the gathered information in a central server. ParticipAct is a MCS platform in which some of the author of this paper have been working on.

Instead, PeaaS and, concretely, its implantation nimBees exploit the smartphones to gather their owners' contextual information, to store it and to compute it in the device itself. Then, the smartphone provides this information as a service to other entities (users or companies). Therefore, PeaaS is a crowd sensing paradigm based on a mobile-centric architectural style.

2.1 ParticipAct

ParticipAct is a MCS platform which primary goal is realizing an efficient and easy-to-use playground to identify the most suitable MCS policies depending on context and application requirements. ParticipAct is a complete supporting infrastructure that allows local actions and the client collection of data, is in charge of transferring sensed data to the ParticipAct backend support, and takes over not only data harvesting, but also post-processing, mining, and maintenance.

The nature of ParticipAct is based on a client-server architecture. The ParticipAct client is the component that takes care of receiving sensing tasks, asking users whether they want to run them, managing data collection, and uploading results to the server. Functionally, ParticipAct client comprises two main components: the task management component and the sensing management component.

The server side of ParticipAct provides management, storage, and analysis of crowd sensed data. The backend consists of three macro-components: Networking, Post-processor, and Data Mining. Networking includes the Data Receiver component that receives data while ensuring authenticity, integrity, and confidentiality. The Data Receiver acknowledges each data packet that it receives, thus allowing the client to delete the data from its local database. Data is then cleaned up and prepared for long-term storage by the Post-processor. Finally, the Data Mining component builds a profile for each user that is used to identify users that are more likely to successfully execute a task based on their recorded behavior

Finally, ParticipAct has an administrative module allowing administrators to man-age the users' profile, design and assign tasks, review the status of each campaign or sensing task, and data review.

2.2 People as a Service and Nimbees

PeaaS is a mobile crowd sensing platform focused on exploiting the ever increasing computing and storage capacity of smartphone with the aim of enabling new communications methods between devices based on the gathered information, reducing the data transfer and the operational cost of cloud environments. Concretely, PeaaS proposes to use the smartphone's sensors to gather the user's contextual information. This information is kept on the device, and securely provided as a service to third parties directly from the smartphone, allowing owners to keep their virtual identity under their own control. At the same time, the consumers of such services are allowed to get real-time information.

The PeaaS model considers four principles. First, the users' smartphones are taken to be the virtual interfaces of their owners. Second, the smartphones' sensors are used to collect information about their owners. This information is used to create the users' virtual sociological profiles. Third, different services can be deployed in the smartphones to provide the stored and inferred information and to process collective sociological information. Users can control which services can be deployed in their devices. Finally, all the stored information is kept exclusively on the smartphone. Thus, users can strengthen their privacy by managing who can access their information, what information can be accessed, and when it can be accessed.

The PeaaS model is currently implemented as a MCS platform denominated nimBees. In addition, nimBees provides push notifications with advanced segmentation capabilities based on the users' sociological profile.

Although, nimBees is a mobile-centric platform, it is composed by a client-side and a server-side. The client-side is an API that can be included in any mobile application. This API is in charge of gathering, computing and storing the contextual information of the user. The API also manages the reception of segmented push notifications, indicating whom, or under with circumstances, the notification should be shown to the user. So that, once a push notification reaches the mobile device, the API decides (based on the owner's profile) whether that owner is an appropriate recipient of the message. Only when the owner is selected the smartphone shows the push notification. Otherwise, everything is left as if nothing happened. All these transitions are transparent to the device's owner. This notification could contain a simple message to the user or some more advance activity that should be performed by the API.

The platform also has a server-side that only acts as a DNS (Domain Name System) forwarding the notifications from the senders to the receivers and checking their correct reception.

3 Integration of Both Platforms

The success of MCS campaigns and platforms lies in three fundamental aspects: selecting the right users that should participate in the campaign, capture the required information by the campaign and keep users happy. Therefore, the key aspects that should be taking into account for a successful campaign are:

1. To do a correct segmentation of users. To that end, this activity should consider as many information as possible, from users' abilities and their specific location sometime in the past to their exact location right now, for instance.
2. Tasks and campaigns with different granularity. MCS platforms should allow the execution of different tasks from the gathering of specific and timely information to the coordination and interaction of different users to obtain more complex data.
3. Reduce the user intervention. As far as possible, the activities for gathering information should limit the number of active tasks asked to users, which in the end leads to a better perception of the system.
4. Reduce the resource consumption. MCS platforms have to monitor users (to have more information to segment them) and assign different campaigns to them, but limiting the resource consumption in order to increase their satisfaction.

Most MCS platforms try to meet these aspects as best as possible. However, as indicated above, the architectural style followed implies that there must be a trade-off and balance between them. For example, if server-centric architecture is applied, the freshness and update of the information has to be limited in order to reduce the resource consumption. On the other hand, with a mobile-centric architecture, the number of entities with which to share the captured information must be limited.

In this paper, we present a platform combining both architectural styles in order to take advantage of each one of them, increasing the likelihood of success of crowd sensing campaigns. The resulting platform is divided into two parts: the client-side and the server-side.

Figure 1 shows the server-side architecture. The backend is in charge of managing and coordinating the sensing tasks, and storing and computing the sensed data in order to obtain high-level information. It is divided into two main components: Data Manager and Crowd Sensing Manager.

The Data Manager component stores and analyses all the gathered information. It consists of three modules: Data Receiver, Post processor, and Data Processor. The Data Receiver module obtains the required data from the smartphones via a Representational State Transfer (REST) API. All received data is then cleaned up and prepared for long-term storage by the Post-Processor module. The Interpolation submodule improves data collection by filling in missing data points that can be inferred with sufficient accuracy. The Integration submodule aims at aggregating data in time and space, collapsing all collected data in the same five minutes window in a single row to enable time-based indexing of all sensed data. It also aggregates data in space by creating a geographical view of sensed data. The Data Processor module takes advantage of those time-based and space-based views in order to segment users when a new sensing task or campaign has to be executed.

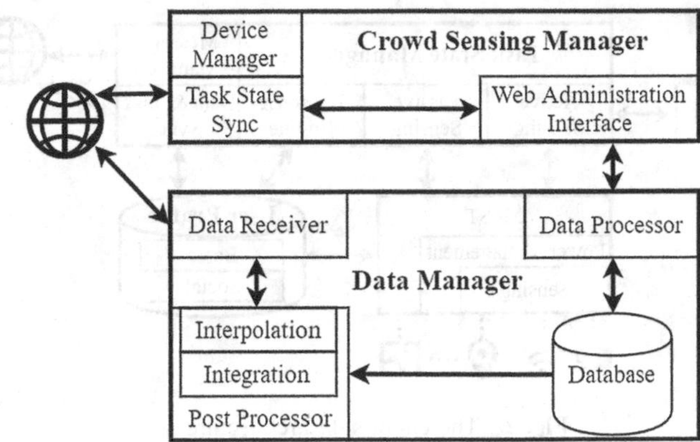

Fig. 1. The server-side architecture.

The Crowd Sensing Manager component is in charge of coordinating the sensing tasks and the users involved in each campaign. The Web Administration Interface allows managers to interact with the system, the design of campaign, the assignment of tasks to users, the visualization of the results of each campaign, etc. The Task Sate Sync, instead, keeps track of the state of each task. Finally, the Device Manager controls all the devices connected with the MCS platforms, allowing the forwarding of tasks and messages to them and providing support to the direct interaction between users for the coordination and orchestration of complex sensing tasks in which more than one user are involved.

Figure 2 shows the client-side architecture. It is responsible for managing the complete lifecycle of the sensing activities and tasks. Its architecture is divided into three main component: MoST, User Profile and Task State Manager.

MoST (Mobile Sensing Technology) is the module providing an uniform access layer to all physical and logical sensors. MoST consists of two main subsystems: the Sensing subsystem and the Power Management subsystem. The Sensing subsystem manages all aspects of sensing, from accessing sensors, to wrapping the detected information into easy-to-manage local objects. The Power Management subsystem controls how the different sensing tasks interacts with the sensors and the frequency at which the information is gathered in order to minimize the power consumption.

The User Profile is the database where all the gathered information is stored in order to compose a comprehensive virtual profile of the user. This information is divided into two subset: Basic and Social Profile. The Basic subset contains the dated raw information gathered from the sensors. All this information is stored in a single timeline in order to better trace the sequence of the information and the events. The Social Profile stores the results of high-level inferences performed over the Basic Profile.

The Task State Manager component manages the lifecycle of the sensing activities and tasks. It is divided into five different packages responsible for different parts of the sensing activities lifecycle. The Message Manager package allows the reception and the delivery of push notifications and tasks and the direct interaction with other users.

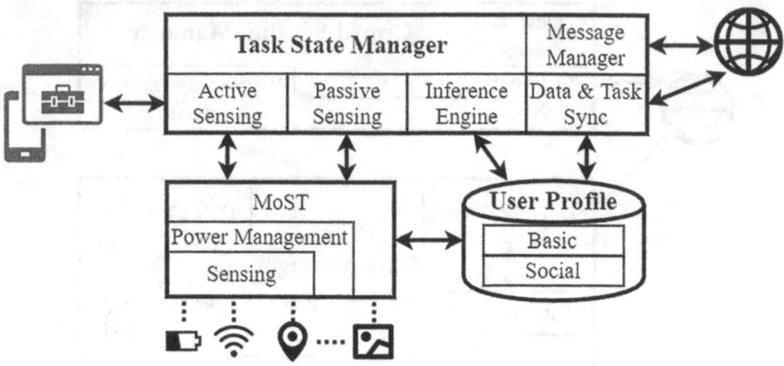

Fig. 2. The client-side architecture

These messages can contain filters to further segment users, allowing a second step of the user segmentation based on the information stored in their smartphone. For the users meeting the filters, the system will ask them whether they want to accept or refuse the sensing task. Each task can be active or passive. Once the user accept it, if it is a passive task, the Passive Sensing module starts gathering all the requirement information. If it is an active task, the Active Sensing package interacts with the user asking her to provide the required data. In addition, some of the assigned tasks may be aimed at obtaining information from the users so that a more complete profile can be obtained to better segment them, reducing the user load and increasing her satisfaction. The Inference Engine module is in charge of further processing all the raw data gathered from the smartphone's sensors in order to obtain high-level information (so that for some tasks the amount of information to be transmitted to the server is reduced). Finally, the Data and Task Sync module manages the synchronization of information between the client-side and the server-side. In order to reduce the resource consumption, this module identifies the best situation to upload the information regarding the freshness of the required information and the device's status (i.e. plugged in, connected to a Wifi, etc.), balancing between uploading the information as soon as possible and minimizing the impact on the resources.

As can be seen, the integration of both platforms allows us to: improve the users segmentation; provide a two-step segmentation process (first, in the server-side and, subsequently, in the smartphone); improve the coordination between users in order to increase the granularity of the task that can be assigned; and, finally, reduce the resource consumption (both in terms of network traffic and energy consumption), processing part of the information directly in the smartphone.

4 Case Studies

In order to evaluate the feasibility of the defined MCS platform, it has been applied in two different scenarios: for monitoring employees to increase the efficiency in smart factories and for the monitoring the well-being of cognitive disable elders.

4.1 Reducing Bottlenecks in a Smart Factory

Nowadays, factories are facing a new revolution to increase the flexibility of their production lines, and to increase the automation and self-organization between equipment and employees to better meet the business objectives [5]. To achieve these goals, factories have to become smarter. So that, they have to be completely equipped with sensors, actuators and autonomous systems managing the whole environment. In this situation, the MCS platforms can improve the self-coordination of people and equipment according to the contextual information.

The platform presented in this paper can be applied in this environment to assess how it would help to resize the number of employees and the tasks they have to do in a particular production line in order to better meet the demand. To that end, each employee is provided with a smartphone for managing the interactions among the employees, the machinery and the MCS platform. The smartphone also manages the employee's virtual profile (storing her abilities, her working hours, the equipment she used, etc.) and monitors her production rate.

The MCS platform is used to allocate production tasks to the employees in real-time depending on their profile and their workload. This is completely done by the platform's backend with the information stored on its database. Once assigned, the smartphone, through the Task State Manager module, communicates the assigned task to the operator. Subsequently, through the Passive Sensing, Active Sensing and MoST modules, the production rate is detected either by interacting directly with the machineries and the surrounding sensors or by asking the operator to actively indicate such information. Thus, we can obtain an estimation on whether the individual production goal will be met in the allocated time.

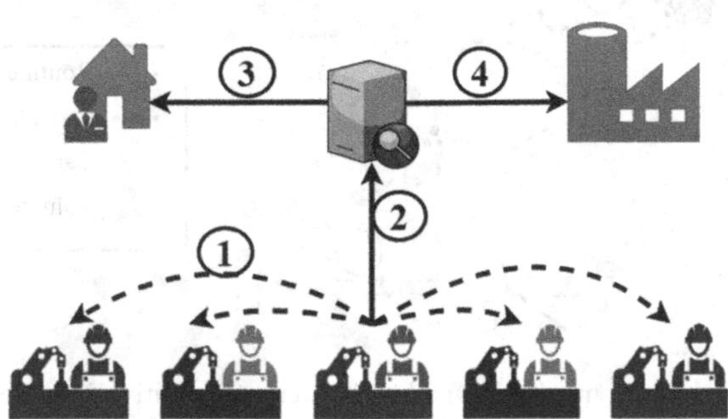

Fig. 3. Application of the combined platform in a smart factory. (Color figure online)

Figure 3 shows the self-organization process when the MCS platforms detected that one employee is overloaded and that her production goal will not be achieved. The figure shows different employees with different colours depending on their workload (from blue to red indicating from low to high overload). When the smartphone identifies that the employee is overloaded, it broadcasts a message (using the Message Manager and the LTE direct technology) to the surrounding devices searching for other employees with the required profile and time to unload on them part of the assigned task (flow number 1 Fig. 3). The requirements to do the task are attached to the broadcasted message and evaluated by the targeted smartphone.

If it were not possible to find an employee who can take care of the task, the situation would be communicated to the server in order to extend the search area to other parts of the factory (number 2). To do this search, the information stored in the backend is used. If, likewise, it could not find any employees in the factory with the right profile, the systems would try to find any employee with the needed skills but outside their working hours, in order to know if they would be willing to accept the task in exchange for the appropriate salary (number 3).

Finally, if the platform does not find any suitable employee accepting the task, the system would delegate the overload to another factory or would indicate the delay for the delivery of the products (number 4).

4.2 Monitoring the Elderly in an Early Stage of Cognitive Impairment

During the last few years, the ParticipAct and the PeaaS platforms have been applied in the care of elderly and cognitive impaired people [3, 15], allowing us to improve their quality of life. However, the resource consumption and the correct coordination of the users' interactions are very sensitive areas in this environment. In this case study, we evaluate how the combined platform improves these issues. This platform has been used to monitor elderly, caregivers, and volunteers who, because of their knowledge, skills and context can provide assistance when an alarm is raised.

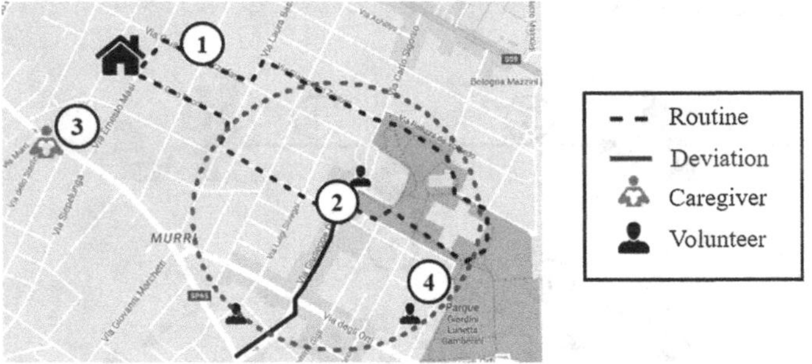

Fig. 4. Improving the monitoring and assistance of cognitive impaired patients.

Patients are constantly monitored in order to identify in real time their routines and any deviation from them. Different information is monitored (such as their location, their pulse, the weather, etc.). This monitoring is done by assigning passive tasks to their smartphone and inferring from the gathered data the routines and the deviations. The routines are uploaded to the server when the device is plugged in. Instead, the deviations raise alarms that can be sent to the caregivers or the volunteers.

Caregivers and volunteers are also monitored with passive tasks, but with less frequency, only to identify their usual routines and their position when an alarm is raised. In addition, periodically, the platform's backend (throughout the Data Processor Module) process the elders and the volunteers routines in order to identify what volunteers are more likely to help each elder in a danger situation (see Sect. 5), sending them encrypted patient's sensitive data. That information is disclosed only when an alarm is generated and if the control plan requires to access it.

Figure 4 shows a situation controlled by the platform. When the patient is performing a certain routine (number 1) and, at some point, she disorients (number 2) the smartphone immediately detects it and triggers an alarm to the caregiver (number 3) through the Message Manager module. If she is too far to assist her in a reasonable time, the smartphones tries to contact the volunteers in the area. To that end, it broadcasts a message (using Bluetooth, Wifi direct or LTE direct [2]) with the key to decrypt the sensitive information about how to take care of her in that situation (number 4). Once received the broadcasted message, if the volunteers accept the task, all the required information is shown to her, indicating how she has to proceed.

Thanks to these different levels of caregivers and volunteers and to the better segmentation of users, we can achieve a lower resources consumption and a more efficient treatment of the deviations.

5 Experimental Results

This section details the outcomes obtained from the application of the above detailed platform in the cognitive impaired elders case study. We only present these results because we are working on involving a higher number of factories in the smart factory case study in order to get highly representative data.

This experiment was carried out with 170 students in the Emilia Romagna region (Italy) simulating the different roles involved in the case study: elders, caregivers and volunteers. The experiments participants were equipped with Android smartphones with an application implementing the client-side of the platform. The experiment lasted one moth, June 2017. We divided it into two different phases, each one lasting two weeks.

During the first phase, the mobile app was constantly executing sensing tasks to gather the contextual information of the users in order to build their virtual profiles. These tasks were especially aimed at identifying the users' abilities and tracking their location in order to be able to infer their movement patterns. All the sensed and inferred information was uploaded to the backend when the device was plugged in and connected to a Wifi network.

The uploaded information was then used to identify the probabilities of the presence of any student-volunteer in a given area to help a student-patient whenever an alarm is raised. Thus, a device-to-device message/alarm can be sent to the nearby student-volunteers in order to disclose the sensitive data, activate the control plan and, thus, provide a faster care. To obtain this information, we divided the evaluation area into 50 clusters of about 300 m $\times$ 300 m. Subsequently, we analysed the volunteers' traces to identify the probabilities of each users of being in each cluster. In addition, we took into account different ranges of separation from the centre of the cluster of 50 m, 100 m, and 150 m. These ranges were selected to identify the probabilities of success of the device-to-device messages when it is sent using different communication technologies (i.e. Bluetooth, Wifi direct or LTE direct, respectively).

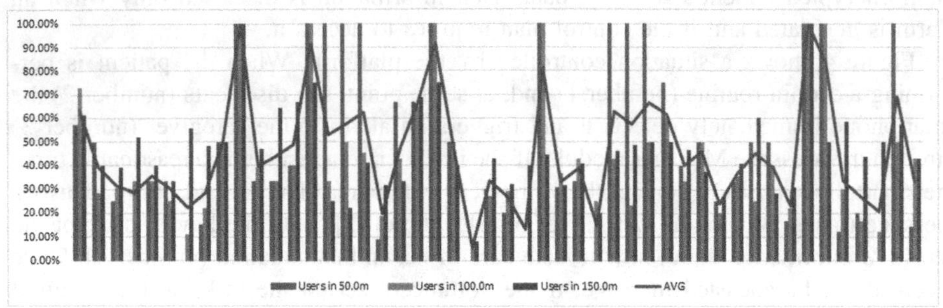

Fig. 5. Probabilities of finding users in each cluster throughout the day.

Figure 5 shows a bar chart with the results obtained for each cluster. As can be seen, there are thirteen clusters with a 100% of probabilities of a user crossing them throughout the day. Normally, these are sub-areas with a large concentration of inhabitants (and, therefore, of users) or crossed by important streets of the city. Therefore, these would be secure zones in which the alarms would have a high probability to be received by a volunteer by mean of a device-to-device communication. For the other clusters, it would be necessary to complement this interaction with others actions to ensure that the alarm is taken care of in a reasonable time (for instance, warning at the patient's caregiver at same time independently of her location).

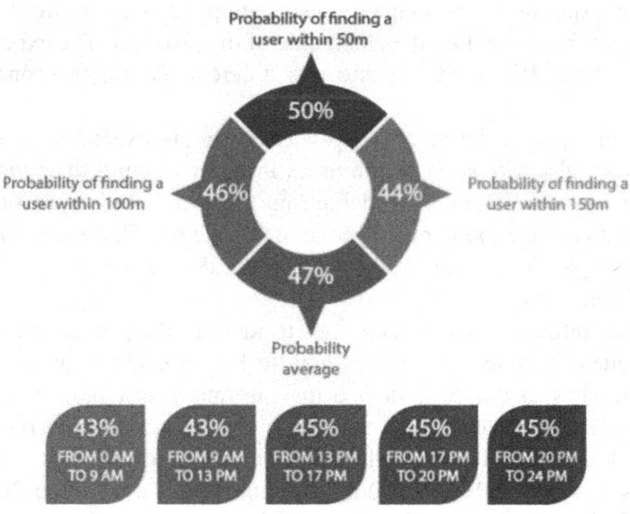

Fig. 6. Aggregated average values over all clusters and in five temporal frames.

The volunteers traces were also analysed taking into account specific time intervals. They are not shown in this paper due to space restrictions. Nevertheless, Fig. 6 shows the aggregated average values over all clusters in five different temporal frames: 9.00AM to 1PM, 1PM to 5PM, 5PM to 8PM, 8PM to 00PM and 00PM to 9.00AM. As can be seen, the average probability of finding a volunteer in any of the clusters is 47%, being from 1PM to 00PM the time intervals with the higher probabilities of finding one.

During the second phase of the experiment, we used the previously detailed data to establish and follow the most appropriate procedure to raise an alarm. To that end, the student-patients simulated different danger situations in different clusters. When an alarm occurred in a secure cluster, a device-to-device communication/alarm was first raised in order to contact the nearby student-volunteers and to assign them a care task. Then, only if the task was not receiver or accepted, the patient's caregiver would be warned. On the other hand, in the other clusters, both volunteers and caregivers were warned, achieving greater effectiveness.

The results of this experiment allowed us to demonstrate the feasibility and the benefits of the MCS platform combining both the server-centric and the mobile-centric architectural styles. These results demonstrate that this platform improves the users segmentation (since, for example, both the information of their routines and their real-time location can be used to improve the care of cognitive impaired people in a dangerous situation). In addition, it also allows us to execute tasks with different granularity depending on the context and, thus, reducing the smartphone resource consumption.

6 Related Work

During the last few years, there has been an increasing interest on crowd sensing platforms. Different architectural styles have been applied to develop these platforms, some of them more oriented to store the sensed data in a central server and others focused on storing it in the user's device.

VITA [9], for example, is a system providing different crowd sensing services that can be integrated with different mobile applications in order to facilitate the sensing of information and the assignment of tasks. This system relies on the smartphone to execute the sensing tasks and to control and monitor the task assigned to them. All the sensed information is then uploaded to the server-side in order to further process it. In addition, the platform's backend manages the coordination of the sensing tasks through the execution of different Business Processes. Nonetheless, part of the users' segmentation is also done in client-side by means of the "Social Vector" module. This module allows developers to improve the assignment of tasks to users depending on their smartphone's capabilities (such as battery, computation power, similar tasks previously executed, etc.). However, it does not keep a complete sociological profile of the user, so that the smartphone's segmentation capabilities are not fully exploited.

In [13], authors state that the sensing and collection of large volume of data could be expensive and not everyone has the financial and computational resources to deal with them. Therefore, they propose a scalable energy-efficient data analytics platform for on-demand distributed mobile crowd sensing called C-MOSDEN. This platform includes different sensing capabilities, but also the Activity-Aware and Location-Aware modules. These modules are used to activate or deactivate specific tasks depending on the user's context. As results, authors indicate that the context-aware capabilities were able to save cost in terms of CPU, battery, memory and network usage. Nevertheless, their context were limited to the activity and the location of the user.

Finally, in [10], authors indicate that existing mobile crowd sensing applications focused primarily on the centralized server-client architecture, but that the increasing computation and storage capabilities of smartphone is leading to a rise in platforms applying a Peer-to-Peer architecture. They indicate that the P2P mobile crowd sensing architecture can effectively reduce the operational cost in the centralized server. Nevertheless, they state that this architecture requires new methods to incentive the sharing of data. Therefore, they propose a data market and a generic pricing scheme for data sharing among sensing users and users requesting that information.

This paper details a platform combining both architectural styles, being able to segment users and to activate/deactivate the sensing tasks using a complete profile at both sides (the client and the server). Thus, a greater reduction in the resources consumption and a better segmentation of users is achieved.

7 Conclusion

User satisfaction is essential for the acceptance of any mobile application and even more so for the participation in crowd sensing campaigns. That satisfaction is greatly influenced both by the resource consumption of the MCS platforms and by the adequacy of the assigned sensing tasks to their profile.

This paper presented a platform that improves both aspects. First, it succeeds in reducing the consumption of the MCS platform by taking advantage of the specific characteristics of the different architectural styles. Secondly, it allows a two-step segmentation process, in which users participating in a campaign can be, first, preselected in the server-side depending on the stored information and, then, finally selected in the client-side based on their specific contextual situation at that time. We are currently evaluating this platform for two case studies. This paper shows the results of one of the case studies to prove the feasibility and benefits of the platform.

Currently, we work on comparing the resource consumption using the different platforms alone and the combined platform, in order to identify when the characteristics of each platform should be used. In addition, we are working on involving a higher number of factories in order to evaluate the platform in a real environment.

Acknowledgements. This research was supported by the Sacher project (no. J32I16000120009) funded by the POR-FESR 2014-20 program through CIRI, by 4IE project (0045-4IE-4-P) funded by the POCTEP program, by the project TIN2015-69957-R (MINECO/FEDER), by the Department of Economy and Infrastructure of the Government of Extremadura (GR15098). The authors would also like to thank Leo Gioia for his help in capturing and processing the data during the realization of the case studies.

References

1. AVG Technologies: Android App Performance Report (2014). http://now.avg.com/wp-content/uploads/2015/02/avg_android_app_performance_report_q4_2014.pdf
2. Bellavista, P., Benedetto, J.D., Rolt, C.R.D., Foschini, L., Montanari, R.: LTE proximity discovery for supporting participatory mobile health communities. In: IEEE International Conference on Communications (2017)
3. Berrocal, J., Garcia-Alonso, J., Murillo, J.M., Canal, C.: Rich contextual information for monitoring the elderly in an early stage of cognitive impairment. Pervasive Mob. Comput. **34**, 106–125 (2017)
4. Berrocal, J., Garcia-Alonso, J., Vicente-Chicote, C., Hernández, J., Mikkonen, T., Canal, C., Murillo, J.M.: Early analysis of resource consumption patterns in mobile applications. Pervasive Mob. Comput. **35**, 32–50 (2017)
5. Brettel, M., Friederichsen, N., Keller, M., Rosenberg, M.: How virtualization, decentralization and network building change the manufacturing landscape: an industry 4.0 perspective. J. Mech. Aerosp. Ind. Mechatron. Eng. **8**(1), 37–44 (2014)
6. Cardone, G., Corradi, A., Foschini, L., Ianniello, R.: ParticipAct: a large-scale crowdsensing platform. IEEE Trans. Emerg. Top. Comput. **4**(1), 21–32 (2016)
7. Ganti, R.K., Ye, F., Lei, H.: Mobile crowdsensing: current state and future challenges. IEEE Commun. Mag. **49**(11), 32–39 (2011)
8. Guillén, J., Miranda, J., Berrocal, J., Garcia-Alonso, J., Murillo, J.M., Canal, C.: People as a service: a mobile-centric model for providing collective sociological profiles. IEEE Softw. **31**(2), 48–53 (2014)
9. Hu, X., Chu, T.H.S., Chan, H.C.B., Leung, V.C.M.: Vita: a crowdsensing-oriented mobile cyber-physical system. IEEE Trans. Emerg. Top. Comput. **1**(1), 148–165 (2013)
10. Jiang, C., Gao, L., Duan, L., Huang, J.: Economics of peer-to-peer mobile crowdsensing. In: 2015 IEEE Global Communications Conference (GLOBECOM), pp. 1–6 (2015)
11. Lee, K., Lee, J., Yi, Y., Rhee, I., Chong, S.: Mobile data offloading: how much can wifi deliver? IEEE/ACM Trans. Netw. **21**(2), 536–550 (2013)
12. nimBees. http://www.nimbees.com
13. Perera, C., Talagala, D.S., Liu, C.H., Estrella, J.C.: Energy-efficient location and activity-aware on-demand mobile distributed sensing platform for sensing as a service in IoT clouds. IEEE Trans. Comput. Soc. Syst. **2**(4), 171–181 (2015)
14. Qian, H., Andresen, D.: Extending mobile device's battery life by offloading computation to cloud. In: Abadi, A., Dig, D., Dubinsky, Y. (eds.) 2015 2nd ACM International Conference on Mobile Software Engineering and Systems, pp. 50–151 (2015)
15. Rolt, C.R.D., Montanari, R., Brocardo, M.L., Foschini, L., Dias, J.D.S.: COLLEGA middleware for the management of participatory mobile health communities. In: IEEE Symposium on Computers and Communication, pp. 999–1005 (2016)
16. Wu, X., Zhu, X., Wu, G.Q., Ding, W.: Data mining with Big Data. IEEE Trans. Knowl. Data Eng. **26**(1), 97–107 (2014)
17. Zhang, Y., Chen, M., Mao, S., Hu, L., Leung, V.C.M.: CAP: community activity prediction based on big data analysis. IEEE Netw. **28**(4), 52–57 (2014)

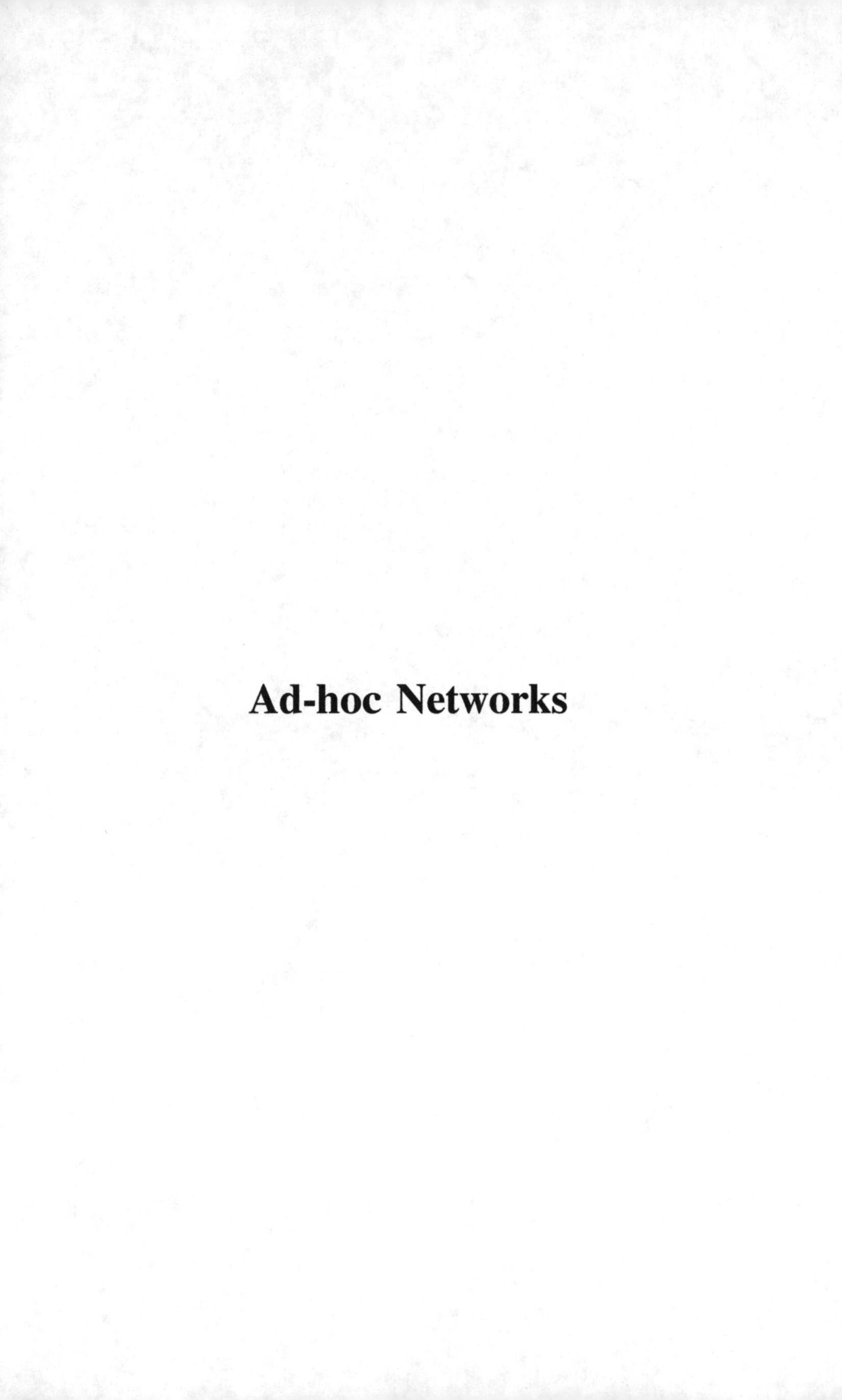

Ad-hoc Networks

SVM-MUSIC Algorithm for Spectrum Sensing in Cognitive Radio Ad-Hoc Networks

Soumaya El Barrak[1(✉)], Abdelouahid Lyhyaoui[1], Amina El Gonnouni[1], Antonio Puliafito[2], and Salvatore Serrano[2]

[1] LTI Laboratory, ENSA-Tangier, University of Abdelmalek Essaâdi, BP 1818 Tanger Principal, Tangier, Morocco
elbarrak.soumaya@gmail.com, lyhyaoui@gmail.com, amina_elgo@yahoo.fr

[2] Engineering Department, University of Messina, Contrada di Dio, 98166 Messina, Italy
{Apuliafito, sserrano}@unime.it

Abstract. Adopting accurate and efficient spectrum sensing policy is crucial in allowing cognitive radio users to be aware of the surrounding parameters related to the radio environment characteristics. Especially, in Ad hoc networks scenario, where dynamic spectrum access is highly required, since the most of the spectrum is already assigned statistically, and the unlicensed bands are becoming overcrowded. This is due to the multiplicity of wireless communication technologies that operate in those bands, and the increasing number of connected devices. In this paper, a spectrum sensing algorithm that combines the Support Vector Machines (SVM) supervised learning technique with the Multiple Signal Characterization (MUSIC) subspace method is used. Our ultimate objective is detecting the presence of primary users (technology signals) in the band of interest. The node's receivers which make up the network collect samples from the radio environment, estimate the number of primary user signals and the corresponding carrier frequencies. Simulations are conducted to demonstrate the efficiency of the proposed SVM based algorithm in detecting the presence of primary users based on lost-detection and false alarm probabilities evaluation.

Keywords: Spectrum Sensing (SS) · Cognitive Radio Ad Hoc Networks (CRAHNs) · Support Vector Machines (SVM) · Minimum Variance Distortionless Response (MVDR) · Multiple Signal Characterization (MUSIC)

1 Introduction

Cognitive radio technology has been proposed to deal with the problem of spectrum scarcity and its inefficient usage, since current wireless networks are based on a fixed spectrum assignment policy, regulated by governmental agencies. This problem becomes more critical and requires more care when it is addressed to ad hoc network scenarios, due to their distributed multihop architecture, node mobility, and spatio-temporal variance in spectrum availability [1]. In Cognitive Radio Ad hoc Networks (CRAHNs), cognitive radio users need to be aware of the surrounding radio

A. Puliafito et al. (Eds.): ADHOC-NOW 2017, LNCS 10517, pp. 161–170, 2017.
DOI: 10.1007/978-3-319-67910-5_13

environment parameters, in order to choose the best frequency channel among the available bands, and dynamically adapting transmission parameters based on the activity of the licensed users. In such an architecture, the primary network and the CR network components can be classified in two groups, as shown in Fig. 1 [2].

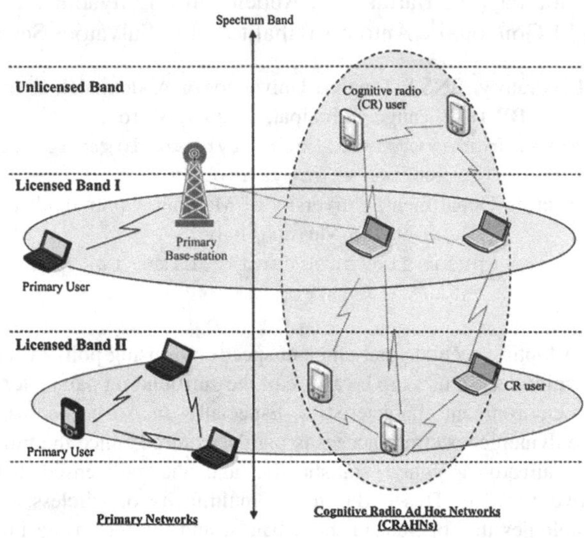

Fig. 1. Co-existence between CRAHNs and primary networks components

The primary network is referred to as an existing network, where the primary users (PUs) have a license to operate in a certain spectrum band. If primary networks have an infrastructure support, the operations of the PUs are controlled through primary base stations. PUs should not be affected by unlicensed users. In the other hand, The CR network (or secondary network) does not have a license to operate in a desired band. Thus, the CR user must sense correctly the channel before transmitting, and vacate it when a primary network activity is detected. In the case of Ad hoc architecture, CR users are assumed to function as stand-alone networks, which do not have direct communication channels with the primary networks. Thus, every action in CR networks depends on their local observations. For this reason, an efficient spectrum sensing mechanism is highly demanded.

Thus, we are interested in this article, in the spectrum sensing feature for cognitive radio ad hoc communication. We will focus on a Support Vector Machines (SVM) estimation technique. More specifically, we will explore a new formulation that takes advantages of the high-resolution of the Multiple Signal Characterization (MUSIC) algorithm [3], and the robustness against overfitting of Support Vector Machines (SVM) supervised learning technique, where few samples are available for estimation. This formulation was developed in [4] for Direction of Arrival (DOA) detection applications. We aim at comparing the performances obtained in terms of primary users' detection, based on the evaluation of lost-detection and false-alarm probabilities. This paper will be organized according to the following schedule. In the second session we will provide a global description of spectral estimation techniques and the formulation of SVM based algorithms. In the third section we will focus on the experimental scenario, simulation results and performance evaluation. The fourth section will be dedicated to the conclusion and future research activities.

2 Spectral Estimation Techniques and SVM Algorithm Formulation

2.1 Main Categories of Spectral Estimation Methods

Spectral estimation has been considered as an established discipline more than a century ago. It was adopted to deal with the problem of determining the spectral content of a time series from a finite set of measurements, by means of either non-parametric or parametric techniques. Spectral estimation finds applications in many diverse fields such as vibration monitoring, speech analysis, radar and sonar systems, medicine, seismology and others.

In cognitive radio systems, there is a resurging interest in spectral estimation methods as a means of characterizing the dynamical behavior of a given primary network in terms of radio spectrum occupancy. Those methods can be divided in three main categories, knowing as:

- The *Beamforming* techniques that were among the first methods used for processing spatio-temporal samples received from a network of antennas [5–7]. They are used to estimate incoming signals, the directions of arrival (DOA), along with removing interferences. The basic idea of those methods is performing spatial filtering by weighting the signals received with suitably calculated coefficients. Linear prediction method [8] and Minimum Variance Distortionless Response (MVDR) [7] are included in this category. The disadvantages of these methods, linked to the restriction of their visibility area, are motivating new research advances.
- The *Maximum Likelihood (ML)* techniques have also been applied first to estimate the direction of arrival [9]. They are very efficient for estimating complex amplitudes or noise's standard deviation, especially in low signal noise ratio (SNR) scenarios. However, they suffer from the high computational load.
- The *Subspaces* category of techniques appeared in the 1970s in the fields of underwater acoustics, seismic and radio astronomy [10, 11]. They are known as high-resolution techniques, and they are based on an algebraic approach that exploits the decomposition of the covariance matrix of data in sub-spaces. The main techniques that were developed are the Multiple Signal Characterization (MUSIC) [3], Min-Norm [12], and Estimation of Signal Parameters Via Rotational Invariance Techniques (ESPRIT) [13]. This category exhibits performances close to the *ML* one, while presenting low computational complexity.

2.2 Spectrum Estimation Based on SVM and Subspace Technique

Support Vector Machines (SVM) have been proposed as a technique in time series prediction [14]. It was originally proposed as an efficient method for pattern recognition and classification [15]. The key characteristic of SVM is that a nonlinear function is learned by a linear learning machine in a kernel induced feature space while the capacity of the system is controlled by a parameter that does not depend on the dimensionality of the space.

We have experienced in our previous study, which was developed in [16], the use of MVDR and MUSIC without resorting to SVM. The development of SVM based MUSIC algorithm was conducted through the use of the power minimization criterion in the Minimum Variance Distortionless Response (MVDR) method [17]. This allows reformulating the MUSIC algorithm from the MVDR, leading to the derivation of the SVM-MVDR and SVM-MUSIC algorithms [4].

SVM-MVDR Estimator. The SVM version is obtained by combining the formula-tion of the MVDR with the complex SVM formulation [18].

We consider $x[n] = [x_1[n], \ldots, x_L[n]]^T$ as a set of L samples of received signals. The power spectrum of $x[n]$ is estimated through a bank of filters, with length L, centred at frequencies $\{w\}_k$. The output power of each filter is:

$$S_x(k) = \mathrm{E}\ [w_k^H x[n] x^H[n] w_k] \tag{1}$$

Where $\mathrm{E}[.]$ is the expectation operator, T represents the transpose and H represents the conjugate transpose. A sample-based approximation of this power estimation is:

$$S_x(k) \approx \frac{1}{N} w_k^H \sum\nolimits_n x[n] x^H[n] w_k = w_k^H R w_k \tag{2}$$

Where $\boldsymbol{R}$ has the expression:

$$R = \frac{1}{N} \sum\nolimits_n x[n] x^H[n] \tag{3}$$

Next step considers a linear estimator of the frequency that can be expressed as:

$$y_k[n] = w_k^H x[n] \tag{4}$$

A set of J complex signals $u_k[n]$ are defined so that their desired array outputs are $r_k[n]$. The optimization must minimize the errors $e_{k,n} = r_k[n] - w_k^H u_k[n]$ over all signals $u_k[n]$. The corresponding functional L_p must contain the following three terms to be jointly minimized: the filter output power, a regularization term consisting on the norm of the filter parameters, and the cost function applied to all errors over $u_k[n]$.

$$L_p = \frac{1}{2} w_k^H R w_k + \frac{v}{2} \|w_k\|^2 + \sum\nolimits_{\mathrm{n}} \left[\mathcal{L}_R\left(\xi_{n,k} + \xi'_{n,k}\right) + \mathcal{L}_R(\zeta_{n,k} + \zeta'_{n,k}) \right] \tag{5}$$

The following constraints are considered, which are equivalent to those applied to the classical MVDR:

$$\begin{aligned}
&\Re\left(r_k[n] - w_k^H u_k[n]\right) \leq \varepsilon - \xi_{n,k} \\
&\Im\left(r_k[n] - w_k^H u_k[n]\right) \leq \varepsilon - \zeta_{n,k} \\
&\Re\left(-r_k[n] + w_k^H u_k[n]\right) \leq \varepsilon - \xi'_{n,k} \\
&\Im\left(-r_k[n] + w_k^H u_k[n]\right) \leq \varepsilon - \zeta'_{n,k}
\end{aligned} \quad (6)$$

These constraints account for real (positive and negative) and imaginary (positive and negative) parts of errors $e_{k,n}$, where ε plays the role of the error tolerance and $\xi_{n,k}$, $\xi'_{n,k}$, $\zeta_{n,k}$, $\zeta'_{n,k}$ are the so-called slack variables, or the part of the error to be minimized.

Next step consists on defining Lagrange multipliers $\alpha_{n,k}, \beta_{n,k}, \alpha'_{n,k}, \beta'_{n,k}$ for real positive, real negative, imaginary positive, and imaginary negative constraints (6) in order to write a Lagrange functional with these constraints and functional L_p (5). An optimization of the Lagrange functional with respect to w_k which is assumed to be complex-valued gives the following result:

$$w_k = R^{-1} U_k \psi_k \quad (7)$$

Where Lagrange multipliers have been grouped as $\psi_k = \alpha_{n,k} + j\beta_{n,k} - \alpha'_{n,k} - j\beta'_{n,k}$ and $U_k = [u_k[1], \ldots, u_k[J]]$, with J being the number of constraints. Plugging result (7) into (2), the estimated spectrum at the frequency is then:

$$S_k = \psi_k^H U_k^H R^{-1} U_k \psi_k \quad (8)$$

SVM-MUSIC Estimator. Assuming that the signal and noise subspaces are known. MUSIC estimator that was developed in [4] is written as:

$$S_x^{MUSIC}(k) = w_k^H V \begin{bmatrix} \alpha \mathrm{I_n} & 0 \\ 0 & \beta \mathrm{I_s} \end{bmatrix} V^H w_k \quad (9)$$

The parameters composing the estimator formula (9) are derived from the following decomposition of the autocorrelation matrix **R,** which has been reconstructed in order to derive a relationship between MVDR and MUSIC:

$$R = V \Lambda V^H \quad (10)$$

Where $\boldsymbol{V}$ is the matrix of eigenvectors of $\boldsymbol{R}$ and Λ is a diagonal matrix containing the corresponding eigenvalues. Let us define $\boldsymbol{V_s}$ and $\boldsymbol{V_n}$ as L_s and L_n eigenvectors of the signal and noise subspaces, respectively. $\boldsymbol{V}$ can be grouped in signal and noise elements, as follows:

$$V = [\,V_s \quad V_n\,] \quad (11)$$

Thus, I_s and I_n are $L_s \times L_s$ and $L_n \times L_n$ identity matrices. α and β are eigenvalues of the signal and noise subspaces, respectively.

The functional (12) can be written from (5) by substituting the autocorrelation matrix $\boldsymbol{R}$ by the matrix (9) under the same constraints as in (6). We obtain:

$$L_p = \frac{1}{2} w_k^H V \begin{bmatrix} \alpha \mathrm{I_n} & 0 \\ 0 & \beta \mathrm{I_s} \end{bmatrix} V^H w_k + \sum\nolimits_n \left[\mathcal{L}_R\left(\xi_{n,k} + \xi'_{n,k}\right) + \mathcal{L}_R(\zeta_{n,k} + \zeta'_{n,k}) \right] \tag{12}$$

The following parameters are obtained $w_k = Q^{-1} U_k \psi_k$, where $Q^{-1} = \alpha^{-1} V_n V_n^H$ Thus, the SVM-MUSIC estimator can be written as follow:

$$S_k = \psi_k^H U_k^H V_n V_n^H U_k \psi_k \tag{13}$$

3 Experimental Results and Performance Evaluation

3.1 Experimental Setup

In our radio environment scenario, we consider the presence of four existing primary networks to be detected by a single node who tries to access opportunistically to the unused frequency channel. Three Bluetooth transmitters with 1 MHz bandwidth are involved, along with one 10 MHz CCK WiFi modulation transmitter. The central frequencies of Bluetooth signals are 10, 20 and 30 MHz. The WiFi signal is centered at 50 MHz. The prior spectrum includes 5 WiFi signals at 10, 30, 50, 70 and 90 MHz, and 9 Bluetooth signals from 10 to 90 MHz as shown in Fig. 2. For all signal sources, the pulse shape is a raised cosine with roll-off factor of 0.5. Signals are sampled at 10 samples per symbol and received with equal power. The SNR of the incoming signals is 1 dB.

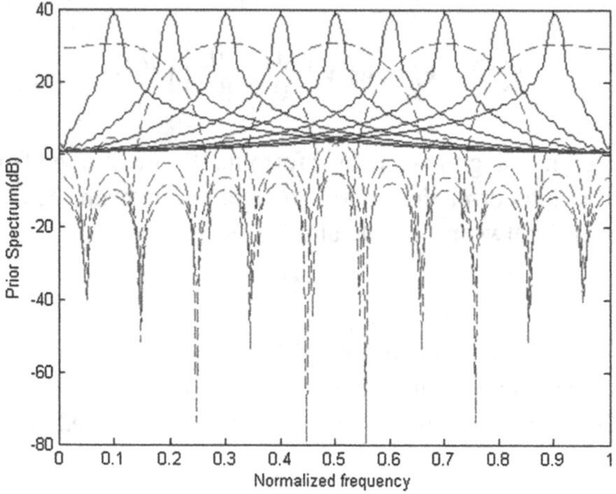

Fig. 2. Prior spectrum: dashed lines are WiFi signals. Continued lines are Bluetooth signals

To assess the frequency estimation of MUSIC, SVM-MVDR and SVM-MUSIC algorithms, we used 100 Monte Carlo iterations.

3.2 Spectrum Sensing Results

The Fig. 3 shows detected spectrum by applying MUSIC technique in our Scenario.

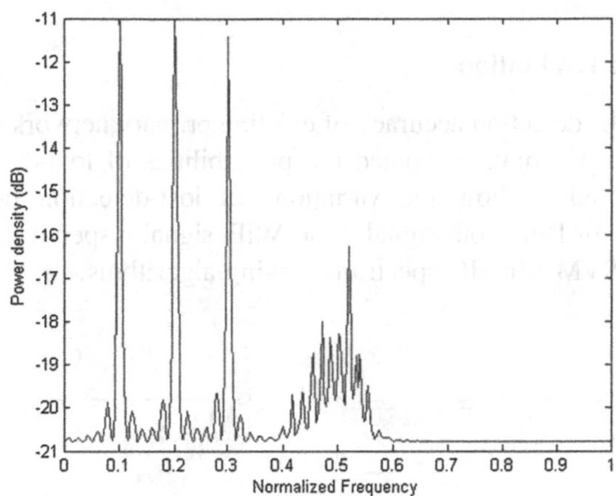

Fig. 3. Spectrum estimation by MUSIC method

Obtained results by applying the SVM version of MUSIC and MVDR are shown in Figs. 4 and 5.

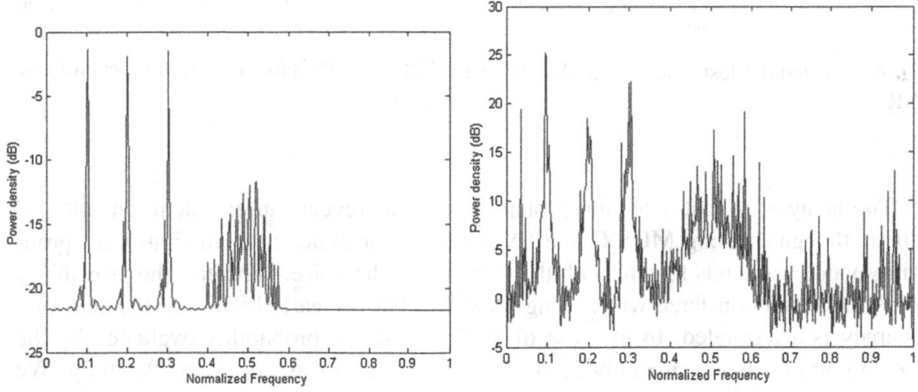

Fig. 4. Spectrum estimation by SVM-MUSIC **Fig. 5.** Spectrum estimation by SVM-MVDR

By analyzing the detection behavior of the three techniques, the following remarks are sealed:

MUSIC and SVM-MUSIC are able to detect the occupied carrier frequencies with high resolution for Bluetooth signals and gives a good estimation of WiFi signal but the resolution is deteriorated. Whereas, SVM-MVDR technique demonstrates low capacity in both, signal detection and peaks resolution. Thus the developed techniques which are based on the sub-space MUSIC algorithm are more efficient and stable in comparison with MVDR.

3.3 Performance Evaluation

To better evaluate the detection accuracy of existing primary network signals, using the previous algorithms, we have evaluated the probabilities of lost-detections and false alarms. Figures 6 and 7 show the variations of lost-detection probability versus Signal-Noise-Ratio for Bluetooth signals, and WiFi signal respectively, using MUSIC, SVM-MVDR and SVM-MUSIC spectrum sensing algorithms.

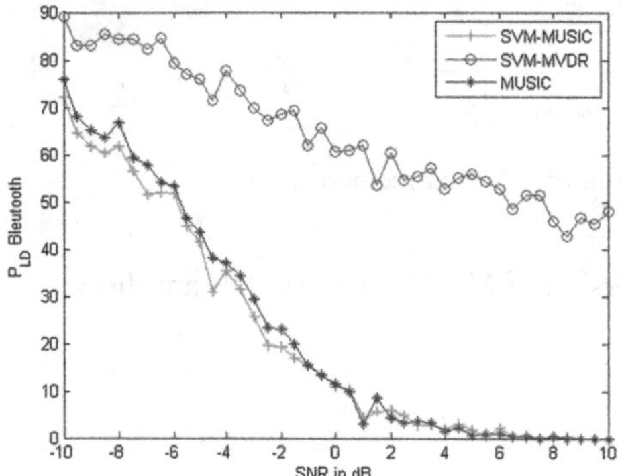

Fig. 6. Bluetooth's lost-detection probability vs. SNR

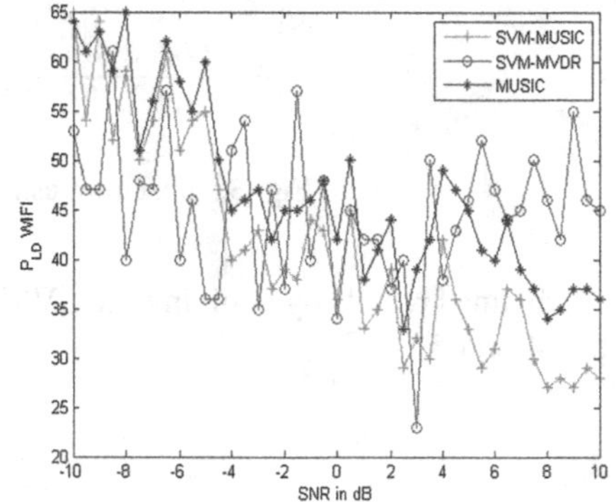

Fig. 7. WiFi's lost-detection probability vs. SNR

The analysis of lost-detection probability graphs reveals an excellent detection of Bluetooth signals using MUSIC and SVM-MUSIC techniques, with significant probability value that tends to zero with the decrease of the noise. Whereas, the probability of lost-detection registered while using SVM-MVDR is clearly high, thus the detection accuracy is deteriorated. In the case of WiFi signal, the probability evaluated by the three algorithms is considerably high. Thus, the detection is not done efficiently. We notice that SVM-MVDR and MUSIC are less accurate comparing to SVM-MUSIC. The only difference between the four primary network signals is the frequency bandwidth of the signals. We deduce that in the presence of narrow bandwidth signals, as in the case of Bluetooth signals, MUSIC and SVM-MUSIC techniques give excellent and similar detection performance.

The global False alarm probability variation, for both Bluetooth and WiFi signals, using the three algorithms is plotted in Fig. 8.

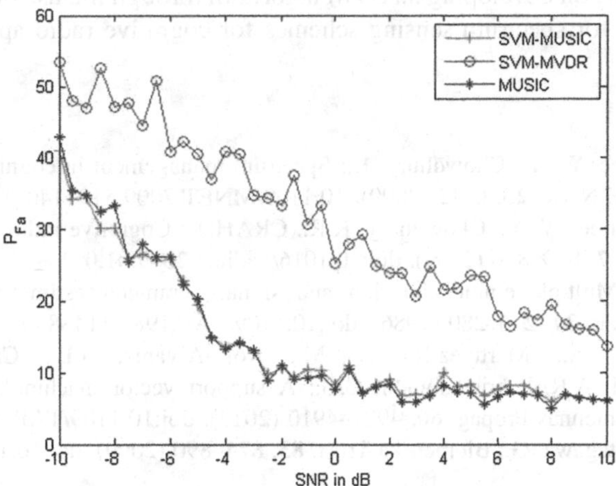

Fig. 8. Global false alarm probability vs. SNR

The probability value of global false alarm detected signals tends to 0.6 while applying MUSIC and SVM-MUSIC. This value is considered as good performance indication with respect to our radio environment scenario constraints, since it gathers both, Bluetooth and WiFi signals false detections. Whereas, the probability of false alarm signals generated by SVM-MVDR is not tolerated.

The over-all detection results are similar for both MUSIC and SVM-MUSIC. That leads to wonder what additional performances are gained while developing the SVM version. In fact, the objective of the work presented in this paper is the implementation of SVM-MUSIC version, so as to evaluate the detection trade-off error, and compare the results with those of MUSIC for our spectrum sensing scenario. Although, our ultimate goal is experiencing other strength points highlighted in [4], that can allow us to deal with the problem of coherent signals detection, and multipath fading effects, which cause harmful interference, especially in the ad hoc architecture conditions.

4 Conclusion

In this work, we have undertaken the spectrum sensing feature, by evaluating the efficiency of Support Vector Machines (SVM) spectrum estimation-based techniques in the detection of present primary signals in a CRAHNs scenario. In the simulation results, we have taken into consideration the three algorithms MUSIC, SVM-MVDR and SVM-MUSIC. MUSIC subspace method is explored because it is able to estimate frequencies with high resolution. Support Vector Machines technique was exploited to obtain a new robust formulation of spectrum sensing, which is SVM-MUSIC. The use of SVM-MVDR seems necessary because it was the base of developing SVM-MUSIC mathematical model. This latter was presented along with the experimental scenario.

The obtained results showed the detection efficiency offered by each technique, so as to determine the spectrum holes. The detection error trade-off was evaluated for the three techniques and proved the accuracy of Support Vector Machines technique. Our future activities will focus on developing an SVM algorithm through the use of kernels, which will be dedicated for spectrum sensing schemes for cognitive radio applications.

References

1. Akyildiz, I., Lee, W.-Y., Chowdhury, K.: Spectrum management in cognitive radio ad hoc networks. IEEE Netw. **23**, 6–12 (2009). 10.1109/MNET.2009.5191140
2. Akyildiz, I.F., Lee, W.-Y., Chowdhury, K.R.: CRAHNs: Cognitive radio ad hoc networks. Ad Hoc Netw. **7**, 810–836 (2009). doi:10.1016/j.adhoc.2009.01.001
3. Schmidt, R.: Multiple emitter location and signal parameter estimation. IEEE Trans. Antennas Propag. **34**, 276–280 (1986). doi:10.1109/TAP.1986.1143830
4. El Gonnouni, A., Martinez-Ramon, M., Rojo-Alvarez, J.L., Camps-Valls, G., Figueiras-Vidal, A.R., Christodoulou, C.G.: A support vector machine music algorithm. IEEE Trans. Antennas Propag. **60**, 4901–4910 (2012). doi:10.1109/TAP.2012.2209195
5. Konishi, S., Kitagawa, G.: Biometrika Trust. **83**, 875–890 (2009). doi:10.1093/biomet/81.3.425
6. Haardt, M., Zoltowski, M.D., Mathews, C.P., Nossek, J.: 2D unitary ESPRIT for efficient 2D parameter estimation. In: 1995 International Conference Acoustics, Speech, Signal Process, vol. 3, pp. 2096–2099 (1995). doi:10.1109/ICASSP.1995.478488
7. Capon, J.: High-resolution frequency-wavenumber spectrum analysis. Proc. IEEE **57**, 1408–1418 (1969). doi:10.1109/PROC.1969.7278
8. Gabriel, W.F.: Spectral Analysis and Adaptive Army Superresolution Techniques aglortihm. Proc. IEEE **68**, 654–666 (1978). doi:10.1109/PROC.1980.11719
9. Ziskind, I., Wax, M.: Maximum likelihood localization of multiple sources by alternating projection. IEEE Trans. Acoust. **36**, 1553–1560 (1988). doi:10.1109/29.7543
10. Pisarenko, V.F.: Some applications of the maximum likelihood method in seismology. Geophys. J. Roy. Astron. Soc. **21**, 307–322 (1970)
11. El-Behery, I.N., Macphie, R.H.: Maximum likelihood estimation of the number, directions and strengths of point radio sources from variable baseline interferometer data. IEEE Trans. Antennas Propag. **75**, 1928–1932 (1978)
12. Reddi, S.S.: Multiple source location-a digital approach. IEEE Trans. Aerosp. Electron. Syst. **15**, 95–105 (1979)
13. Roy, R., Kailath, T.: ESPRIT - estimation of signal parameters via rotational invaraince techniques. IEEE Trans. Acoust. **37**, 984–995 (1989)
14. Mukherjee, S., Osuna, E., Girosi, F.: Nonlinear Prediction of Chaotic Time Series Using Support Vector Machine. IEEE NNSP (1997)
15. Vapnik, V.: The Nature of Statistical Learning Theory. Springer, New York (1995)
16. El Barrak, S., Lyhyaoui, A., El Gonnouni, A., Puliafito, A., Serrano, S.: Application of MVDR and MUSIC spectrum sensing techniques with implementation of node's prototype for cognitive radio ad hoc networks. Presented at ICSDE 2017 (2017)
17. Bienvenu, G., Kopp, L.: Optimality of high resolution array processing using the eigensystem approach. IEEE Trans. Acoust. **31**, 1235–1248 (1983)
18. Ramon, M.M., Xu, N., Christodoulou, C.G.: Beamforming using support vector machines. IEEE Antennas Wirel. Propag. Lett. **4**, 439–442 (2005)

Optimization of a Modular Ad Hoc Land Wireless System via Distributed Joint Source-Network Coding for Correlated Sensors

Dina Chaal[1](✉), Asaad Chahboun[1], Frédéric Lehmann[2], and Abdelouahid Lyhyaoui[1]

[1] LTI Laboratory, ENSA-Tangier, Tangier, Morocco
dina.chaal@gmail.com, lyhyaoui@gmail.com
[2] SAMOVAR Laboratory, Télécom SudParis, Évry, France
frederic.lehmann@telecom-sudparis.eu

Abstract. This paper outlines a proposition of a framework optimizing the communication scheme from the physical part of the transmitters to the data gathered at the receiver in a wireless sensor network. We propose a coding scheme able to take into account the correlation between measurements obtained by the sensors. This scheme consists of joint distributed source encoding and linear network coding. Several coding strategies are compared. (1) linear source coding, based on the low density generator matrix (LDGM) codes, where the compression process is performed by every sensor independently (2) distributed source coding, where the correlation between sources is taken into account, without cooperation between sensors. The compression ratios costs are evaluated analytically for each strategy, with respect to the communication costs between the nodes of the network. The results show a significant improvement in terms of compression rate and distortion compared to linear source encoding.

Keywords: Wireless Sensor Networks · Modular Ad-hoc networks · Lossless compression · Source and network coding · Distributed coding · Random Linear Network Coding (RLNC)

1 Introduction

Satellite Remote Sensing (RS) is an important and evolving technology for monitoring the world global environmental change. Scientific community has made extensive use of satellite image data for mapping land cover and estimating geophysical and biophysical characteristics such as Land Surface Temperature (LST). The classical LST in-situ validation method consists on the use of radiometers and data loggers along predefined transect. Measurements are performed on fixed location, or in points uniformly distributed on the target field to account for the spatial heterogeneity. Emerging field of Wireless Sensor Network (WSN) combines sensing, computation and communication into a single tiny

A. Puliafito et al. (Eds.): ADHOC-NOW 2017, LNCS 10517, pp. 171–183, 2017.
DOI: 10.1007/978-3-319-67910-5_14

device (called mote or node). The mesh networking connectivity seeks out and exploits any possible communication path by hopping data from node to node in search of its destination. The main objective of this low cost adopted method is to provide an alternative and operative method for in-situ LST validation measurements. These wireless modular networks of interconnected and calibrated sensors are made up of a multitude of small autonomous devices, capable of self-organizing to communicate with each other, thus working together to collect information about their environment, all without human intervention. The size of the transmitted data gathered, and the cost effectiveness performance expected from such systems require high levels of data flow optimization, therefore, source coding and routing techniques are primordial in the communication scheme of such structures. The application demands for robust, scalable, low-cost and easy to deploy networks are perfectly met by the developed WSN. This paper is an extension of a previous work [1] where we proposed a data processing system using LDGM codes and network coding, to further improve the compression performances previously obtained, and take advantage of the data correlation between the sensors, we present an upgrade of this system using the distributed source coding at the transmitter and network coding as a routing process, then compare the results obtained from both strategies. In Sect. 3 we describe the proposed global scheme for data compression and routing. Section 3.1 will deal with the theory of distributed coding. The illustration of the network coding process will be detailed in Sect. 3.2. Section 4 will be reserved for the results of the experimental simulations obtained from the real scenarios, we provide a comparison of the proposed scheme with the previous method of compression by evaluating the compression ratio of each, finally conclusion will be the subject of the last paragraph.

2 Related Work

Distributed Source Coding (DSC) [2] refers to the case of highly correlated signals that are separately encoded at the transmitter and jointly decoded at the receiver. This kind of techniques has recently been studied as a potential solution for data compression in applications requiring simple encoders, since the computation complexity is switched to the receiver which is generally a base station with high processing capabilities. From a theoretical point of view, the DSC relies on the Slepian-Wolf theorem [3] established in 1973 and the Wyner-Ziv [4] theorem in 1976, as well as on some necessary tools to implement solutions based on this coding which are the LDPC error correcting codes that are part of our proposed solution. Slepian-Wolf's work on DSC was intended for lossless source coding of discrete sources. However, lossy distributed compression approach was a foundational work of Wyner-Ziv [4,5]. Wyner's achievements included channel coding, due to the strong connection between DSC and channel coding concepts [6]. Therefore, the majority of DSC applications joint channel coding, such as Turbo codes [7] and LDPCs (Low Density Parity Codes) [8]. [7] exploited punctured Turbo codes for compression of correlated binary sources, unlikely the implementation of the method was not efficient because of the absence of theoretical link between the design of the two codes (Slepian-Wolf and Turbo code). [9], [10] processed distributed source coding using syndromes (DISCUS). [11] exploited a novel approach of distributed adaptive signal processing framework and algorithm to reduce energy consumption in sensor networks, their approach consisted on the blind data compression of each sensor without any communication between the nodes of the network. [6] exposed a continuation to [10] results. Through relevant research efforts, Slepian and Wolf 1973 [3]; [5] concluded that Slepian-Wolf source coding and Wyner-Ziv coding are actually source-channel coding problems. [12] achieved high performances in rate transmission by taking advantages of data correlation among the sensor nodes, to do so he implemented an approximated version of Slepian-Wolf coding called localized Slepian-Wolf coding, however, this approach wasn't effective since it estimates static link capacity, moreover it becomes constraining and not practical when the neighborhood size is not scalable. In recent work, [13] illustrated the robustness performance of a DSC-based system (slotted ALOHA) in terms of throughput, delay, and energy efficiency, he approximated the average traffic contained in each time interval and deduced a closed-form expression for average rate, then exploited Markov chain analysis to provide the average delay and energy consumption. Researches on DSC for WSN derive from the Slepian-Wolf theorem, hence, all proposed DSC algorithms require prior knowledge of the data correlations at different sensors.

3 The Proposed System

The proposed system is composed of three main blocs as shown in Fig. 1. A source bloc which collects the physical data and transmits it via a distributed source encoder, a network bloc which optimizes the information flow, and a receiver bloc which reconstructs the original signal via a network and a joint decoder.

3.1 The Distributed Source Coding

While conventional source coding exploits the redundancies in the source in order to perform compression, distributed source coding takes advantage of the correlation available between multiple sources to optimize the compression ratio for each source without the necessity to identify the neighboring information. This shifts the complexity of the problem to the decoder, which is convergent in many applications that generally uses a low complex device at the source and a more robust device at the receiver. We adapted a channel coding approach for compression, using Algebraic Binning as shown in the Fig. 2. The used algorithm is described below:

- Distribute all the possible realizations of x (of length n) the source information into bins.
- Each bin is a coset of an (n, k) linear channel code C, with parity-check matrix H of size $(n, n-k)$ indexed by a syndrome s.

- For an input x, the encoder forms a syndrome
$$s = xH^T.$$
- The decoder interprets the side information as a noisy version (output of virtual communication channel called correlation channel) of x.
- Find a code word of the coset indexed by s closest to y by performing conventional channel decoding.

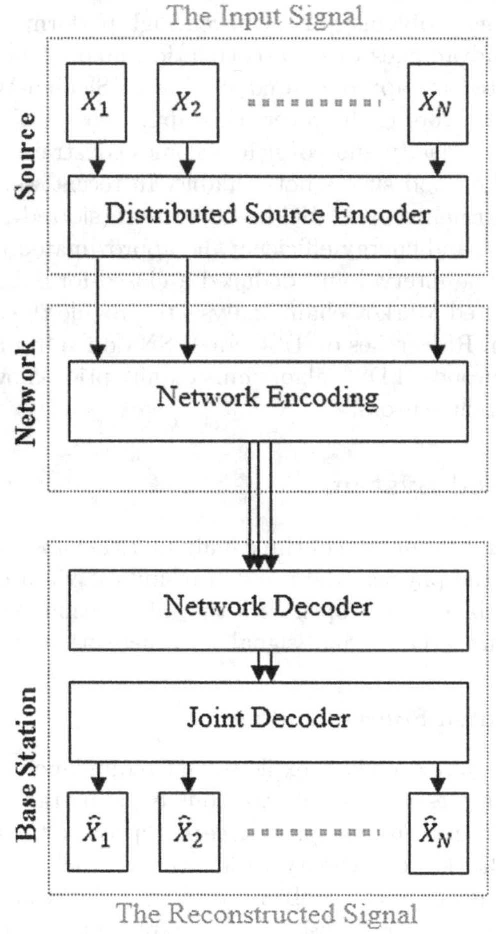

Fig. 1. Structure of the proposed framework

3.2 The Network Coding Processor

Network coding is a concept recently introduced by [14] to improve the throughput of multicast information transmissions in a network formed by one or more

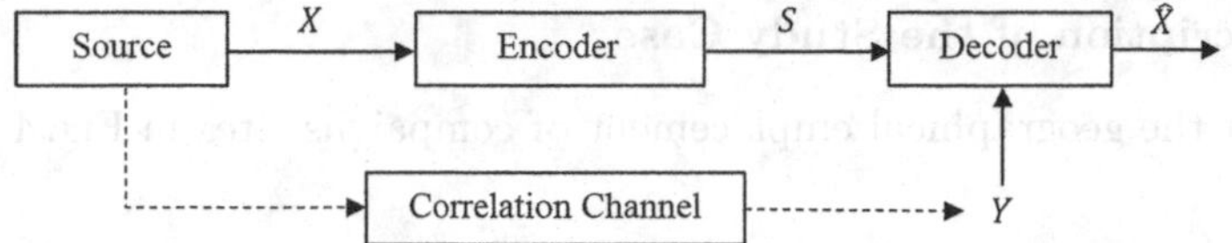

Fig. 2. Distributed source coding scheme

sources and one or more receivers. The principle of this mechanism is to allow intermediate nodes to perform encoding operations on the data of incoming packets instead of simply sending them as in the conventional routing strategy. Each link in the network transfers the error-free packets of M symbols into a given finite field F_q. During each transmission, the source node sends n information vectors of $X_1, X_2......X_n$, (packets) each with the size $1 \times M$. The packet coming out of a transmission node is obtained as a random linear combination (with finite field values F_q) of all incoming packets received. The receiver can obtain the set of transmitted packets $X_1, X_2......X_n$ from the received set of packets $Y_1, Y_2......Y_n$ by simply solving the set of linear equations $Y = AX$, where A is a matrix of $N \times n$ which corresponds to the linear transformation system applied by the network (Fig. 3).

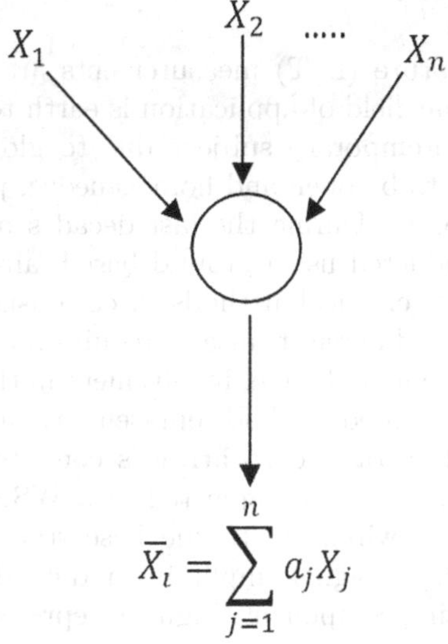

Fig. 3. Transmission packets process

3.3 Description of the Study Case

We present the geographical emplacement of compaigns sites in Fig. 4 below.

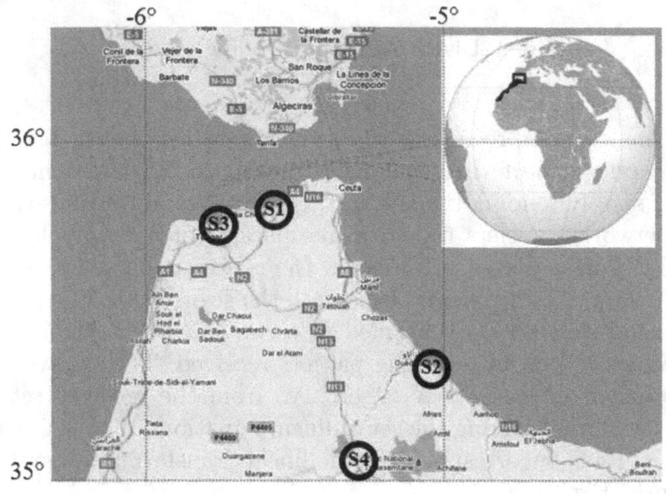

Fig. 4. Geographical emplacement of campaigns sites

Land surface temperature (LST) measurements are used for remote sensing (Fig. 5). One particular field of application is earth temperature monitoring that became a more contemporary subject due to global warming concerns. The observed area needs to be open and homogeneous, particular methods and instruments are used [15,16]. During the last decades many earth observation campaigns had been conducted using ground based, air-borne, ship-borne and satellite experiments, these classical methods witch consist of the use of radiometers and data loggers along different transects require generally expensive instruments and/or physical or remote human involvement in the gathering or processing phase of data. The proposed method represents the measurement system as a set of modules of node arrays, each array is consisted by intelligent nodes deployed in the observed area and connected in a WSN topology, the data is processed and transmitted wirelessly to the base station. Each node is composed by (*i*) TIR module, (*ii*) signal amplifying and conditioning stage and (*iii*) processing and transmitting component. Figure 6 represents the bloc diagram of the node system.

The TIR used in our system is $OSM101$, a very low cost battery powered and portable infra-red to analog converter module, its range of measure is $-18\,°$C to $538\,°$C, it has a laser cycle indicating the optical FOV, a 3.5 VDC Lithium battery and an SMP male-to-male K-type retractable cable. The TRI sensor $OSM101$ provides a very low output voltage (3.3 mV at $80\,°$C), this voltage level is not sufficient to connect it to the wireless sensor node and therefore an

Site		Latitude	Longitude	Time Period	Sample	
Kasr-Seghir	(S1)	+35° 51' 28.08"	-5° 32' 1.35"	16th of May 2009	Bare-Soil	(BS1)
					Bare-Soil	(BS2)
					Sparse-Vegetation	(SV)
					Gravels	(G)
					Mineral-Soil	(MS)
					Water	(W1)
Targha	(S2)	+35° 24' 4.49"	-5° 1' 0.05"	4th to 5th July 2009	Bare-Soil	(BS3)
					Water	(W2)
				8th to 10th August 2009	Bare-Soil	(BS3)
Tangier	(S3)	+35° 44' 56.43"	-5° 50' 48.12"	17th to 18th August 2009	Water	(W3)
Chefchaouen	(S4)	+35° 10' 0.07"	-5° 16' 5.02"	28th to 29th August 2009	Water	(W4)

Fig. 5. Validation sites characteristics

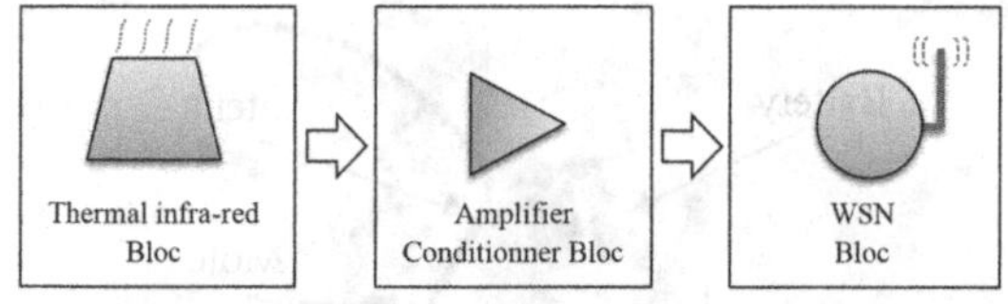

Fig. 6. The block diagram of the node system

Fig. 7. Location of the sensors in "Ksar Sghir" site

$AD595$ amplifier is added for conditioning the sensor's signal, thus the output of the amplifier will be 10 mV/°C. For the WSN module, a Mica2dot mote ($MPR510CA, CROSSBOW, U.S$) is selected mainly for its size, price and range.

Figure 8 presents the overall structure and components of a sensor node. The modules are deployed in four distant and heterogeneous sites located in northern morocco (Kasr-Seghir, Targha, Tangier and chefchaouen). In each module targets as detailed in Figs. 4 and 5. The case study is focused on the “Ksar Sghir” site ($S1$) since it contains the maximum number of sensors. The site consists of 6 sensors, two of which are located in bare soil environment (labeled as 01 and 03), the rest is respectively deployed in the following environments: water (04), gravels (05), vegetation (02) and mineral soil (06). In addition to that the site includes a base station (G) which is located in the center of the network, and serves as a terminal to process the collected data from the sensors. The layout and locations of the sensors and the base station are illustrated in Fig. 7.

The sensors are spatially correlated even with the difference in their environments nature. Neighboring sensors collect similar values, Table 1 presents the distance matrix between the sensors.

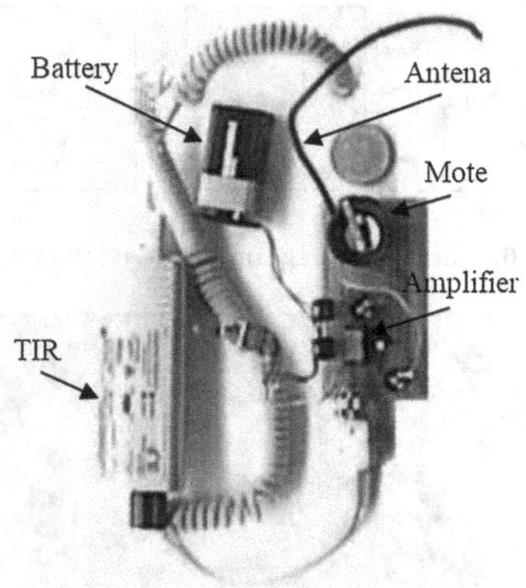

Fig. 8. Structure and components of a sensor node

Table 1. Distance matrix between the sensors

Distance (m)	Bare soil 2	Vegetation	Bare soil 1	Water	Gravel	Mineral soil
Bare soil 2	–	32	67	45	54	45
Vegetation	32	–	48	22	25	19
Bare soil 1	67	48	–	25	38	67
Water	45	22	25	–	22	38
Gravel	54	25	38	22	–	29
Mineral soil	45	19	67	38	29	–

4 Simulation Results

The database is formed by measurements of 6 sensors, each sensor collects a temperature value every 10 s for a given period of time. The maximum difference between two neighboring data is 2, 5 °C, which means for a precision of 0, 1 °C and for a given neighboring value, we will have 50 possible values and therefor a maximum difference of 6 bits between two neighboring sources. In our previous work [1] we used an LDGM encoder which focused on the redundancy within the information itself, providing us with an entropy close to 0, 85 for our case study and therefore a compression rate of 0.85 and a main error equal to 0.4 °C. While using the correlation described above the side information provided us with a joint entropy of 0, 65 and a main error of 0 °C since the compression is lossless. We will present a graphical comparison of the reconstructed signal between the LDGM coding and the distributed coding for three of the studied sensors. The data of each sensor is devised into 20 packets and then simultaneously injected to the network. The emitted message from a certain node is a linear combination of the node's own packets along with the received useful packets. The decoding step is triggered once there is enough data to reconstruct the initial packets. The main delay time in the system's initial performance was 0, 6 s, along with 33% rate loss in packet transfers. The implementation of our system ensured a smaller delay and unmistakably a bandwidth enhancement. The results showed significant improvement in comparison with our previous work [1] and some existing related algorithms [17,18] (Figs. 9, 10, 11, 12, 13 and 14).

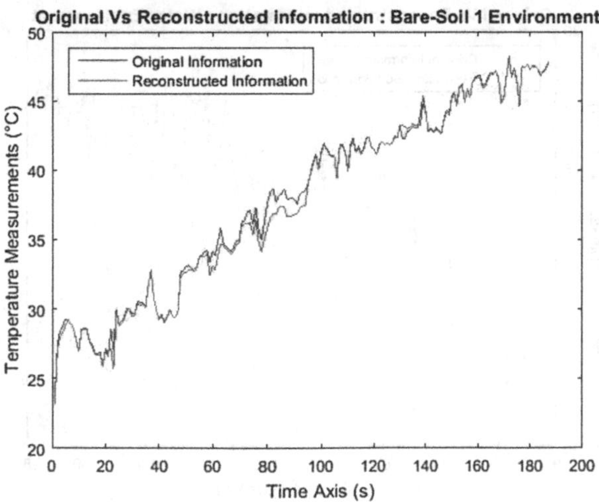

Fig. 9. Input data vs. reconstructed data for sensor bare-soil 1 with LDGM, R = 85%

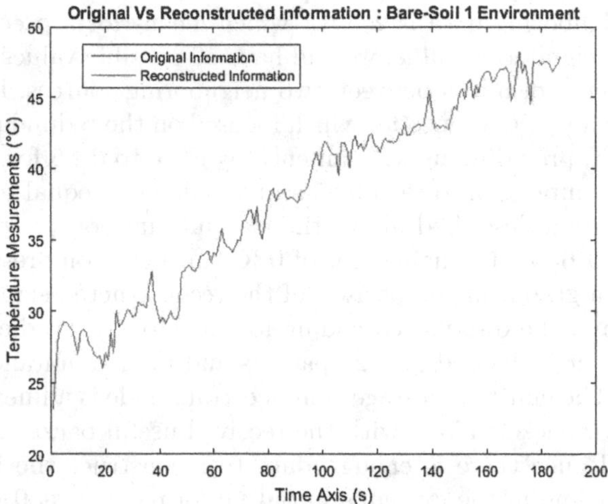

Fig. 10. Input data vs. reconstructed data for sensor bare-soil 1, with DSC, R = 69%

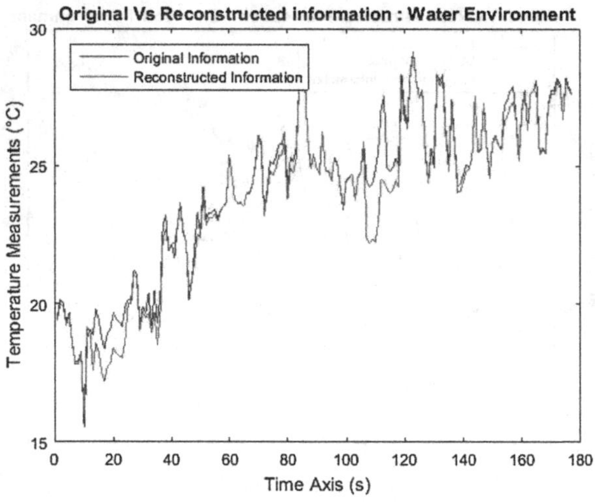

Fig. 11. Input data vs. reconstructed data for sensor water with LDGM, R = 85%

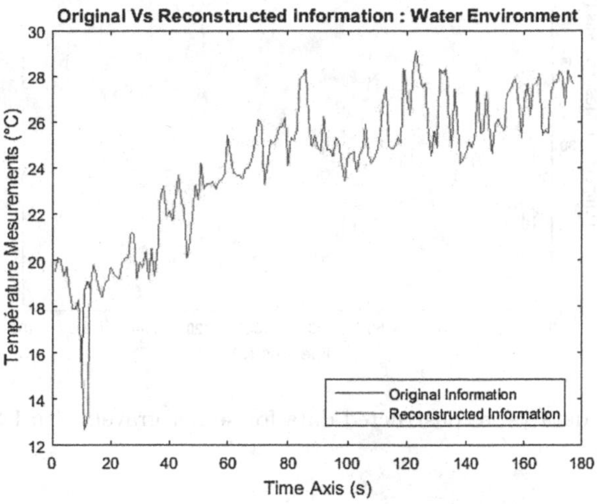

Fig. 12. Input data vs. reconstructed data for sensor water, with DSC, R = 69%

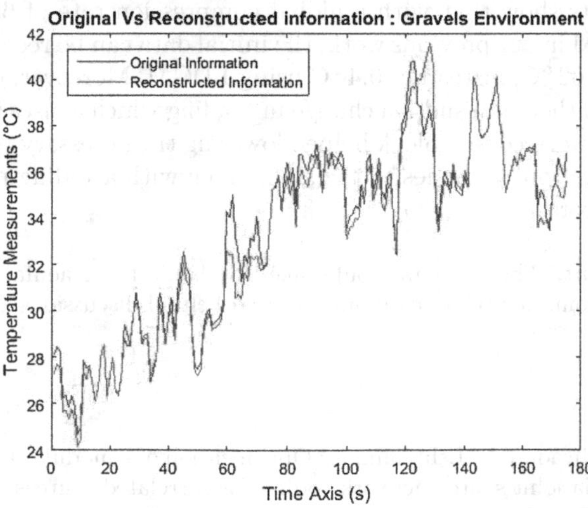

Fig. 13. Input data vs. reconstructed data for sensor gravel with LDGM, R = 85%

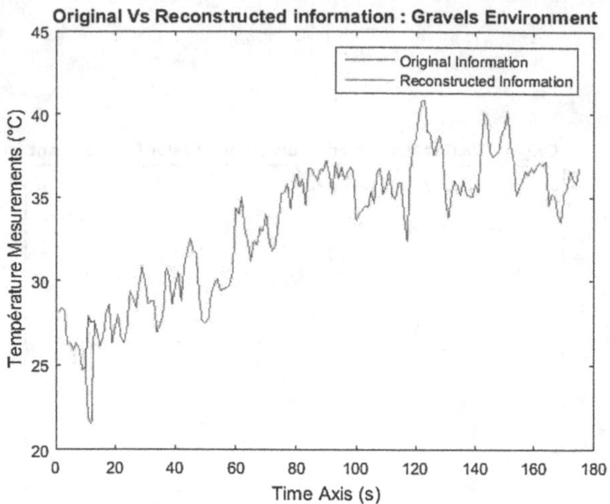

Fig. 14. Input data vs. reconstructed data for sensor gravel, with DSC, R = 69%

5 Conclusion

At the source, distributed source coding significantly lowered the number of bits required to be sent in the channel, therefore enhancing the bandwidth. The simulation results show that with a global compression rate of 69%, instead of 85% using LDGM in our previous work, the initial data can be reconstructed with a distortion of 0.02 °C, instead of 0.4 °C using LDGM. Moreover, distortion only takes place when there is a sudden change in reading which causes inexact values. Whereas the network coding block helped lowering the necessary processing time in packet routing from all nodes to the sink, along with avoiding packet loss and data transfer errors.

Acknowledgment. The authors would like to thank Pr. Naoufal Raissouni and Pr. Asaad Chahboun, for their collaboration, and helpful discussions.

References

1. Chaal, D., Lyhyaoui, A., Lehmann, F.: Optimization of a modular ad hoc land wireless system via joint source-network coding for correlated sources. In: Proceedings of Engineering and Technology, vol. 20, pp. 21–24 (2017)
2. Dragotti, P.L., Gastpar, M.: Distributed Source Coding, 1st edn., January 2009
3. Slepian, D., Wolf, J.: Noiseless coding of correlated information sources. IEEE Trans. Inf. Theory **19**(4), 471–480 (2006)
4. Wyner, A.D.: The rate-distortion function for source coding with side information at the decoder\3-II: general sources. Inf. Control **38**(1), 60–80 (1978)

5. Kaspi, A., Berger, T.: Rate-distortion for correlated sources with partially separated encoders. IEEE Trans. Inf. Theory **28**(6), 828–840 (1982)
6. Liveris, X., Cheng, S.: Distributed source coding for sensor networks. IEEE Sign. Process. **21**(5), 80–94 (2004)
7. Garcia-Frias, J.: Compression of correlated binary sources using turbo codes. IEEE Commun. Lett. **5**(10), 417–419 (2001)
8. Liveris, A.D., Xiong, Z., Georghiades, C.N.: Compression of binary sources with side information at the decoder using LDPC codes. IEEE Commun. Lett. **6**(10), 440–442 (2002)
9. Pradhan, S.S., Ramchandran, K.: Distributed source coding: symmetric rates and applications to sensor networks. In: Proceedings of Data Compression Conference, DCC 2000, pp. 363–372 (2000)
10. Pradhan, S.S., Kusuma, J., Ramchandran, K.: Distributed compression in a dense microsensor network. IEEE Sign. Process. Mag. **19**(2), 51–60 (2002)
11. Chou, J., Petrovic, D., Ramachandran, K.: A distributed and adaptive signal processing approach to reducing energy consumption in sensor networks. In: Twenty-Second Annual Joint Conference of the IEEE Computer and Communications Societies, IEEE INFOCOM 2003, vol. 2, pp. 1054–1062, March 2003. (IEEE Cat. No. 03CH37428)
12. Yuen, K., Liang, B., Li, B.: A distributed framework for correlated data gathering in sensor networks. IEEE Trans. Veh. Technol. **57**(1), 578–593 (2008)
13. Hong, Y.W.P., Tsai, Y.R., Liao, Y.Y., Lin, C.H., Yang, K.J.: On the throughput, delay, and energy efficiency of distributed source coding in random access sensor networks. IEEE Trans. Wireless Commun. **9**(6), 1965–1975 (2010)
14. Ahlswede, R., Cai, N., Li, S.Y.R., Yeung, R.W.: Network information flow. IEEE Trans. Inf. Theory **46**(4), 1204–1216 (2006)
15. Wan, Z., Zhang, Y., Zhang, Q., Li, Z.: Validation of the land-surface temperature products retrieved from Terra Moderate Resolution Imaging Spectroradiometer data. Remote Sens. Environ. **83**(1), 163–180 (2002)
16. Dash, P., Olesen, F.-S., Prata, A.J.: Optimal land surface temperature validation site in Europe for MSG
17. Regalia, P.A.: A modified belief propagation algorithm for code word quantization. IEEE Trans. Commun. **57**(12), 3513–3517 (2009)
18. Pulikkoonattu, R.: A source coding scheme using sparse graphs: Modern Coding Theory Course Exam 2008 (2008)

AdhocInfra Toggle: Opportunistic Auto-configuration of Wireless Interface for Maintaining Data Sessions in WiFi Networks

Anurag Sewak[1(✉)], Prakhar Mehrotra[2], Bhaskar Jha[1], Mayank Pandey[1], and Manoj Madhava Gore[1]

[1] Computer Science and Engineering Department, Motilal Nehru National Institute of Technology Allahabad, Allahabad, India
{anuragsewak_2013rcs51,mayankpandey,gore}@mnnit.ac.in, bhaskaron9@gmail.com

[2] Adobe Systems Incorporated, Noida, India
prakhar1194@gmail.com

Abstract. Mobile device users can utilize IEEE 802.11 WiFi networks to facilitate device-to-device local communications using applications like video streaming, P2P file sharing, network gaming, etc. The wireless access points in these networks are sometimes not accessible either due to weak signal strength or due to the constraint to support a limited number of users. So, the ad hoc modes of wireless interface cards can also be used to provide device-to-device connectivity. The objective of this work is to enable toggling between the two modes (infrastructure and ad hoc) of wireless interfaces of these devices, so as to opportunistically switch between these modes depending on the availability of an access point. This provides seamless connectivity within the network, without making any changes in the existing hardware or in the protocol stack. We have developed application-level scripts to enable this mode toggling feature in WiFi equipped scenarios having less dynamism. This toggling feature can also be used to implement application-level routing between two or more ad hoc network clusters via access points, so as to provide intermittent connectivity between them. The results obtained from our experimental testbed show that our scripts work well for toggling between the two modes, and data transfer gives satisfactory performance.

Keywords: WiFi · Infrastructure · Ad hoc · Toggling · Opportunistic switching · Application-level routing

1 Introduction

The proliferation of portable and mobile devices like laptops, tablets and mobile phones have accelerated the use of various applications, like browsing, messaging, file sharing, multimedia streaming, gaming, social networking, etc. In this paper, we focus on less dynamic (low mobility and low churn) scenarios having

A. Puliafito et al. (Eds.): ADHOC-NOW 2017, LNCS 10517, pp. 184–198, 2017.
DOI: 10.1007/978-3-319-67910-5_15

WiFi infrastructure, like airports, railway stations, workplaces, university campuses, etc. Mobile device users can easily get hooked to the WiFi network at these places for getting access to services on Internet. The WiFi infrastructure can also be utilized to establish local communication among mobile device users. Therefore, distributed applications, like file sharing, become relevant for these users where Internet connectivity is not required for their execution.

WiFi access points periodically broadcast beacon frames within their transmission range (also known as WiFi zone) [5]. Devices that receive these beacon frames, attach themselves to an access point for getting WiFi access. However in such scenarios, there are certain connectivity issues that result in disruption of any ongoing data session between devices. These issues relate to mobility, signal impairments and limitations of access points. The major ones are listed below:

- Users may loose connectivity when they move between WiFi zones.
- Certain zones (known as *dead zones*) may exist in WiFi equipped scenarios where the signal strength of access points is either very low or absent.
- Commercial access points can provide access to limited number of users (usually 20–25) [4]; therefore some users may be denied WiFi access.
- Which is the best AP for association among all APs transmitting beacon frames in the range?

In order to handle the above stated issues, ad hoc connectivity between nodes can alleviate the problem in some situations. But, this requires changing the mode of WiFi interface cards of nodes from infrastructure to ad hoc. This raises a demand for opportunistic auto-configuration of WiFi interface of these devices, which shall automatically switch between the modes of WiFi interface depending upon the situation. As a result, some ad hoc network clusters may start existing in isolation at specific places; thus posing a challenge to interconnect them.

In this paper we have devised a solution, where mobile devices opportunistically connect to a WiFi infrastructure network or form an ad hoc network depending on the availability of an access point. However, some security issues may arise while making such connections, which we have not considered in this work. Our contribution is manifold-

- We have developed a script that enables the feature of opportunistic auto-configuration of WiFi modes of mobile devices.
- Another script has been developed that selects the access point with best signal strength and quality for attachment.
- Our final contribution is to intermittently connect two isolated ad hoc network clusters via WiFi infrastructure network using special bridging devices.

Video file transfer and streaming have been used to test the working of these scripts on suitable network topologies. The performance results show that our scripts help in maintaining data sessions, even when communicating nodes move across different WiFi zones.

The remainder of the paper is structured as follows. Section 2 highlights the possible solutions existing in the literature, that handle connectivity and coverage issues in WiFi networks. Our approach to solve the connectivity issues is given in Sect. 3. The experimental evaluation of our scripts and results obtained are discussed in Sect. 4. Finally, the conclusion of our work and future research direction are given in Sect. 5.

2 Related Work

The WiFi standard defines two operational modes for the network interface cards - infrastructure and ad hoc. WiFi interface cards can be set to work in either of these modes, but not in both simultaneously [2]. In this section, we try to highlight those approaches that have used both the infrastructure and ad hoc modes (not simultaneously) of WiFi interfaces of mobile devices for establishing network connectivity. Solutions for expanding the wireless network coverage and selecting the best signal strength are also mentioned in this section. Efforts have been made to devise solutions at different levels of the protocol stack.

Juggler [1] is a link-layer implementation of a virtual 802.11 network layer that toggles between the two modes. Juggler achieves switching times of approximately 3 ms, and less than 400 μs in certain conditions, but it switches between the channels, rather than switching between the modes. WiFi cards have most of their functionality of IEEE 802.11 protocol in the card firmware, rather than in software device driver. Therefore, use of a software defined radio (SDR) can also help in toggling between modes at the physical layer [3].

Efforts have also been made to connect a device to multiple networks at the same time. This either requires multiple network interface cards in devices, or a change in the network protocol stack to support connectivity to multiple networks. Multiple network interfaces require multiple physical radios, which leads to high power consumption. Hence, a software based approach using a single interface is preferable. MultiNet [12] establishes simultaneous connections to multiple networks by virtualizing a single wireless card. The virtualization is achieved by introducing an intermediate layer below IP, which continuously switches the card across multiple networks. This is further improved by FatVAP [8] in terms of throughput and switching efficiency. FatVAP proposed an 802.11 driver, that can facilitate connections with multiple access points by aggregating their bandwidth and performing load balancing between them.

Efforts have also been made to reduce interference in order to expand the coverage area of wireless networks. The radio waves distribution problem has been resolved in [11], so as to expand indoor coverage. In our approach, network expansion has been done by interconnecting various ad hoc clusters via WiFi infrastructure network.

Selection of the best AP (access point) out of signals received from multiple APs has always been a challenging task. WiFi-Reports [7] has been developed as a service that helps in AP selection by leveraging historical information (related to performance and application support) about APs contributed by users.

In contrast to above stated approaches, mode switching is performed at application-level in our approach. Therefore, no changes are required in the device drivers, WiFi cards, or even in the protocol stack. Our approach does not focus on interference issues, and the selection of best AP is done using parameters like signal strength and quality.

3 AdhocInfra Toggle

In this section, we present our approach to maintain uninterrupted data sessions between mobile devices by switching opportunistically between the ad hoc and infrastructure modes of wireless NICs of those devices. We have termed this feature as AdhocInfra Toggle or WiFi Mode Toggle. We have written application-level Python scripts for enabling this feature. Kernel-level commands have been used in these scripts to gather information about available wireless networks, and also to toggle between the infrastructure and ad hoc modes of wireless interface cards. Following Linux-based *CLI utilities* have been used for fetching network parameters required in toggling function: (1) *iw* - a new nl80211 based configuration utility that is used to show/manipulate wireless devices and their configuration [9]; (2) *nmcli* - a tool for controlling Network Manager and getting its status [10]. The commands based on these utilities that have been used in the mode toggling scripts are given in Table 1. The contributions made to address the connectivity issues mentioned in Sect. 1 are given in the following subsections.

Table 1. Commands used in the scripts

Command	Function
***cmd₁**: nmcli d wifi connect {ssid} password {p/w} iface {devname}*	Connects a specific wireless interface to a specific access point
***cmd₂**: nmcli con up id {profilename}*	Uses the profile name of an access point for making connection
***cmd₃**: nmcli con down id {profilename}*	Disassociates from an already connected access point or ad hoc profile by using its profile name
***cmd₄**: iwlist [interface] scan*	Returns a list of access points and ad hoc cells in range, along with their details like ESSID, Quality, Frequency, Mode, Signal Level, etc
***cmd₅**: nmcli con status*	Returns a list of all the connected networks' SSIDs

3.1 Opportunistic WiFi Mode Toggling

In this subsection, we deal with cases when an access point becomes available or unavailable for mobile devices engaged in some ongoing local data session. In order to maintain these data sessions, an auto-configuration of WiFi interface of these devices is required. This auto-configuration is done by opportunistically switching between the WiFi modes of these devices, which avoids restarting the data sessions all over again. These data sessions are mostly TCP sessions which do not time out till a specific TCP wait time [6], even if the underlying network connectivity ceases to exist. This gives sufficient scope for the WiFi interface cards to switch between their modes.

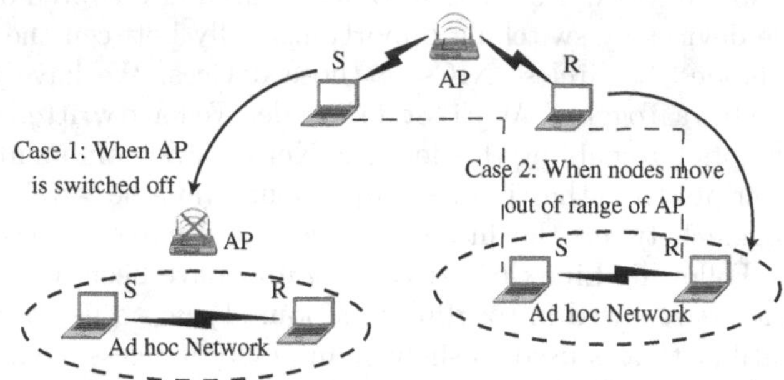

Fig. 1. WiFi mode toggling (when AP goes off or when nodes move out of range)

In Fig. 1, we present a scenario where two mobile devices (S and R) are initially associated with an access point (AP), as their WiFi interface cards are in infrastructure mode (by default). When the AP becomes unavailable - either the AP is switched off (Case 1) or both devices move away from the transmission range of AP (Case 2), our mode toggling feature starts searching for any available APs in the range. If no AP is found, both devices switch their modes from infrastructure to ad hoc and connect with each other (if present in the transmission range of each other). Later on, if they come in range of some AP, they can switch back to infrastructure mode in order to connect with AP. This action is event-driven, and is triggered as soon as the devices start receiving beacon frames from the AP. Since infrastructure networks offer better bandwidth and connectivity [2,5], infrastructure mode should be preferred over ad hoc mode.

Algorithm 1 shows the working of this opportunistic mode toggling script. The Service Set Identifiers (SSIDs) of infrastructure (AP) and ad hoc networks along with a time, t are taken as input. The time, t is passed to a sleep() function, which is used to run the script periodically on the devices. The *IsAvailable()* function uses $\boldsymbol{cmd_4}$ (see Table 1 for commands) to check if the node is receiving signal from a particular network or not. The *IsConnected()* function uses $\boldsymbol{cmd_5}$ to check if the node is connected to a particular network or not. The *connectInfra()* and *connectAdHoc()* functions use $\boldsymbol{cmd_2}$ for connecting to specific infrastructure and ad hoc networks respectively. The *disconnectAdHoc()* function uses $\boldsymbol{cmd_3}$ for disconnecting from an ad hoc profile. This script checks if an AP (infrastructure network) is available or not [line no. 2]. When an AP is available, it connects with the AP; otherwise it associates with some ad hoc profile of other devices (if available) in its range [line nos. 3–12]. This feature

Algorithm 1. Opportunistic WiFi Mode Toggling

```
INPUT: t, SSIDInfra, SSIDAdHoc
while true do
    if IsAvailable(SSIDInfra) then
        if not IsConnected(SSIDInfra) then
            if IsConnected(SSIDAdHoc) then
                disconnectAdHoc()
            end if
            connectInfra(SSIDInfra)
        end if
    else
        if not IsConnected(SSIDAdHoc) then
            connectAdHoc(SSIDAdHoc)
        end if
    end if
    sleep(t)
end while
```

of mode toggling can be considered as an opportunity to maintain any ongoing data session, when devices loose connectivity with the infrastructure network while moving from one location to the other.

3.2 Selection of Best AP

In this subsection, we deal with cases when mobile devices need to connect to an access point that has the best signal strength, or when devices lie in a *dead zone*. This situation arises when the transmission range of adjacent access points in a WiFi infrastructure network overlap with each other, and mobile devices lie at locations where they receive signals from multiple access points. The mobile devices may also pass through some *dead zone* while they are in transit.

Figure 2 depicts a scenario, where node S receives beacon frames of different signal strengths and qualities from two different APs (AP_1 and AP_2), and node R receives beacon frames from AP_2 and AP_3. A Python-based script runs periodically on these nodes in the background, in order to select an access point with best signal strength and quality among AP_1, AP_2 and AP_3. Nodes S and R both associate themselves to the nearest AP (i.e. AP_3), whose received signal strength and quality are better than that received from other APs. Our script uses a Linux command (***cmd₄*** specified in Table 1) that gives a list of all the networks/signals available in the range, along with various signal characteristics of each network. Following signal characteristics [5] are utilized by our script to select the best AP among all available APs in the range:

1. *Mode:* It signifies the type of network (connectivity mode) offered by the signal received at wireless interface of the mobile node. It's value is *Master* for infrastructure networks, and *Ad Hoc* for ad hoc networks.

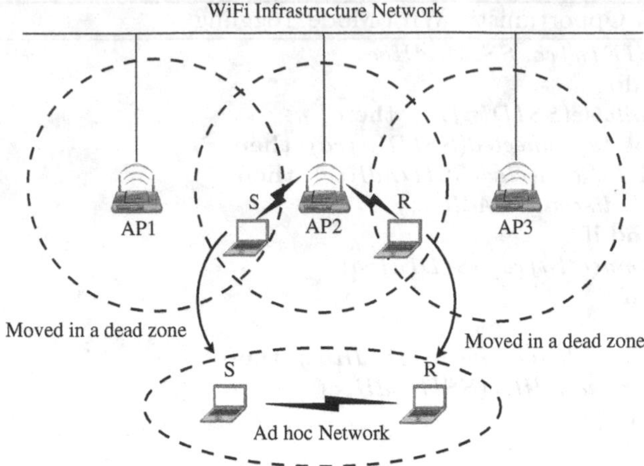

Fig. 2. Best AP selection and WiFi mode toggle based on signal characteristics

2. *Quality:* This parameter signifies the interference or noise level present between the AP and the device's decoder. It reflects the goodness of the signal that the wireless node is receiving. Noise is low if the quality is high, and increases on degradation of quality.
3. *Signal Level:* This represents strength of the signal that the mobile node receives. The closer the node is to the AP transmitter, the better signal strength it receives from it.

Algorithm 2. Best AP Selection and Mode Toggle in Dead Zones

```
INPUT: t, SSIDPool
networks ← intialiseSSIDPool(SSIDPool)
while true do
    params ← fetchParameters(networks)
    infraSSID ← allNodesWith(params.mode = 'Master' & params.signalLevel ≥ −70dbm)
    if infraSSID ≠ NULL then
        profile ← max(infraSSID.quality)
        connectInfra(profile)
    else
        adhocSSID ← allNodesWith(params.mode =' AdHoc')
        profile ← max(adhocSSID.Quality)
        connectAdHoc(profile)
    end if
    sleep(t)
end while
```

The working of our script for best AP selection is given in Algorithm 2. We have kept the minimum acceptable signal level of APs for connection to be −70 dBm. The script accepts a pool of SSIDs for all APs in the WiFi network (including existing ad hoc profiles) and a time, t for making the script periodic using sleep() function. The function *initialiseSSIDPool()* initializes the SSIDs for all APs [line no. 1]. The initialized SSID pool in parameter *networks* is passed to the function *fetchParameters()*, which uses $\boldsymbol{cmd_4}$ along with *grep* utility of Linux to fetch the required signal characteristics of the available networks [line no. 2]. Now, all infrastructure networks, i.e. APs with *mode 'Master'* and having *signal level* ≥ -70 dBm are selected using function *allNodesWith()* [line no. 4]. Out of these APs, the one which has maximum value for *quality* is chosen for connection [line nos. 5–7]. If no AP is found, network profiles with *mode 'Ad Hoc'* are searched, and the device switches its *mode* to ad hoc for making connection with the ad hoc profile having the best *quality* [line nos. 9–11]. The *connectInfra()* and *connectAdHoc()* functions use $\boldsymbol{cmd_2}$ for connecting to respective networks. This script runs periodically on devices and keeps searching for AP with best signal quality amongst all scanned APs that have minimum acceptable signal strength. In situations when devices lie in a *dead zone*, the mode toggle feature is used to connect to some ad hoc network profile for maintaining ongoing data sessions.

3.3 Application-Level Routing for Intermittent Connectivity Between Isolated Ad Hoc Network Clusters

As discussed in Sect. 1, ad hoc network clusters may exist in isolation at different locations of the application scenario. In order to connect these isolated ad hoc clusters, the WiFi infrastructure at these locations can be utilized. For this, at least one device of each ad hoc cluster must be in the transmission range of some AP. This device is chosen as an *anchor node* [13], which participates in both the infrastructure and ad hoc networks by using a periodic WiFi mode toggling feature. There may be many candidates for an *anchor node* in an ad hoc cluster, out of which one is chosen depending upon one or more criteria (position, good signal strength, large dwell time, good power backup, high processing power and large memory, no. of neighbours, etc.). These *anchor nodes* can either be one of the existing mobile devices in the scenario, or they can be pre-designated nodes that have been strategically placed at specific locations [13]. These *anchor nodes* behave as routing entities, as they bridge between ad hoc network clusters and the WiFi infrastructure network (Fig. 3). We have developed a script that executes continuously on all anchor nodes by periodically switching between the two modes (ad hoc and infrastructure). This enables the *anchor nodes* to participate in both the networks on a periodic basis. Each *anchor node* belonging to an ad hoc network cluster, maintains an application-level routing table in which the entries of other *anchor nodes* of other ad hoc clusters are present. This routing table is used by the *anchor node* of the sender's ad hoc cluster to resolve the *anchor node* of the receiver's ad hoc cluster.

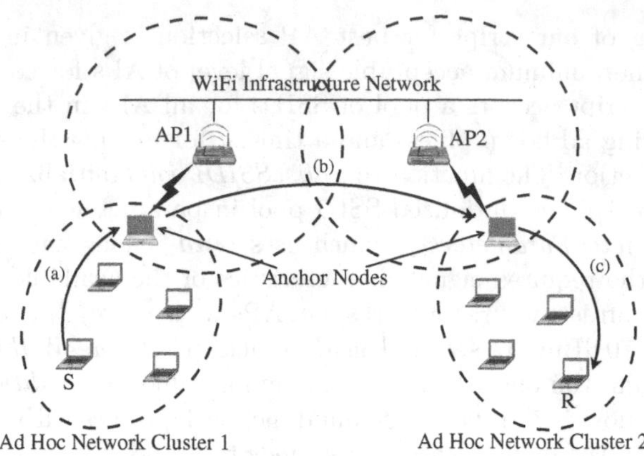

Fig. 3. Application-level routing between isolated ad hoc network clusters

Algorithm 3 describes the working of our script which comprises of three functions - *Switching*, *ReceiverSide* and *SenderSide*. These functions run as separate threads on sender, receiver and *anchor nodes*. The *Main* function initializes the flags s_1, s_2 and *eofFlag* stored in shared memory, and starts all three threads [line nos. 1–8]. The *Switching* function provisions the establishment of two concurrent data connections on *anchor nodes* by periodically switching (using sleep() function) between the ad hoc and infrastructure modes of their WiFi interface [line nos. 10–17]. *ReceiverSide* and *SenderSide* functions are used to run two data connections concurrently on each *anchor node*. Step (a) in Fig. 3 represents data connection between *SenderSide* thread running on sender, S and *ReceiverSide* thread running on *anchor node* at sender's side. Step (b) represents data connection between *SenderSide* and *ReceiverSide* threads running at *anchor nodes* of sender and receiver sides, respectively. Finally, step (c) represents data connection between *SenderSide* thread running on *anchor node* at receiver's side and *ReceiverSide* thread running on receiver, R. *ReceiverSide* thread uses a function *startServer()* to start a server process, sets the flag s_1 to 1, and waits for the *SenderSide* thread to set s_2 to 1 using *wait()* function [line nos. 18–21]. On the other hand, the *SenderSide* thread uses a function *connectToServer()* to establish a TCP connection with the server process, sets the flag s_2 to 1, and waits for the *ReceiverSide* thread to set s_1 to 1 [line nos. 28–31].

Data transfer will start only after $s_1 = 1$ and $s_2 = 1$, i.e. both the data connections are established on an *anchor node*. The *SenderSide* thread reads data bytes from a file till EOF (end of file) using *readFromFile()* function, and sends them to the server process running as *ReceiverSide* thread using *sendToServer()* function [line nos. 32–35]. The *ReceiverSide* thread receives data bytes from the client process running as *SenderSide* thread using *receiveFromClient()* function, and writes the bytes received to file using *writeOnFile()* function [line nos. 22–25]. The *anchor node* on sender's side would use data connection of step (a) to receive incoming bytes of the file from one ad hoc cluster. All these incoming bytes are

Algorithm 3. Algorithm for Interconnecting Ad hoc Clusters using Wifi Mode Toggling at Anchor Nodes

```
INPUT: t
function MAIN
    s1 ← 0
    s2 ← 0
    eofFlag ← 0
    startThread(Switching)
    startThread(ReceiverSide)
    startThread(SenderSide)
end function
function SENDERSIDE
    connectToServer()
    s2 ← 1
    wait(s1)
    while eofFlag ≠ 1 do
        readFromFile(bytes)
        sendToServer(bytes)
    end while
end function
function RECEIVERSIDE
    startServer()
    s1 ← 1
    wait(s2)
    while bytes ≠ NULL do
        receiveFromClient(bytes)
        writeOnFile(bytes)
    end while
    eofFlag ← 1
end function
function SWITCHING
    while true do
        connectAdHoc(SSIDInfra)
        sleep(t)
        connectInfra(SSIDAdHoc)
        sleep(t)
    end while
end function
```

stored in a buffer in the form of a *temporary file.* The data connection of step (b) reads bytes from the same file and sends it across to the *anchor node* on receiver's end. Eventually, the data connection of step (c) transfers the contents of the file from the *anchor node* to the receiver in other ad hoc cluster. s_1, s_2 and *eofFlag* flags have been used to avoid concurrency problems by creating following provisions - (1) sender and receiver threads wait for both connections to get established, before initiating any file transfer; (2) *eofFlag* prevents the file read descriptor to move ahead of the file write descriptor.

Our script provides intermittent connectivity between isolated ad hoc clusters, if anchor nodes exist in these clusters. However, it may experience a conflict in periodic switching time of two different systems due to difference in adapter properties, system specifications, etc. Because of this conflict, anchor nodes may switch in opposite order at some stage, thereby loosing the infrastructure connectivity temporarily.

4 Experimental Evaluation

We have written Python-based scripts for the algorithms proposed in previous section. The execution of these scripts is either periodic or event-based depending upon the scenario and topology. The complete experimental setup, along with the results obtained and inferences drawn are presented in this section.

4.1 Experimental Setup

An experimental testbed of IEEE 802.11 wireless devices (specifications given in Table 2) is setup for testing and evaluation of the scripts. The current implementation is based on Linux, however it would be interesting to investigate same provisions for other operating systems also. Separate scripts are written for all three algorithms of Sect. 3, and are tested on topologies of Figs. 1, 2 and 3, respectively.

4.2 Results and Discussion

We have used a file transfer application (using *vsftpd daemon*) to transfer video files of different sizes (20 MB to 50 MB), in order to evaluate the scripts. We have also tested the scripts in case of live video streaming, for which we used *Sarxos API in JAVA* that helps in detecting the web camera. We have considered smaller file sizes, as the connectivity issues mentioned in the paper occur for very small durations of time. So, in order to evaluate our mode toggling scripts, lesser duration of data transfer was required.

Table 2. Wireless device specifications

Parameter	Value
Operating system	Ubuntu 14.04 (Linux)
Kernel	4.4.0-71-generic
Scripting language	Python 2.7.6
Wireless NIC adapter type in nodes	802.11b/g/n
Connection type	Infrastructure priority
Infrastructure AP specifications	Frequency: 2.462 GHz, 11 Channels

The performance results of all the three scripts have been calculated based on file transfer traces captured using packet analysis tool, *Wireshark* [14]. Network latency, throughput and switching overhead have been used as performance metrics. Latency and throughput values have been recorded for each file transfer in all the experiments, and then average values have been computed. The *average latency* is a weighted average of latency values for different file sizes; whereas *average throughput* has been calculated by normal averaging method. *Switching overhead* has been calculated in terms of difference between latency obtained with mode toggling feature and latency obtained in ideal scenario.

Opportunistic WiFi Mode Toggling Feature. The topology of Fig. 1 was used to evaluate our mode toggling script given in Algorithm 1. In the first phase, both the nodes (S and R) were kept initially in infrastructure mode and then the AP was switched off and on once, twice and thrice (this is similar to nodes moving out of range of AP and then coming back into the range).

The performance metrics for file transfer were recorded under three scenarios, specially created for comparative analysis of experimental results (see Table 3):

- *Ideal Case:* When there was no movement of nodes and the file transfer completes with continuous availability of AP.
- *Movement without toggle script:* When nodes move out of range of AP with no script running, and then return later to complete the file transfer.
- *Movement with toggle script:* When nodes move out of range of AP with toggle script running, and then return later to complete the file transfer.

Table 3. Nodes in infrastructure mode moving between APs (with and without toggle script)

Metric	Scenario						
	Ideal case (no movement)	No script			Toggle using script		
		Once	Twice	Thrice	Once	Twice	Thrice
Avg. latency (s)	17.24	96.96	150.15	199.93	73.03	54.23	70.90
Avg. throughput (Mbps)	17.35	3.14	1.99	1.52	4.15	5.50	4.13
Switching overhead (s)	0	N/A	N/A	N/A	55.79	36.99	53.66

In Table 3, the values of *average latency* and *average throughput* obtained in case of movement of nodes with our toggle script, lie between the values obtained in ideal case and the values obtained when nodes moved without any script. It can also be observed that when mode is toggled twice, the values of performance metrics are better than the values obtained in case of toggling once and thrice. This is due to the fact that toggling the mode twice brings back the device in infrastructure mode, thus increasing the performance (as infrastructure mode offers better bandwidth and throughput in comparison to the ad hoc mode).

In the second phase, the nodes were kept initially in ad hoc mode and above experiments were repeated under all three scenarios, with nodes moving out of range of each other instead of moving away from an access point, and then coming back in range. The values of performance metrics have been shown in Table 4. In this case also, the values of *average latency* and *average throughput* in case of our toggle script lie in between the ideal case and the case when no script was used. However the overall performance drops, as ad hoc connection supports lower bandwidth and throughput as compared to infrastructure connectivity. Here, when mode is toggled once or thrice by the script, the values of metrics are better than the values in case of toggling the mode twice. The reason is still the same, that toggling the mode once or thrice brings the device in infrastructure mode; thus yielding a better performance.

The *switching overhead* in both the phases (Tables 3 and 4) is mostly due to the time taken in toggling the modes, which shall decrease for file transfers

Table 4. Nodes in ad hoc mode (with and without toggle script)

Metric	Scenario						
	Ideal case (no movement)	Nodes move out of range (no script)			Mode toggle with movement (using script)		
		Once	Twice	Thrice	Once	Twice	Thrice
Avg. latency (s)	24.25	102.61	157.82	206.81	40.36	87.87	64.88
Avg. throughput (Mbps)	13.08	2.76	1.81	1.39	7.48	3.28	4.53
Switching overhead (s)	0	N/A	N/A	N/A	16.13	63.64	40.65

of longer duration. This overhead can significantly be reduced, if toggling feature can be designed at the MAC level instead of application level. But, this application-level implementation serves as a trade off between the switching overhead and opportunistic connectivity provided, without making any changes in the interface card or protocol stack. Hence, it is evident that our mode toggle script works well in maintaining data sessions, when nodes are under movement in and around the coverage area.

Best AP Selection. The topology of Fig. 2 was used to evaluate our best AP selection script given in Algorithm 2. The nodes were made to move across multiple APs in the WiFi network, and results were recorded with and without the best AP selection script. The results were recorded for all the three scenarios (ideal case, without script and with script) when handoff between APs occurred once, twice and thrice. The values of performance metrics (given in Table 5) for handoffs with our best AP script are better than the values in case of normal handoffs without script. Hence it is evident that, using our best AP selection script improves the file transfer performance. This happens because our script facilitates quick handoff to a better access point than normal handoffs (when signal level falls below the threshold level), thus giving improved throughput and latency. Therefore, our script works well on Linux-based devices for selecting the best AP for attachment based on signal strength and quality, when exposed to signals from multiple access points.

Table 5. Nodes performing handoff across APs for best AP selection (with and without script)

Metric	Scenario						
	Ideal case (no movement)	No script			Best AP script		
		Once	Twice	Thrice	Once	Twice	Thrice
Avg. latency (s)	17.24	27.76	34.00	44.63	20.97	22.69	26.20
Avg. throughput (Mbps)	17.35	10.79	8.67	6.66	14.17	12.95	11.29

Application-Level Routing for Intermittent Connectivity between Isolated Ad hoc Clusters. The topology of Fig. 3 was used to evaluate our intermittent connectivity script given in Algorithm 3. The multi-threaded script was deployed on sender, receiver and *anchor nodes*, and network latency values of 155.55 s, 316.87 s, 485.68 s and 685.31 s were recorded for file transfer of sizes 20 MB, 30 MB, 40 MB and 50 MB respectively. This drastic increase in latency with increase in file size is due to the reason that, when *anchor nodes* periodically switch between their modes they loose synchronization with each other due to their individual system clocks. But, after a certain period of time they regain synchronization, again due to the difference in their system clocks.

Table 6. Application-level communication between isolated ad hoc clusters

Metric	Scenario		
	Ideal case (infrastructure mode)	Ideal case (ad hoc mode)	Intermittently connected ad hoc clusters
Avg. latency (s)	17.24	24.25	473.64
Avg. throughput (Mbps)	17.35	13.08	0.76
Switching overhead (s)	0	0	452.9

The *average latency*, *average throughput* and *switching overhead* values are given in Table 6. The performance depicted in these values is much low in comparison to ideal situations of file transfer using infrastructure and ad hoc connectivity, still the biggest achievement of this script is to find connectivity between the isolated ad hoc clusters. The performance will significantly be improved, once a solution is devised to handle the synchronization issue between the system clocks of anchor nodes. The idea of having *anchor nodes* (with an extra overhead) to connect ad hoc clusters is still better than the case of loosing connection with one AP and attaching to another AP (if more APs are installed in order to avoid ad hoc connections), as this may interrupt any ongoing TCP session for a longer time period. This interruption leads to multiple retransmission attempts with exponentially increasing retransmission times, which increases the recovery time and worsens the situation. However, the extra overhead incurred by the *anchor nodes* is detrimental to battery life of nodes. We are currently investigating to include those nodes as *anchor nodes*, which are either attached to power sources for charging or have large power backup.

5 Conclusion and Future Work

The application-level scripts that we have developed in this paper work well in the case of less dynamic scenarios. The results demonstrate that these scripts can efficiently handle the connectivity issues of IEEE 802.11 networks. The proposed approach is practical and feasible, as it is based on application-level control, in contrast to the existing approaches that depend on changes in the hardware, device drivers, or in the network protocol stack. Apart from this, the overhead incurred in switching modes trades off well against the loss of connectivity and disruption in data transfer.

The scripts proposed in our work execute on a periodic basis, thus raising a synchronization issue, especially in the case of data transfer between two ad hoc network clusters. We are planning to propose a synchronization mechanism, so that the anchor nodes remain in the same mode while performing data transfer.

References

1. Nicholson, A.J., Wolchok, S., Noble, B.D.: Juggler: virtual networks for fun and profit. IEEE Trans. Mob. Comput. **9**(1), 31–43 (2010)
2. Basagni, S., Conti, M., Giordano, S., Stojmenovic, I.: IEEE 802.11 AD HOC Networks: Protocols, Performance, and Open Issues, pp. 69–116. Wiley-IEEE Press (2004)
3. Cai, X., Zhou, M., Huang, X.: Model-based design for software defined radio on an FPGA. IEEE Access **5**, 8276–8283 (2017)
4. Cisco: Approximating maximum clients per access point
5. Crow, B.P., Widjaja, I., Kim, J.G., Sakai, P.T.: IEEE 802.11 wireless local area networks. IEEE Commun. Mag. **35**(9), 116–126 (1997)
6. Faber, T., Touch, J.D., Yue, W.: The TIME-WAIT state in TCP and its effect on busy servers. In: Proceedings of IEEE INFOCOM 1999, The Conference on Computer Communications, Eighteenth Annual Joint Conference of the IEEE Computer and Communications Societies, The Future Is Now, New York, NY, USA, 21–25 March 1999, pp. 1573–1583. IEEE (1999)
7. Pang, J., Greenstein, B., Kaminsky, M., McCoy, D., Seshan, S.: WiFi-reports: improving wireless network selection with collaboration. IEEE Trans. Mob. Comput. **9**(12), 1713–1731 (2010)
8. Kandula, S., Lin, K.C.J., Badirkhanli, T., Katabi, D.: FatVAP: aggregating AP backhaul capacity to maximize throughput. In: 5th USENIX Symposium on Networked Systems Design and Implementation, San Francisco, CA, April 2008
9. Linux: iwlist(8) - linux man page. https://linux.die.net/man/8/iwlist
10. Linux: nmcli(1) - linux man page. https://linux.die.net/man/1/nmcli
11. Ranveer Chandra, P.B., Bahl, P.: Expansion of wireless networks of standard mimo indoors. In: EUROCON 2009. IEEE, July 2009
12. Ranveer Chandra, P.B., Bahl, P.: MultiNet: connecting to multiple IEEE 802.11 networks using a single wireless card. In: IEEE INFOCOM, pp. 882–893, September 2004
13. Sewak, A., Pandey, M., Gore, M.M.: Forming structured p2p overlays over disjoint MANET clusters. In: 2016 IEEE International Conference on Advanced Networks and Telecommunications Systems (ANTS), pp. 1–6, November 2016
14. Wireshark. https://www.wireshark.org/

Simulation of AdHoc Networks Including Clustering and Mobility

J.R. Emiliano Leite, Edson L. Ursini, and Paulo S. Martins(✉)

School of Technology, University of Campinas (UNICAMP), R. Paschoal Marmo, 1888, Limeira, Sao Paulo 13484, Brazil
paulo@ft.unicamp.br

Abstract. We introduce a discrete-event simulation model of an AdHoc network considering the presence of clusters and node mobility. The main goal is to study the volume of traffic in relatively large networks of sensor systems and Internet of Things considering fading and network connectivity. We also evaluate relevant parameters such as the mean CPU utilization and the mean queueing time in each node. The model is relatively general in that it combines the traffic from small devices such as sensors as well as more complex intermediate systems such as gateways and Internet nodes. It is also extensible to other types of scenarios and it allows the evaluation of the network under other performance criteria or evaluation metrics. The results show that the model yields simulation values that could be analytically validated by Jackson networks.

Keywords: Emergency services · MANET · VANET · WSN · IoT · Traffic planning · Network dimensioning · Connectivity · Clustering

1 Introduction

Node mobility is a critical element in the design of stable, scalable and adaptive clusters in mobile AdHoc networks (MANETs). It has a significant impact on network performance and it must be accounted for in the analysis of such networks. Another key architectural component for the self-organization of an AdHoc network is the presence of network clusters. Both features (i.e. mobility and clusterization) are thus essential when planning and dimensioning the next generation AdHoc networks.

In this work, we provide a simulation model based on discrete event (for network traffic) and Random Way Point (RWP) (for mobility) simulation for an AdHoc network that accommodates clusters and the effects of node mobility. The model captures a number of features of complex systems, including gateways, emergency nodes, Internet and IoT traffic. The goal is to investigate through a case study the mean queueing time and the CPU utilization for each node for a given probability of connectivity of the nodes resulting from the mobility. We also aim to estimate the incoming and outgoing traffic for each cluster, considering its connectivity.

A. Puliafito et al. (Eds.): ADHOC-NOW 2017, LNCS 10517, pp. 199–209, 2017.
DOI: 10.1007/978-3-319-67910-5_16

Despite the existence of a body of work on AdHoc networks, none of the studies cited and surveyed in the literature (Sect. 2) tackles the features of the AdHoc under consideration in our work, e.g. the issue of functions to increase network robustness (emergency clusters). Additionally, they do not address a more comprehensive simulation model for traffic performance evaluation - with the adoption of discrete event simulation (with Arena simulation software), dynamic mobility model (with MATLAB), and the extension of traffic analysis - allowing the inclusion of clusters.

The remainder of this paper is organized as follows: In Sect. 2 we review previous work. The proposed approach is introduced in Sect. 3. A case study illustrating the application of the model is shown in Sect. 4. In Sect. 5 the results are discussed. We summarize and present our conclusions in Sect. 6.

2 Related Work

Amis *et al.* arrange the network in clusters using the Max-min algorithm [1]. In our work, clusters are formed according to the application under consideration, i.e. nodes are grouped according to their functional similarity. Furthermore, our major goal is to study the traffic load flowing through the clusters for the purpose of dimensioning these devices and their links. Pramanik *et al.* [7] use MATLAB to evaluate the connectivity of an AdHoc network by means of the Random Way Point algorithm. However, the authors neither make use of discrete event simulation of network traffic nor arrange the network in clusters. Nassef [5] shows the influence of fading and mobility in network node connectivity. This is incorporated in our work through the adopted mobility model.

Although clustering has traditionally been used for increasing the scalability of networks, the work by Phanish and Coyle also opts for multi-level clustering large-scale AdHoc and sensor networks as a means to support energy-efficient strategies for gathering data [6]. Two design variables are influential to the optimality of a multi-level network: (1) The number of levels, and (2) The number of nodes operating at each level. These variables are characterized within a multi-hop, multi-level hierarchical network of variable size that gathers and aggregates data at each level. Unlike the work by Phanish and Coyle [6], our model considers a single level of clustering. As mentioned earlier, our goal is the dimensioning of network traffic.

In the work by Cai *et al.* [2], the authors present a clustering algorithm based on the group mobility and a revised group mobility metric obtained from the speed and direction of nodes. The authors also account for the residual energy of nodes and the number of neighbor nodes in their model. Unlike our work, Cai *et al.* [2] uses the Gauss Markov instead of the Random Way Point mobility model, and their goal is to compose more stable clusters including highly dynamic environments. Our model has more stable clusters due to the sensor networks, and it includes mobility and connectivity to evaluate the total volume of traffic.

Ren *et al.* [8] proposed a new mobility and stability-based clustering algorithm (MSCA) for urban scenarios. The performance of the proposed algorithm was assessed by changing the maximum lane speed and the traffic flow rate. As mentioned earlier, our model considers more stable clusters, as it is formed by sensor networks, targeted to IoT applications, which are not meant to be as mobile as e.g. VANETs, and it takes into consideration mobility and connectivity of the internal nodes for the evaluation of the total traffic volume. This volume is estimated by means of a combination of discrete-event simulation with the Random Way Point mobility model.

Celes *et al.* [3], using discrete-event simulation with NS-3, point out that five widespread vehicular mobility traces present gaps, potentially leading to error prone operation. They present a framework allowing more fine-grained traces, which leads to more trustworthy simulation results. The authors employ clustering algorithms to fill the gaps of real-world traces. They also compare the communication graph of the original and the calibrated traces using network metrics. The results reveal that the gaps lead to unrealistic network topologies. Like our model, their work approaches simulation with a previous trace of node mobility. Our approach also employed simulation with clustering and mobility, and since we also model sensor network nodes, we have a previous knowledge of the approximate location of clusters. Nevertheless, our goal differs from theirs since we aim at estimating the traffic for the purpose of dimensioning and planning network capacity.

3 Network Model and Proposed Approach

A TCP/IP packet is modeled as an entity that arrives to the system and crosses several internal queues in a cluster before its departure (i.e. before it is consumed by an application). The network model is a hierarchy consisting of clusters which contain nodes, which in turn have multiple CPUs, thus allowing several parallel connections. Inherent to each queue is the waiting delay before a packet can be processed by a server. Clearly, both queueing and processing times are subject to statistical distributions. Therefore, a network cluster may be regarded as a set of internal queues (each one associated with an outbound link).

Figure 1 shows the network model with its inputs (packets) and outputs (packets) to each cluster. The details are as follows:

- 7 clusters ($CLT_1....CLT_7$); these are non-mobile and homogenous for the sake of simplicity. However, the model does not restrict the addition of heterogeneous clusters. Each cluster consists of n mobile nodes, where n is a configurable parameter;
- 4 gateways or Internet nodes ($GW_1....GW_4$); Both GW_1 and GW_2 are output gateways; GW_3 is an emergency gateway, i.e. it is used as a backup gateway for GW_1, e.g. when the latter overflows its internal buffers; GW_4 and also GW_2 are protocol converters, i.e. they are used to integrate two subnets;
- 7 Inputs: the inputs model data packets generated by IoT sensors;
- 3 Internet outputs: they model the flow of IP packets outbound;
- Node mobility: the Distributed Dynamic Routing algorithm for mobile AdHoc networks is employed;

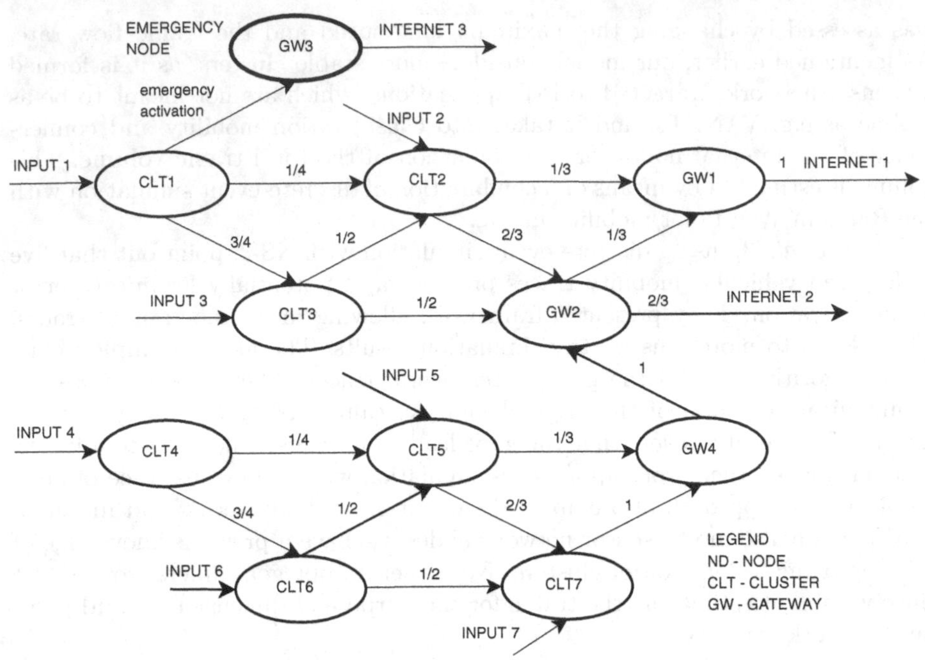

Fig. 1. Mobile AdHoc network model

- Input variables: data arrival and service time distributions in a node;
- Control variables: probability of node connectivity in a cluster. This probability is provided by the Random Way Point algorithm (which depends on a range of variables (Table 2);
- Output variables: mean queue time and mean CPU utilization on each cluster for a given position of the nodes within the cluster.

Each cluster contains several nodes (Fig. 2) which in turn have internally one or more CPUs (only CPUs for nodes 5 and 9 are shown to avoid overcrowding the figure, and because they are connected to the cluster outputs). The model is dynamic and the illustration is only a snapshot representation of an arbitrary instant t in time. For example, at instant $t + 1$ it might be other nodes that engage in transmission outbound. In addition to that, each cluster has one or more output CPUs which are used for its output channels/links (Table 1). These output CPUs are fixed in our model (without sacrificing the quality of the results), although it is possible to configure them to have some limited degree of mobility as well. Nodes share the output CPUs for relaying outbound traffic, provided that they have connectivity, i.e. they are within the power range of either an output CPUs or an intermediate node.

Each node is modeled as 4 simulation blocks connected in series:

1. *Enter block:* the enter block simulates the arrival of a packet in a cluster. It counts the number of packets entering the cluster;

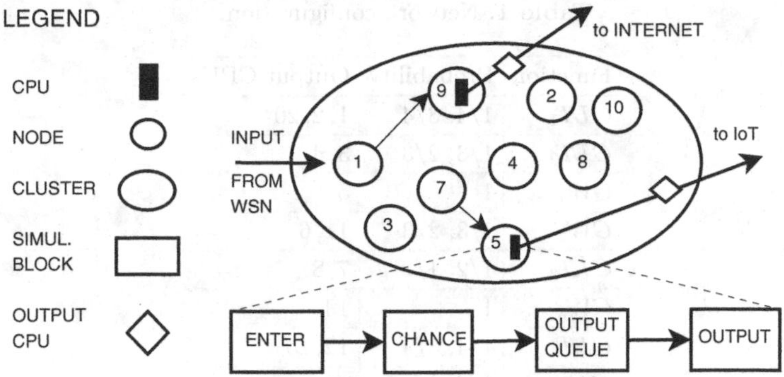

Fig. 2. Cluster organization model (snapshot)

2. *Chance* is an Arena DECIDE block, and it distributes the packets across a set of outgoing lines, where each line is associated with an outgoing queue; An important parameter in this block is the probability of packet loss, and its value was obtained from the case study (Sect. 4). The probabilities of a packet being forwarded to an outgoing link are initially configured as shown in Fig. 1 (e.g. 1/4 from cluster 1 to cluster 2 and 3/4 from cluster 1 to cluster 3);
3. *Output queue* represents the queueing time in the outgoing line;
4. *Output cluster* simulates the output (i.e. forwarding) of packets from the cluster. It is also responsible for counting the number of packets leaving the cluster.

Table 1 shows the relation of cluster/gateways to output CPUs. The column "Probability" is associated to the column "Output CPUs". Each probability is used to define the traffic management of each node according to a given application. These values also indicate the probability of a packet being serviced by the indicated output CPU. For example, the probability that cluster CLT_2 sends a packet to output CPU_3 is 1/3, and this probability is 2/3 for output CPU_4.

Each node receives packets at the input link and forwards them to one of the outbound links using UDP over IP (Datagram). Since the arrival of requests for the AdHoc network can be modeled as a Poisson process, the traffic volume of each individual node can be extended to the traffic volume of a cluster by the simple sum of the rates of Poissonian arrivals. Thus, we sum the rates of each node to form a cluster of ten nodes.

4 Case Study

To evaluate each node independently, a MATLAB routine generates random positions for the ten nodes within the cluster, every one sec (in our case). Table 2 shows the input parameters for the MATLAB algorithm. This case used the Random Waypoint Mobility Model (RWP) to simulate the performance of the network. By changing different parameters, we can either increase or decrease

Table 1. Network configuration.

Function	Probability	Output CPUs
CLT_1	1/4, 3/4	$1, 2, 20$[a]
CLT_2	1/3, 2/3	$3, 4$
GW_1	1	5
GW_2	1/3, 2/3	$11, 6$
CLT_3	1/2, 1/2	$7, 8$
GW_3	1	14
CLT_4	1/4, 3/4	$12, 15$
CLT_5	1/3, 2/3	$16, 13$
CLT_6	1/2, 1/2	$9, 10$
CLT_7	1	18
GW_4	1	17

[a]Output-CPU 20 is used only in an emergency

the connectivity. For example, it is possible to increment the connectivity by increasing (1) the number of user nodes, or (2) the number of gateways (interconnection), or (3) transmission power or else (4) by decreasing the simulation area, or a combination of these factors. Mobility determines the location of each node that selects a random destination, and travels towards it in a straight line at a randomly chosen uniform speed. The distance to connect nodes lies within the range from 200 to 500 m. We used a 1000×1000 m area. The adopted mobility model is the one presented by Pramanik [7].

Two basic propagation models (FS = Free Space and TR = Two-Ray ground propagation model) were considered, which are described by the following equations:

$$P_{t,FS} = \frac{P_t G_t G_r \lambda^2}{(4\pi)^2 d^2 L}; \quad d = \sqrt{\frac{P_t G_t G_r \lambda^2}{(4\pi)^2 P_{t,FS} L}} \tag{1}$$

$$P_{t,TR} = \frac{P_t G_t G_r h_t^2 h_r^2}{d^4 L}; \quad d = \left(\frac{P_t G_t G_r h_t^2 h_r^2}{P_{t,FS} L}\right)^{1/4} \tag{2}$$

where d is the minimum distance (in meters) required for connection between a pair of nodes. Using the values from Table 2, we obtain $d_{FS} = 582$ m and $d_{TR} = 564$ m. Depending on the scenario (indoor, free space) it is possible to switch from one propagation model to another.

In this case study, since both values obtained are close, we adopted the more conservative value of 500 m. By running the model (MATLAB simulation) 20 times, we observed distances larger than 500 m;

The histogram in Fig. 3 illustrates the number of nodes that remained connected per observation (i.e. simulation run). In 15 simulation runs we had 10 nodes connected, 2 runs yielded 9 nodes, 2 runs ended up with 8, and one execution

Table 2. MATLAB input parameters

Input parameters	Values
Receiver threshold	−88 dBm
Area size	1000 × 1000 m
Antenna type	Omnidirectional
Antenna height (ht, hr)	1.5 m
Antenna gain (G_t, G_r)	1.0
System Loss Coefficient (L)	1.0
# Mobile nodes in a cluster	10
Mobility model	Random waypoint
Speed interval	[0.2–2.2] m/s
Pause interval	[0–1] s
Walk interval (walk time)	[2–6] s
Direction interval	[−180 to +180]°
Transmission frequency	5.8 GHz
Transmission power (Pt)	15 dBm
Simulation time	306 s

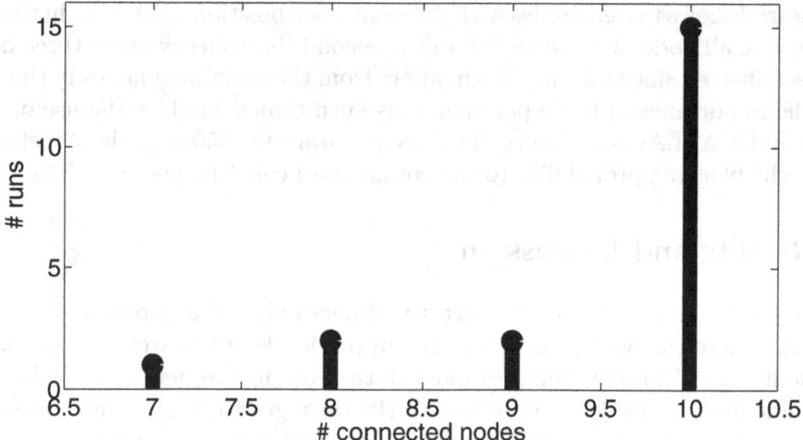

Fig. 3. Number of active nodes in a cluster.

resulted in 7 nodes (in a total of 20 rounds). Thus, the probability of no connection P_f is given by:

$$P_f = 1 - (15 \times 10 + 2 \times 9 + 2 \times 8 + 1 \times 7)/200 = 191/200 = 0.495(4.95\%). \quad (3)$$

i.e. we had 9 transmissions without success at 200 possible transmissions (20 × 10 nodes), or = 0.045 or 4.5%. The P_f value was used in an Arena Chance Block to represent the "Disconnected State", or the probability of lost connection.

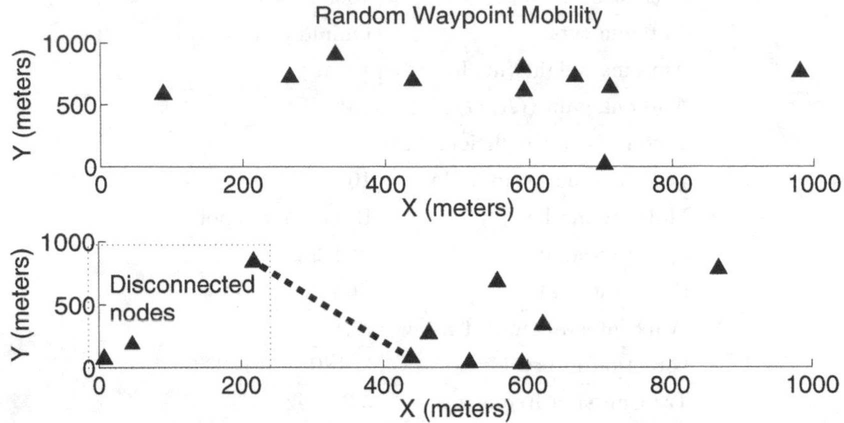

Fig. 4. Node connectivity due to distance.

Figure 4 shows two examples with different node positions. Whereas in the first (top) graph all nodes are connected, in the second (bottom) example three nodes are disconnected since they are 500 m apart from the remaining nodes in the cluster. The importance of these parameters is such that it justifies the use of both Arena and MATLAB simulators. If we use a distance of 250 m (or less) instead of 500 m, the blocking probability (or no connection) could be less than 50%.

5 Results and Discussion

This work is part of a major effort to dimensioning the capacity of an IoT network, where the bottleneck lies in the upper levels of the architecture, e.g. at the mediator, which concentrates most data flows in the network. Notice that the traffic under consideration in this work converges to the IoT mediator.

The discrete-event network simulation model allows an approximate placement of the sensor nodes in a cluster, as well as their approximate traffic load, but it does not express their mobility. The mobility of nodes within a cluster is given by the RWP algorithm. A degradation in connectivity for some nodes may cause the overall reduction of the traffic in the upper layers. It may also increase the traffic in the surrounding nodes. In the latter case, the possibility of using emergency gateway nodes is crucial for many types of applications, as a measure to counteract the performance degradation in the affected nodes.

The most important result of this work is the analysis and estimation of the total traffic in the network of clusters considering the effect of mobility and fading, i.e. the traffic volume under dynamic conditions.

We have executed two runs, described as follows:

- *First Run*: In the first run, the inter-arrival times for a cluster are EXPO(0.2), i.e. one packet is received each 0.2 s ($1/0.2=5$), thus generating an arrival rate of 5 packets/s. We used the service rate of 1/0.33, meaning that the service rate is 3, i.e. three packets are processed each second. Some nodes were unstable since the service rate was less than the arrival rate and their utilization rate was close to 100%. This caused excessive delays and we could see the behavior (i.e. activation) of the emergency node. Thus, in the first run a situation was forced in which the network presented instability, mainly to verify the performance of the emergency node. By running the complete model (1.000 s) in the simulation model, it was possible to extract the following traffic load at the Gateways:
(1) Traffic at the Internet Gateway (GW_1): 5.2550 packets/s; (2) Traffic at the Internet Gateway (GW_2): 2.2350 packets/s; (3) Traffic at the Emergency Gateway (GW_3): 0.3910 packets/s.
- *Second Run:* In the second run, the inter-arrival times for a cluster are EXPO(0.6), i.e. one packet each 0.6 s (the arrival rate is $1/0.6=1.67$ packets/s, i.e. 10 nodes $\times$ 0.167 packets/s per node). We used the service rate of 1/0.1, meaning that the service rate is 10, i.e. ten packets are processed each second (a threefold speed increase in relation to the first run). The increase in the service rate was enough to stabilize the system, as shown in Fig. 5. Therefore, the stable network caused the emergency node (output-CPU-20) to remain mostly inactive, i.e. only a few packets flowed through it. Recall also that in the actual system, the processing delay is different for each network element.
By running the complete model (5.000 s) for the second run, it was possible to extract the following traffic load at the Gateways: (1) Traffic at the Internet Gateway (GW_1): 4.3828 packets/s; (2) Traffic at the Internet Gateway (GW_2): 6.3736 packets/s, and (3) Traffic at the Emergency Gateway (GW_3): 0.0030 packets/s.
In the second run, we also used the Jackson's network analytical model [4], which was calculated as a Markov chain to validate the simulation model under the exponential distribution for both arrival and service distributions. Therefore, note that the simulation model is not limited to the use of the Poissonian distribution initially assumed in this work. We have adopted this type of distribution since it allowed the validation of this model. However, once validated, it was possible to evaluate other conditions not allowed by the analytical model, such as different distributions other than the exponential, the inclusion of the loss of connectivity, and the probability distributions regarding the traffic between clusters. Moreover, the proposed model is flexible in that it is not restricted to the use of Random Waypoint mobility model, and other types of mobility models may be used and compared.

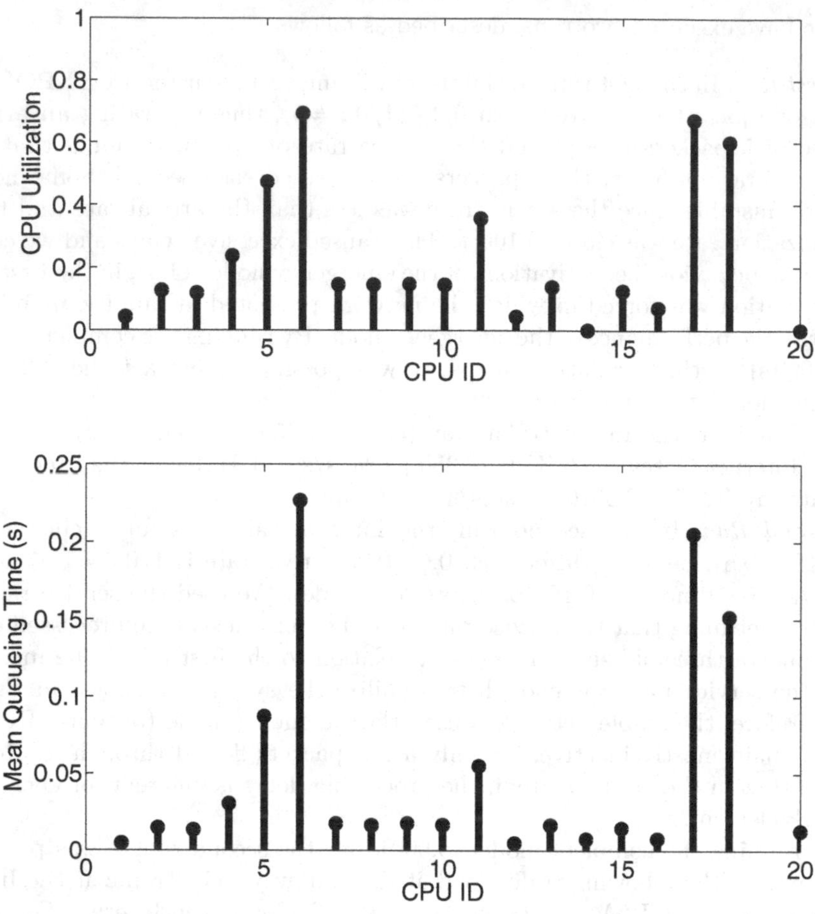

Fig. 5. Mean output-CPU utilization (upper) and mean queueing time (lower) (for second run)

6 Conclusion

In this work, we presented a simulation model based on both discrete event and Random-Way Point simulation for an AdHoc network that accommodates clusters and the effects of node mobility. The model captures a number of features of complex systems, including gateways, emergency nodes, Internet and IoT traffic. We analyzed through a case study the mean queueing time and the output-CPU utilization for each node for a given probability of connectivity of the nodes resulting from the mobility. Through the model it was possible to estimate the incoming and outgoing traffic for each cluster, considering its connectivity.

Ongoing work attempts to include fuzzy logic for decision making regarding emergency cases. This fuzzification would allow a fully dynamic model, which reacts by triggering the emergency node based on the actual traffic, instead of using the current procedure that is based on the probability distribution of traffic volume. We intend to use this model for dimensioning and planning the IoT network mediator. Future work may also contemplate specific applications as well as configurations regarding clustering and mobility/connectivity. Another possibility is to apply prediction methods to evaluate network traffic in order to improve future dimensioning.

References

1. Amis, A.D., Prakash, R.T., Vuong, H., Huynh, D.: Max-min cluster formation in wireless ad hoc networks. In: Proceedings of Ad Hoc Infocom 2000 (2000)
2. Cai, M., Rui, L., Liu, D., Huang, H., Qiu, X.: Group mobility based clustering algorithm for mobile ad hoc networks. In: 2015 17th Asia-Pacific Network Operations and Management Symposium (APNOMS), pp. 340–343, August 2015
3. Celes, C., Silva, F., Boukerche, A., Andrade, R., Loureiro, A.: Improving VANET simulation with calibrated vehicular mobility traces. IEEE Trans. Mob. Comput. **PP**(99), 1 (2017)
4. Jackson, J.R.: Networks of waiting lines. Oper. Res. **5**(4), 518–521 (1957)
5. Nassef, L.: On the effects of fading and mobility in on-demand routing protocols. Egypt. Inform. J. **11**, 67–74 (2010)
6. Phanish, D., Coyle, E.J.: Application-based optimization of multi-level clustering in ad hoc and sensor networks. IEEE Trans. Wireless Commun. **PP**(99), 1 (2017)
7. Pramanik, A., Choudhury, B., Choudhury, T.S., Arif, W., Mehedi, J.: Simulative study of random waypoint mobility model for mobile ad hoc networks. In: IEEE Proceedings Global Conference on Communication Technologies (GCCT 2015), pp. 112–116 (2015)
8. Ren, M., Khoukhi, L., Labiod, H., Zhang, J., Veque, V.: A new mobility-based clustering algorithm for vehicular ad hoc networks (VANETs). In: 2016 IEEE/IFIP Network Operations and Management Symposium, NOMS 2016, pp. 1203–1208, April 2016

Implementations and Validations

Highlighting Some Shortcomings of the CoCoA+ Congestion Control Algorithm

Simone Bolettieri, Carlo Vallati(✉), Giacomo Tanganelli, and Enzo Mingozzi

Department of Information Engineering, University of Pisa, Pisa, Italy
{simone.bolettieri, carlo.vallati, giacomo.tanganelli, enzo.mingozzi}@iet.unipi.it

Abstract. The Constrained Application Protocol (CoAP) is expected to be the de-facto standard application protocol for the future Internet of Things (IoT). Future IoT devices will be interconnected by networks characterized by high packet error rates and low throughput. For this reason, congestion control will be crucial to ensure proper and timed communication in these networks. In this context, CoCoA+, an advanced congestion control for CoAP, is currently under standardization. In this work, we present a critical analysis of CoCoA+ by means of simulation, and highlight some of its shortcomings and pitfalls. We considered a typical scenario with an increasing traffic load due to an increasing number of CoAP requests. We show how CoCoA+ may be characterized by many spurious retransmissions at some offered loads close to congestion.

Keywords: CoAP · CoCoA · CoCoA+ · Congestion control

1 Introduction

The Internet of Things (IoT) is getting momentum both in academia, as a very active research topic, and in industry, with smart objects that are meant to thoroughly affect our lives. IoT systems are built on top of IoT devices that collect data and interact with the physical environment. IoT devices are usually cheap, battery-powered devices, with limited computation capabilities. Those devices are equipped with low power transceivers that, however, allow forming Low Power Lossy Networks (LLN), characterized by possibly large packet error rates, and low throughput. Due to these characteristics, new communication protocols tailored for constrained devices operating in lossy networks are required. To this aim, IETF formed the CoRE WG that defined a communication protocol stack for IoT devices. Such communication stack includes at the top the Constrained Application Protocol (CoAP) [1], which allows applications to exchange information with IoT devices. Considering the limited capabilities of nodes and the limited amount of bandwidth available in such networks, congestion control is crucial to ensure proper and timely delivery of data. CoAP employs the User Datagram Protocol (UDP), which, however, does not provide any congestion control mechanism. For this reason, CoAP implements optional reliable data delivery through retransmissions and regulates the amount of data transmitted through a simple congestion avoidance mechanism. The definition of the latter, however, is not trivial: the shared

A. Puliafito et al. (Eds.): ADHOC-NOW 2017, LNCS 10517, pp. 213–220, 2017.
DOI: 10.1007/978-3-319-67910-5_17

nature of the wireless medium results in collisions of transmissions, and the limited buffer size of IoT devices often produces buffer overflows. Both phenomena can cause frequent packet losses that bring to additional messages transmitted, and additional traffic that can eventually lead to congestion.

Since CoAP defines only a basic congestion control mechanism, a new advanced congestion control have been proposed recently in [2], and further extended in [3]. Such algorithm, called CoCoA, is currently under standardization within the CoRE Working Group. CoCoA exploits the measured round-trip time (RTT) between the client and the server to adjust the timeout for retransmissions, and avoid frequent retransmissions. CoCoA has been originally evaluated against the standard CoAP in [2] and [3], showing that it can improve the performance in terms of throughput, packet delivery ratio and average delays. More recent works have focused instead on evaluating its performance per se. In [6], for instance, CoCoA is evaluated in a typical large-scale IoT scenario in which GPRS communications are employed to connect IoT nodes. The authors compared CoCoA against other congestion control mechanisms defined for TCP applications. Results showed that CoCoA performs equally or better than TCP-based algorithms. Similar conclusions are drawn in [7], where the authors compare CoCoA with other TCP-based congestion control mechanisms on an emulated Zigbee network. In [5], the authors evaluate CoCoA on a cloud infrastructure, characterized by high network bandwidth and powerful nodes. The evaluation is carried out through an extensive set of experiments that shown how CoCoA can help in increasing the throughput by more than 100% as compared to the default congestion control. Finally, in [8] the authors propose a modification to CoCoA introducing a more complex RTT estimator to distinguish between wireless losses and congestion losses. Results show an improvement of the performance in networks where the loss rate is particularly high.

In this work, we evaluate the CoCoA congestion control algorithm by means of an extensive set of experiments. An in-depth analysis of simulation results allowed us to highlight some novel critical issues of the algorithm. The rest of the paper is structured as follows. Section 2 presents the methodology adopted for the performance evaluation, while Sect. 3 reports performance evaluation results, and summarizes the main issues identified by their in-depth analysis. Finally, Sect. 4 draws the conclusions.

2 Simulation Methodology

We evaluate the performance of CoCoA+ by simulation. To this aim, the Cooja platform is used, leveraging a ContikiOS implementation of CoCoA+ originally obtained from its authors, and then updated to be in line with the latest CoCoA+ specification [3]. Cooja motes are used to emulate wireless sensor devices. The relevant simulation parameters are summarized in Table 1. In all scenarios, we consider a grid of 6×6 nodes, as reported in Fig. 1. In the network, one node is the RPL border router (node ID 1), one node act as a CoAP server (node ID 3), and, finally, the remaining nodes are CoAP clients. The linear distance between two nodes in a row is 10 m.

Table 1. Simulation parameters.

Parameter	Value
Channel model	Unit disk GM, tx range = 10 m, interference range = 20 m
MAC buffer size	8 packets
MAC level max retransmissions	8
CoAP base ACK-TIMEOUT	3 s
CoAP requests buffer size	4

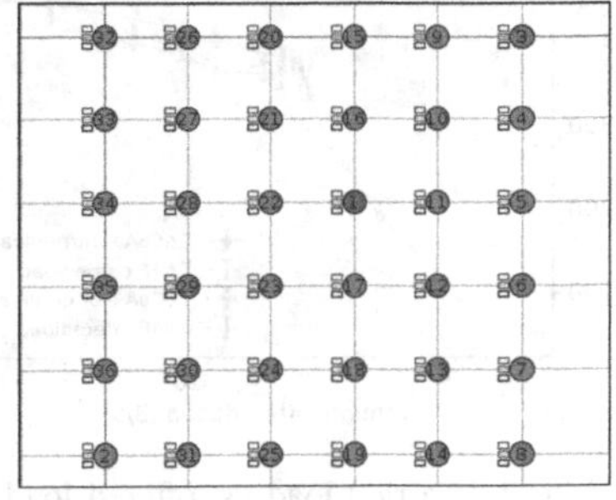

Fig. 1. Network topology.

3 Performance Evaluation

In this section we analyze the results of the performance evaluation. In all experiments, periodic CoAP traffic with CoCoA+ congestion control is considered to emulate a network of sensors that report their measurements at a fixed interval. In particular, all CoAP clients periodically send confirmable POST requests towards the server with ID 3. Each request has a size of 95B. For each successfully received request, the server replies with a response piggybacked onto a CoAP ACK message. All requests are addressed to an IPv6 global address, which forces messages to be routed through the RPL border router in order to reach the server. Before starting to transmit CoAP requests, nodes wait for 60 s to let the RPL topology stabilize, and then for a further random time to avoid synchronization effects. Finally, nodes start sending requests with a common period T, ranging from 70 s to 0.5 s, in order to inject in the network an increasing amount of data. Each experiment has a duration of 800 s. In order to obtain statistically sound results, twelve independent replicas for each scenarios are run, and metrics of interest are then estimated for each scenario along with a 95% confidence interval.

Figure 2 shows the aggregate carried load and actual offered load as a function of the nominal offered load for both CoAP and CoCoA+. The carried load is defined as the overall amount of data successfully delivered at the server. The actual offered load, instead, is defined as the overall amount of generated data that is transmitted at least once, i.e., that is not discarded at the client because of a buffer overflow at the application layer. From Fig. 2 we can first observe that for both CoAP and CoCoA+ the respective carried and actual offered loads are very close to each other even when the carried load stops increasing linearly (at a nominal offered load of approximately 380 B/s), i.e., congestion has started in the network. This means that in both cases the server receives almost all the requests transmitted by clients, and therefore the main bottleneck that causes congestion is located at the client, and not in the network. This is expected, as only one outstanding request for each node is allowed when N_START is set to one. When the offered load increases, the network traffic increases resulting in longer times to complete a transaction (considering that the client retransmits a request until the ACK is received), which, consequently, results in more frequent buffer

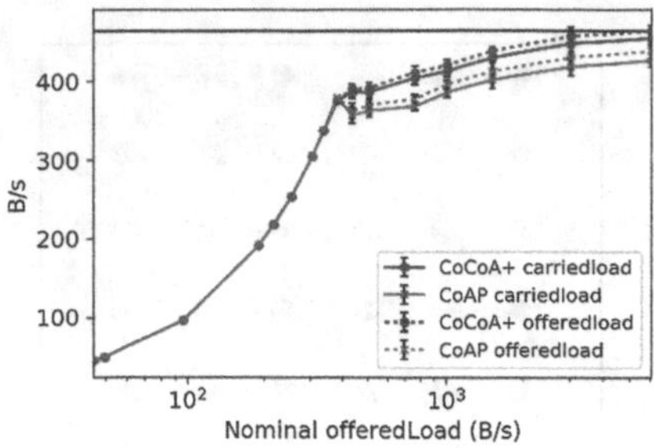

Fig. 2. Carried load vs. offered load.

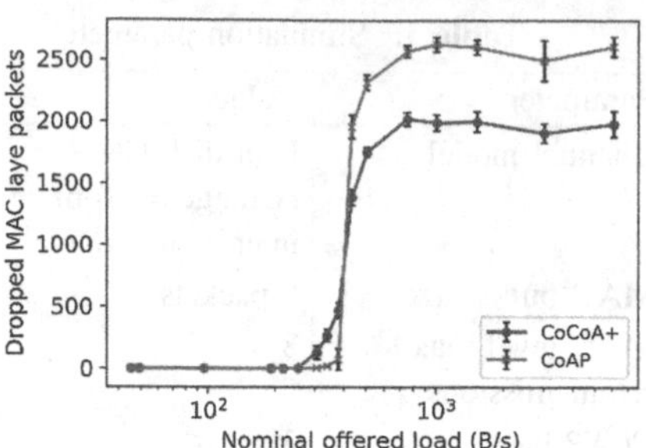

Fig. 3. Overall in-network MAC buffer overflows.

overflows at the application layer. However, if we focus on the results obtained when congestion has started, we can also notice that the carried load continue increasing, though at a much slower rate. This can be explained by considering that congestion is not evenly distributed among all nodes. In particular, nodes farther from the server start discarding requests earlier, while nodes closer to the server, instead, experience shorter delays to complete a transaction, consequently they experience application layer overflows at higher offered loads. Although omitted here for the sake of brevity, this is confirmed by analyzing the carried load per node: the increase of the carried load after congestion has started is mainly due to the three clients closest to the server (nodes 4, 9, and 10).

By comparing CoAP and CoCoA+ carried loads from Fig. 2, we can then observe that their performance differ under congestion: as expected, CoCoA+ always outperforms the CoAP simple congestion control mechanism by dynamically adapting retransmission timeouts. In order to get a better insight into this result, we consider a few additional relevant metrics as a function of the nominal offered load. In particular, Fig. 3 shows the overall MAC buffer overflows in the network, i.e., the overall amount of requests discarded at the MAC layer in the network. Figure 4 shows instead the average number of transmissions (at the client) for each successful transaction. Finally, Fig. 5 shows the average number of ACK messages received at the client for each successful transaction (including duplicates). These results highlight that three different offered load intervals can be identified to compare the performance of CoCoA+ and CoAP congestion control: a non-congestion interval, with offered loads up to 250 B/s, a pre-congestion interval, with offered loads ranging between 250 B/s and 380 B/s, and, finally, a congestion interval, with offered loads above 380 B/s.

In particular, in the non-congestion interval, corresponding to one request every 12 s, or more, per node, the network is essentially operating out of congestion, no buffer overflow occurs and therefore the two algorithms behave in the same way. On the other hand, in the congestion interval, corresponding to one request every up to 8 s per node, the network is running under substantial congestion in most of its nodes, and the actual offered load stops increasing linearly with the nominal offered load. In this case, the amount of buffer overflows rapidly increases together with packet delays. A fixed RTO value is therefore not efficient, as its value eventually becomes close or

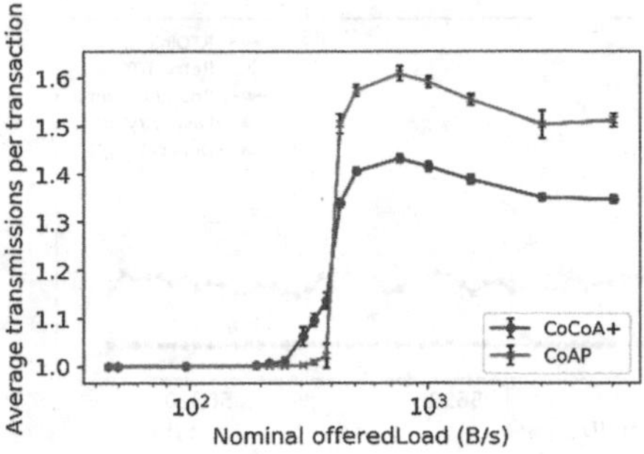

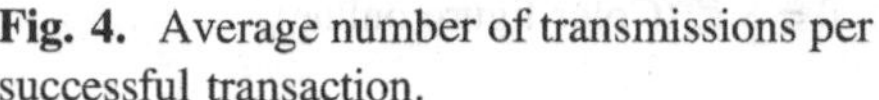

Fig. 4. Average number of transmissions per successful transaction.

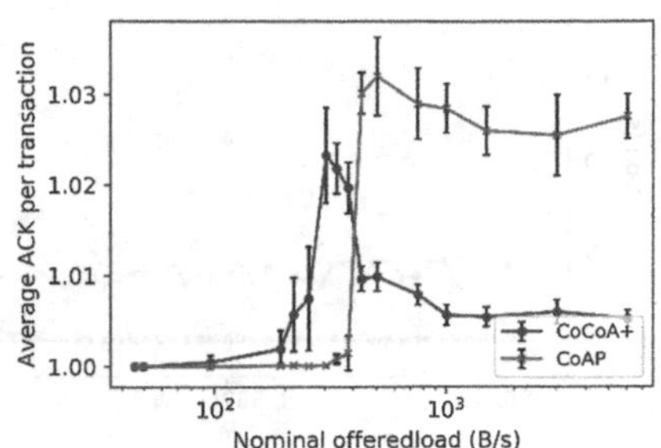

Fig. 5. Average number of ACKs per successful transaction.

even lower than round-trip times. For this reason, the timeout expires more frequently and often unnecessarily, leading to the transmission of additional traffic that further congests the network. CoCoA+ shows instead its effectiveness in adapting to increasing RTTs by using larger RTO values and therefore reducing the amount of retransmissions per transaction (see Fig. 4). The amount of unnecessarily retransmitted requests is also drastically reduced, as shown in Fig. 5.

Finally, we compare the performance of CoCoA+ and CoAP in the pre-congestion interval, which corresponds to request arrival periods per node between 8 s and 12 s. This case is interesting since both CoAP and CoCoA+ achieve the same carried load in this interval, but they clearly behave differently. In particular, we observe from Fig. 4 that CoCoA+ triggers a number of request retransmissions that increases linearly with the offered load, while retransmissions with CoAP are still negligible. However, considering that the carried load is practically the same, such retransmitted requests with CoCoA+ are unnecessary and they only contribute to increase the traffic load in the network. In fact, either such spurious retransmissions are dropped before reaching the server, as shown by Fig. 3, or they produce duplicate ACKs that are sent back to the client, as shown by Fig. 5. In this range of offered loads CoCoA+ requires therefore more network resources than CoAP to achieve the same carried load, or, said

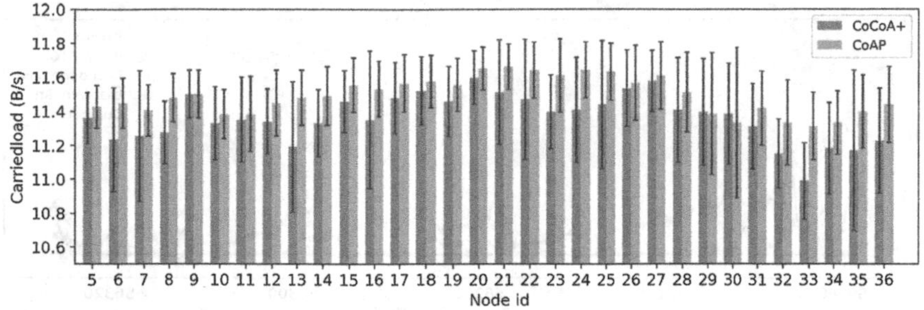

Fig. 6. CoAP vs. CoCoA+ carried load per node, T = 8 s.

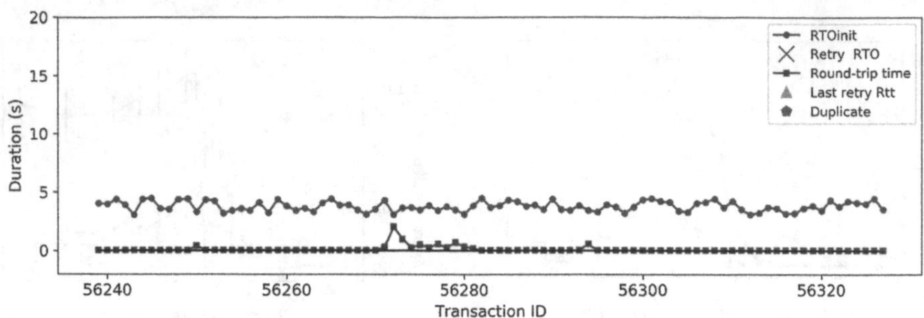

Fig. 7. CoAP operation, node ID 10, T = 8 s. (Color figure online)

alternatively, congestion is reached earlier than CoAP. As a matter of facts, with a request period of 8 s, the carried load with CoCoA+ is even slightly lower than CoAP. Such behavior is consistently exhibited by all nodes in the network, as reported in Fig. 6, which shows the carried load per node for both protocols.

In order to understand this behavior, we provide an insight into the operation of both the algorithms in a selected time interval, in the case of a request period equal to 8 s. Figures 7 and 8 show the values of all the relevant parameters controlling the transmission of requests over subsequent transactions in CoAP and CoCoA+, respectively. Specifically, in both figures values reported with blue circle marks represent the RTO_{init} values, while green square marks represent the measured RTTs. More precisely, in case of a retransmission, the RTT value is calculated as the delay between the first request transmission and the arrival of an ACK (this is due to the RTT ambiguity problem [4]); in this case, the measured delay between the last retransmission and the ACK reception is also reported using the yellow triangle mark. Finally, in both figures we report with blue cross marks the RTO values calculated as a result of a back-off because of a retransmission, and highlight spurious retransmissions with brown small circle marks.

From Fig. 7 we can observe that RTO values randomly chosen by CoAP in the predetermined range are large enough to cope with RTT variations, thus causing a negligible number of unnecessary fluctuations. On the other hand, from Fig. 8 we can observe that, with CoCoA+, when a series of similar RTT values is sampled, the contribution of RTTVAR to the RTO computation vanishes and, consequently RTO

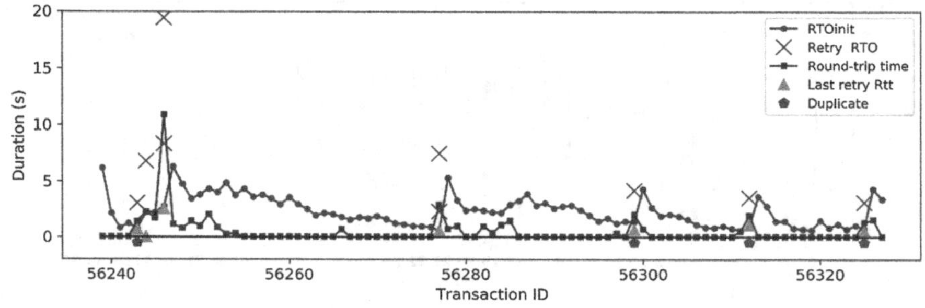

Fig. 8. CoCoA+ operation, node ID 10, T = 8 s. (Color figure online)

values eventually get very close to actual RTTs. When this occurs, small RTT variations cause spurious retransmissions, which, besides being unnecessary, potentially further exacerbate the issue, since they contribute to increase RTTs in the network, and therefore trigger additional retransmissions. Furthermore, from the analysis of the RTO traces, we discover that also a small increase in the RTT can causes a steep increment of the RTO value due to the lack of an aging mechanism for the weak estimator. If the weak estimator has grown to a large value at some time in the past, it will influence the RTO whenever a retransmission occurs (even if the retransmission is spurious), irrespectively of how much time has passed since then.

4 Conclusions

In this paper we presented an in-depth analysis of the CoCoA+ advanced congestion control algorithm. Specifically, we considered a common scenario in which periodic CoAP traffic is exploited to collect sensors' data. The simulation results allowed us to highlight some issues of the algorithm, mostly related to the relationship between RTT estimation and RTO calculation. In particular, we have shown that in certain scenarios CoCoA+ unnecessarily triggers more retransmissions than CoAP. In order to overcome such shortcomings, the CoCoA+ algorithm could be modified considering the mechanisms for congestion control and RTT calculation, in one of the many TCP variants. For instance, to avoid spurious retransmissions, CoCoA+ could be simplified, eliminating the usage of a weak estimator in favor of a precise request/response matching using the transmission counter option defined in CoAP. Moreover, other modifications could be considered, aimed at preventing congestion before it happens. This could be performed introducing a delay between subsequent transmissions, to reduce the transmission rate when required.

Acknowledgment. This work has been partially supported by the project "IoT and Big Data: methodologies and technologies for large scale data gathering and processing" – PRA_2017_37, funded by University of Pisa.

References

1. Shelby, Z., Hartke, K., Bormann, C.: The Constrained Application Protocol (CoAP)
2. Betzler, A., et al.: Congestion control in reliable CoAP communication. In: Proceedings of Modeling, Analysis & Simulation of Wireless and Mobile Systems, ACM MSWIM (2013)
3. Betzler, A., Gomez, C., Demirkol, I., Paradells, J.: CoCoA+ : an advanced congestion control mechanism for CoAP. Ad Hoc Netw. **33**, 126–139 (2015)
4. Karn, P., Partridge, C.: Improving round-trip time estimates in reliable transport protocols. SIGCOMM Comput. Commun. Rev. **17**, 2–7 (1987)
5. Betzler, A., et al.: Congestion control for CoAP cloud services. In: Emerging Technology and Factory Automation (ETFA). IEEE (2014)
6. Betzler, A., et al.: CoAP congestion control for the Internet of Things. IEEE Commun. Mag. **54**(7), 154–160 (2016)

7. Järvinen, I., Laila, D., Markku, K.: Experimental evaluation of alternative congestion control algorithms for Constrained Application Protocol (CoAP). In: Internet of Things (WF-IoT). IEEE (2015)
8. Bhalerao, R., Sridhar, S.S., Joseph, P.: An analysis and improvement of congestion control in the CoAP Internet-of-Things protocol. In: Consumer Communications & Networking Conference (CCNC). IEEE (2016)

Experimental Evaluation of Non-coherent MIMO Grassmannian Signaling Schemes

Jacobo Fanjul(✉), Jesús Ibáñez, Ignacio Santamaria, and Carlos Loucera

Department of Communications Engineering, University of Cantabria, 39005 Santander, Spain
{fanjulj,jesus.ibanez,i.santamaria,carlos.loucera}@unican.es

Abstract. In this paper, we present an over-the-air (OTA) performance analysis of Grassmannian signaling strategies in an orthogonal frequency-division multiplexing (OFDM) single-user multiple-input multiple-output (SU-MIMO) scenario. Specifically, we compare the Grassmannian signaling technique to the differential Alamouti scheme and a novel space-time non-coherent scheme recently proposed in the context of 5G. As a performance benchmark we include in the comparison the coherent Alamouti scheme. We study the practical impairments associated to frequency synchronization mismatches (frequency offsets), as well as the effects of time-varying channels for different spectral efficiencies. The experimental results show that non-coherent techniques are more robust to the aforementioned impairments than the coherent Alamouti approach, while Grassmannian methods are close to the differential Alamouti scheme with 2 transmit antennas.

Keywords: Non-coherent communications · Grassmannian signaling · MIMO testbed · OFDM · Over-the-air (OTA) experiments

1 Introduction

The vast majority of wireless communications systems rely on the use of channel state information (CSI), at least at the receiver end. However, some scenarios might present short coherence times due, for instance, to large Doppler spreads associated to communications with terminals mounted in high-speed vehicles. For very fast time-varying scenarios, channel estimation might not be even feasible. Even if accurate channel estimates can be obtained in fast fading scenarios, the associated overhead implies a significant reduction in terms of throughput. In the context of 5G, the concept of massive multiple-input multiple-output (MIMO) is attracting significant research efforts [7]. Considering the large

J. Fanjul—This work has been supported by the Ministerio de Economía y Competitividad (MINECO) of Spain under grants TEC2013-47141-C4-R (RACHEL), TEC2016-75067-C4-4-R (CARMEN) and FPI grant BES-2014-069786. The software tools to control the USRP devices in this work have been provided by the GTEC Group, University of A Coruña, Spain.

A. Puliafito et al. (Eds.): ADHOC-NOW 2017, LNCS 10517, pp. 221–230, 2017.
DOI: 10.1007/978-3-319-67910-5_18

amount of antennas and, consequently, the associated channel state information to be handled in these scenarios, non-coherent strategies are receiving renewed interest for massive MIMO.

Since the first differential coding schemes for phase-shift keying (PSK) systems were proposed for systems using a single transmit antenna [16], several non-coherent transmission techniques have been developed for MIMO systems. A remarkable approach is the differential Alamouti scheme presented in [15], which retains the nice properties of the popular Alamouti [2] with only a 3 dB loss with respect to the coherent case. This scheme was later generalized to other differential space-time block codes (DSTBC) with more than two transmit antennas in [9], but with a reduction in rate. The design of a rate-2 differential STBC was described in [1].

The study of the capacity of non-coherent MIMO systems in block fading channel models was considered in [12], where the structure of the capacity achieving input distribution was characterized (the capacity achieving transmitted signals are isotropically distributed unitary matrices). Shortly afterwards, the use of differential space-time modulations was proposed in [6,8]. Following a similar line, Zheng and Tse presented in [17] the Grassmannian signaling technique, which relies on the fact that the space spanned by the transmitted matrices is invariant to the channel matrix. In this way, the design of optimal transmit matrices for non-coherent schemes can be posed as a sphere packing problem on the Grassmann manifold. To solve this problem, an alternating projection algorithm is provided in [4]. Further advances on Grassmannian constellation design and detection are presented in [3,5].

In the experimental area, only a few works have been conducted to evaluate the performance of differential STBC schemes over real scenarios (see [13]). However, the experimental performance evaluation of Grassmann-based signaling schemes using over-the-air (OTA) transmissions is still lacking. In this work, we attempt to fill this gap and present an experimental comparison between Grassmannian-based signaling schemes and other well-known non-coherent techniques, namely the differential Alamouti method in [15] and the PSK-based ST scheme proposed in [10].

1.1 Notation

Uppercase and lowercase boldface letters will be used for matrices and column vectors, respectively. $(\cdot)^T$ will represent transpose, whereas $(\cdot)^H$ denotes conjugate transpose (Hermitian). Additionally, $\mathbf{I}$ stands for the identity matrix, $\mathrm{Tr}(\cdot)$ represents the trace operator and $\mathrm{vec}(\cdot)$ is used for vectorization. Finally, the notation $\mathbb{G}(T, M)$ will be used to represent the Grassmann manifold containing all subspaces of dimension M in a T-dimensional ambient space.

2 Grassmannian Signaling Overview

In this section, we present a brief description of Grassmannian signaling for MIMO non-coherent transmissions. Given a transmitter with M antennas, a receiver equipped with N antennas, and a coherence time T within which the channel remains constant, the received signal $\mathbf{Y} \in \mathbb{C}^{N\times d}$ is given by

$$\mathbf{Y} = \mathbf{HX} + \mathbf{W}, \tag{1}$$

where $\mathbf{X} \in \mathbb{C}^{M\times d}$ is the transmitted signal over d time slots, $\mathbf{H} \in \mathbb{C}^{N\times M}$ is the channel matrix, and $\mathbf{W}$ denotes additive white Gaussian noise (AWGN). Also, notice that time indexes have been omitted in this general signal model for the sake of notational simplicity.

It was shown in [6,8,12] that at high-SNR the capacity of a non-coherent block-fading MIMO channel is achieved transmitting isotropically-distributed unitary matrices provided that $T \geq \min(M,N) + M$. The resulting capacity achieving approach has a nice geometric interpretation as a Grassmannian signaling scheme [17], where the transmitted signals are K-dimensional subspaces in $\mathbb{C}^T$, with $K = \min(M, N, \lfloor \frac{T}{2} \rfloor)$. It is also worth mentioning that, as stated in [17], no additional benefit in terms of capacity can be attained by increasing either M or N beyond $T/2$.

The main intuition behind Grassmannian signaling is that the received signal $\mathbf{Y}$ spans the same row space as the transmitted signal $\mathbf{X}$ for any nonsingular channel $\mathbf{H}$. For a given configuration with $M \leq N$ and $T \geq 2M$, we can send up to $M\,(T-M)$ information streams over T time slots by transmitting subspaces from a codebook $\mathcal{X}$ formed by subspaces in the Grassman manifold $\mathbb{G}\,(T,M)^1$. Consider that the transmitted subspace is $\mathbf{X}\,[n] = \mathbf{X}_i$; then, at the receiver side the optimal decoding rule is given by the generalized likelihood ratio test (GLRT)

$$\hat{\mathbf{X}}_i = \arg \max_{X_j \in \mathcal{X}} \left(\operatorname{Tr}\left(\mathbf{Y}\mathbf{X}_j^H\mathbf{X}_j\mathbf{Y}^H\right)\right), \tag{2}$$

with a complexity that grows exponentially with the block-length T.

3 Frame Format and Experimental Setup

This section describes the indoor MIMO testbed that has been used to conduct the over-the-air experiments, as well as the frame format. Figure 1 shows the experimental set-up, whose main points are the following:

- The link is a 2×2 MIMO system, and the Tx-Rx distance is approximately 2 m.
- Both transmitter and receiver functionalities have been implemented with Universal Software Radio Peripheral (USRP) B210 software-defined radio (SDR) devices, which are equipped with Analog Devices AD9361 single-chip direct-conversion transceivers and Spartan6 XC6SLX150 FPGA.

[1] The (complex) dimension of the Grassmann manifold $\mathbb{G}\,(T,M)$ is $\dim(\mathbb{G}\,(T,M)) = M\,(T-M)$, and therefore the multiplexing gain or pre-log factor of the system is $M\,(1-M/T)$.

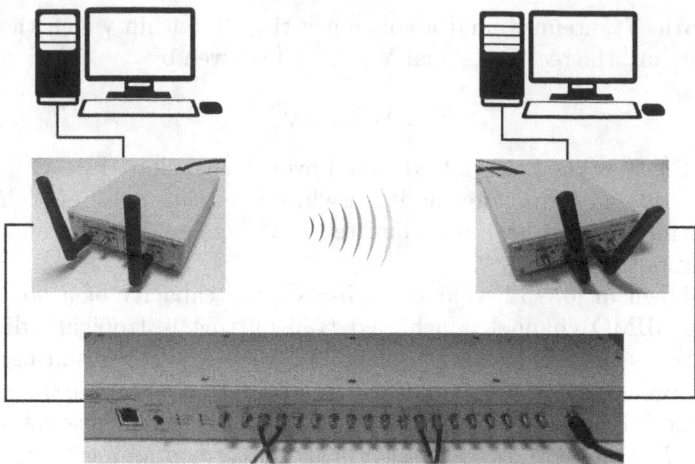

Fig. 1. Experimental set-up. The distance between transmitter and receiver is approximately 2 m.

Standard 802.11a header	Coherent training	Alamouti OFDM symbols	2	+ diff. Alamouti OFDM symbols	Grass. short T OFDM symbols	Grass. long T OFDM symbols	Space-time UL OFDM symbols

Fig. 2. Frame format used in our experiments.

- Additionally, for those experiments requiring precise frequency synchronization, we use the Ettus OctoClock module, which provides a high-accuracy timing reference signal for up to eight nodes.
- Both transmitter and receiver have been configured and controlled using the GTIS software provided by the GTEC Group from University of A Coruña, Spain.

Using this set-up, we transmit frames with the format shown in Fig. 2. Since we consider broadband transmissions over frequency-selective channels, we use the OFDM-based IEEE 802.11a wireless local area network (WLAN) physical-layer standard to construct the frames. Each subcarrier can be viewed as a 2×2 MIMO channel, and the non-coherent schemes are encoded over T consecutive OFDM symbols. The initial block is the common header for 802.11a transmissions (short-training symbols for frame detection and long-training symbols for coarse frequency estimation).

We include in the comparison the coherent Alamouti scheme, which is the first payload data after the channel estimation stage (needed only for this scheme). Then, we append in the frame a number of OFDM symbols for the differential Alamouti, the Grassmannian signaling scheme with two different ambient space configurations, and finally, the non-coherent scheme in [10].

The schemes under study have been evaluated for two different spectral efficiencies, namely, $\eta = 1$ and $\eta = 2$ bps/Hz, which are achieved as follows:

- For $\eta = 1$ bps/Hz, Alamouti-based schemes rely on BPSK symbols. For the Grassmannian signaling we transmit 2-dimensional subspaces in ambient spaces of dimension $T_{short} = 4$ and $T_{long} = 6$. In order to accommodate the corresponding number of bits within each subspace, 16 and 64-element codebooks are used for $T_{short} = 4$ and $T_{long} = 6$, respectively. For the PSK-based non-coherent scheme, we take $L = 4$-symbol codebooks.
- For spectral efficiency $\eta = 2$ bps/Hz, QPSK constellations are transmitted for both coherent and differential Alamouti techniques, the Grassmannian codewords are drawn from 256-element $\mathbb{G}(4, 2)$ and 1024-element $\mathbb{G}(5, 2)$ codebooks, respectively. For the last scheme, an $L = 16$-symbol constellation is used.
- The Grassmannian codebooks used throughout this work have been designed by means of the alternating projection algorithm in [4].

For each experiment, the previous sequences have been transmitted over 48 data subcarriers in the 2.487 GHz band, and spanning a total bandwidth of 8 MHz.

4 Experimental Results

In this section we evaluate the performance of the non-coherent MIMO schemes in realistic wireless scenarios by means of OTA experiments. We have carried out several measurement campaigns to analyze the performance degradation caused by frequency estimation offsets as well as the impact of time-varying channels. For each experiment, the results of 1000 independent transmissions at different transmit power levels have been averaged.

4.1 Frequency Offset

For each spectral efficiency ($\eta = 1$ and $\eta = 2$), we made 1000 OTA experiments with two different configurations:

- A static scenario (coherence time much longer than the OFDM symbol duration) that uses an external frequency synchronization reference signal generated by an OctoClock device. This configuration is associated to the solid lines in Figs. 3 and 4.
- The same static scenario, but removing the high-accuracy timing reference signal. In this case, the frequency offset has been estimated using the long training symbols (LTS) included in the 802.11a standard. Dashed lines in Figs. 3 and 4 represent the results for this scenario.

The aim of this measurement campaign is to determine how a given frequency offset impacts the different transmission techniques. Figures 3 and 4 show the symbol error rate (SER) curves for spectral efficiencies $\eta = 1$ bps/Hz and $\eta = 2$ bps/Hz, respectively. As expected, the MIMO non-coherent schemes are significantly more robust to frequency synchronization errors than the coherent Alamouti approach. For the coherent Alamouti scheme, the offset provokes

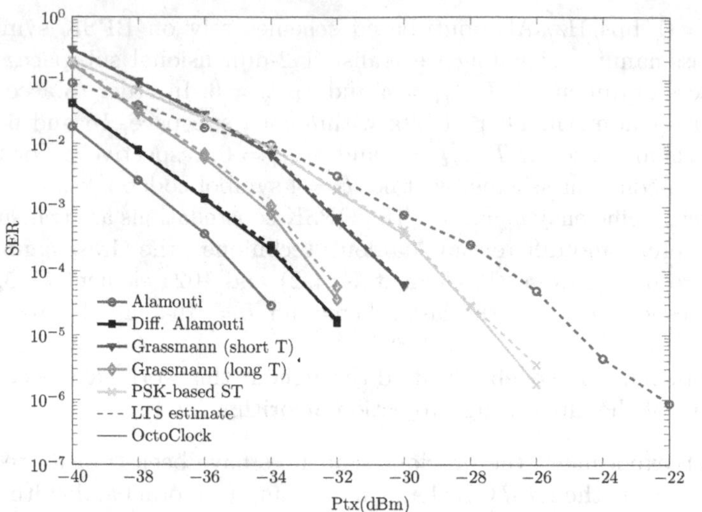

Fig. 3. SER for $\eta = 1$ bps/Hz with different freq. sync. approaches.

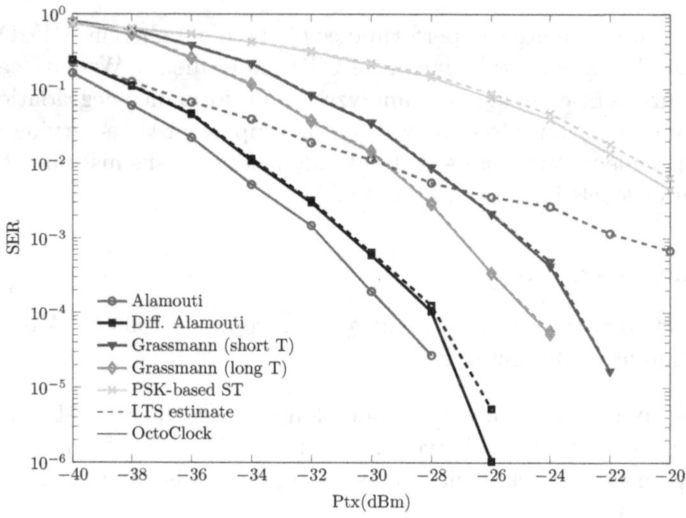

Fig. 4. SER for $\eta = 2$ bps/Hz with different freq. sync. approaches.

a rotation of the data symbols with respect to the channel estimate obtained from the training preamble, which in turn increases the SER especially for those symbols located at the end of the coherent packet. On the other hand, for the non-coherent schemes the effect of the frequency offset is limited to the length of a codeword and it does not accumulate with time. In particular, the frequency offset effect is restricted to $T = 4$ OFDM symbols for the differential Alamouti (2 consecutive ST codewords), either T_{short} or T_{long} OFDM symbols for the

2 Grassmanian signaling schemes under comparison, and only $T = 2$ OFDM symbols for the PSK-based non-coherent scheme. Therefore, the performance degradation due to frequency offsets is much more limited.

Turning now our attention to the comparison among the different non-coherent schemes, we observe that the PSK-based scheme in [10] provides the poorest performance, probably due to the fact that the method does not optimize the pairwise distance between ST codewords. Nevertheless, this method requires a coherence time of only $T = 2$ OFDM symbols, whereas the differential Alamouti requires the channel to remain constant for 2 consecutive blocks ($T = 4$), and for the Grassmannian signaling scheme it has to remain constant during the selected ambient space dimension (either $T = 4$ or $T = 6$ for $\eta = 1$ bps/Hz).

Regarding the Grassmannian signaling scheme, it can be observed that the behaviour improves when the ambient space dimension, T, increases. This is in agreement with theoretical works in [11,17], which indicate that the performance of this non-coherent technique approaches the coherent capacity as the coherence time, T, tends to infinity (static channel).

Finally, it can be concluded from the figures that the best performing method is the differential Alamouti scheme. However, we can notice that the gap between the Grassmannian approaches and the differential Alamouti scheme is decreased when we reduce the spectral efficiency. This fact could be related to the number of elements in each codebook. Recall that, for spectral efficiencies $\eta = 1$ bps/Hz and $\eta = 2$ bps/Hz, the differential Alamouti builds on BPSK and QPSK constellations, respectively; whereas the Grasmannian signaling with T_{long} uses 64 and 1024-element codebooks, respectively. Obviously, for a given ambient dimension a lower number of codewords allows us to increase the minimum distance between symbols and the codebook optimization is easier. Finally, it is worth mentioning that unlike the differential Alamouti, which is limited to 2 transmit antennas, Grassmannian signaling schemes can be applied to more general antenna configurations.

4.2 Time-Varying Channels

In this section, we evaluate the performance of the MIMO non-coherent schemes in fast-fading channels that arise in high-mobility wireless communications. In general, experimental evaluation of wireless technologies in high-mobility scenarios requires expensive equipment and sophisticated software processing [14]. To avoid these costs, in this work we emulate the fast-fading process at the transmitter side, and transmit frames filtered with time-varying channels generated with different Doppler spreads.

To focus only on the time selectivity of the channel, for this set of experiments we use an external clock to ensure frequency synchronization between the nodes. The frame format is the same as in Sect. 4.1, and we consider two different Doppler spreads: 400 Hz and 500 Hz. The SER curves for the schemes under comparison for $\eta = 1$ bps/Hz and $\eta = 2$ bps/Hz are presented in Figs. 5 and 6, respectively.

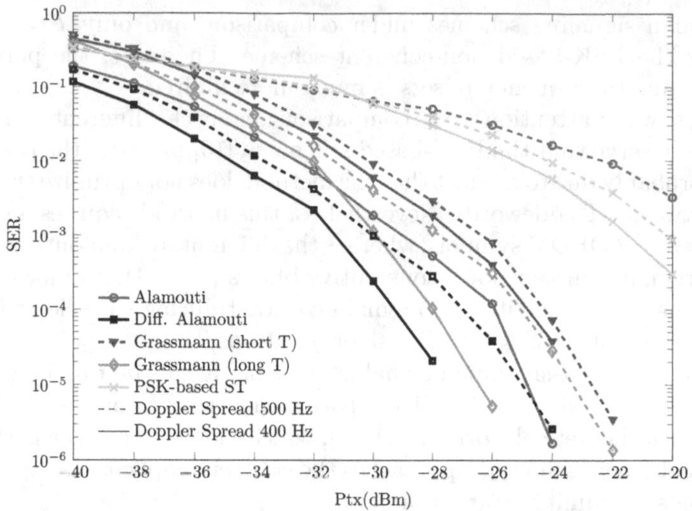

Fig. 5. SER for $\eta = 1$ bps/Hz and different Doppler spreads.

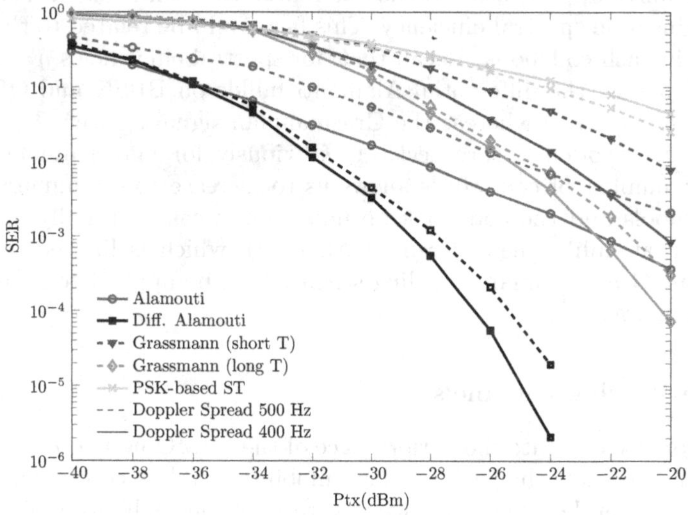

Fig. 6. SER for $\eta = 2$ bps/Hz and different Doppler spreads.

As observed in both figures, with the exception of the scheme in [10], non-coherent techniques are also more robust in fast-fading channels than the coherent scheme. Again, the best performing non-coherent scheme is the differential Alamouti, especially for $\eta = 2$ bps/Hz. Remember also that the differential Alamouti scheme has a very simple optimal decoding rule in comparison to the GLRT detector used for the Grassmanian signaling scheme, whose complexity is exponential in T. Regarding the Grassmaniann signaling schemes, one would expect that increasing the Doppler spread would be more harmful for the scheme with higher ambient dimension T. However, Figs. 5 and 6 show that the codebook size (i.e., the spectral efficiency) also plays an important role here. This aspect requires further theoretical analysis and will be considered in a future work.

5 Conclusion

In this work, we have presented an experimental evaluation of 3 different non-coherent techniques in a wireless 2×2 MIMO-OFDM scenario. In particular, we have compared the performance of the subspace-based signaling technique to the differential Alamouti scheme and a recently proposed non-coherent scheme that uses PSK modulations. We have focused our study in two aspects of practical importance: (i) the impact of frequency offsets between transmitter and receiver, and (ii) the performance of these schemes under fast-fading (emulated) channels. While all non-coherent schemes are clearly more robust than the coherent Alamouti scheme under frequency offsets and time-selective channels, the differential Alamouti scheme seems to be the best performing non-coherent technique. Our study also showed that for the Grasmannian signaling scheme there are some interesting trade-offs between the ambient space dimension (number of channel uses over which the channel should remain constant) and the spectral efficiency (codebook size) that require further theoretical study. Also, it might be of interest to extend this experimental study to scenarios with more transmit antennas, for which existing DSTBC schemes have a penalty in rate [9].

References

1. Al-Dhahir, N.: A new high-rate differential space-time block coding scheme. IEEE Comm. Lett. **11**(11), 540–542 (2003)
2. Alamouti, S.: A simple transmitter diversity scheme for wireless communications. IEEE J. Sel. Areas Commun. **16**, 1451–1458 (1998)
3. Beko, M., Xavier, J., Barroso, V.A.N.: Noncoherent communication in multiple-antenna systems: receiver design and codebook construction. IEEE Trans. Signal Proc. **55**(12), 5703–5715 (2007)
4. Dhillon, I.S., Heath, R.W., Strohmer, T., Tropp, J.A.: Constructing packings in grassmannian manifolds via alternating projection. Exper. Math. **17**(1), 9–35 (2008)
5. Gohary, R.H., Davidson, T.N.: Noncoherent mimo communication: grassmannian constellations and efficient detection. IEEE Trans. Inf. Theory **55**(3), 1176–1205 (2009)
6. Hochwald, B.M., Sweldens, W.: Differential unitary space time modulation. IEEE Trans. Commun. **48**, 2041–2052 (2000)
7. Hoydis, J., Hosseini, K., Brink, S.T., Debbah, M.: Making smart use of excess antennas: massive MIMO, Small Cells, and TDD. Bell Labs Tech. J. **18**(2), 5–21 (2013). http://dx.doi.org/10.1002/bltj.21602
8. Hughes, B.: Differential space-time modulation. IEEE Trans. Inf. Theory **46**(7), 2567–2578 (2000)

9. Jafarkhani, H., Tarok, V.: Multiple transmit antenna differential detection from generalized orthogonal designs. IEEE Trans. Inf. Theory **47**(6), 2626–2631 (2001)
10. Li, G., Gong, F.: Space-time uplink transmission in non-coherent systems with receiver having massive antennas. IEEE Commun. Lett. **21**(4), 929–932 (2016)
11. Li, Y., Nosratinia, A.: Product superposition for MIMO broadcast channels. IEEE Trans. Inf. Theory **58**(11), 6839–6852 (2012)
12. Marzetta, T.L., Hochwald, B.M.: Capacity of a mobile multiple-antenna communication link in Rayleigh flat fading. IEEE Trans. Inf. Theory **45**(1), 139–157 (1999)
13. Ramírez, D., et al.: A comparative study of STBC transmissions at 2.4 GHz over indoor channels using a 2x2 MIMO testbed. Wirel. Commun. Mobile Comput. **8**, 1149–1164 (2007)
14. Suárez-Casal, P., Rodríguez-Piñeiro, J., García-Naya, J.A., Castedo, L.: Experimental evaluation of the WiMAX downlink physical layer in high-mobility scenarios. EURASIP J. Wirel. Commun. Netw. **2015**, 109 (2015)
15. Tarokh, V., Jafarkhani, H.: A differential detection scheme for transmit diversity. IEEE J. Sel. Areas Commun. **18**(7), 1169–1174 (2000)
16. Weber, W.: Differential encoding for multiple amplitude and phase shift keying systems. IEEE Trans. Commun. **26**(3), 385–391 (1978)
17. Zheng, L., Tse, D.N.: Communication on the grassmannian manifold: a geometric approach to the noncoherent multiple-antenna channel. IEEE Trans. Inf. Theory **48**(2), 359–383 (2002)

WEVA: A Complete Solution for Industrial Internet of Things

Giuseppe Campobello(✉), Marco Castano, Agata Fucile, and Antonino Segreto

Department of Engineering, University of Messina, Messina, Italy
gcampobello@unime.it

Abstract. In this paper a new architecture for Wireless Sensors and Actuators Network (WSAN) is proposed, which is specifically suited to the paradigm of the Industrial Internet of Things (IIoT). Starting from a detailed analysis of the requirements of industrial applications, a complete solution for IIoT is devised, which is able to cope with the main challenges of industrial environments. In comparison with other standards and commercial solutions, the proposed architecture offers several advantages: it is based on open source software and communication protocols; it encompasses reliability and discovery mechanisms; it provides a graphic tool for network set-up and maintenance.

Conversely to several other papers, all aspects of the proposed architecture are discussed, from hardware specifications to software applications, communication protocols and implementation issues.

1 Introduction

The main goal for industrial automation is visibility and management improvement in production processes [1]. Today, this can be obtained thanks to low-cost processors and memories and gradual reduction in dimensions and energy consumptions for embedded systems, which allow a large number of devices and machines to be deployed in a same area. These technological progresses are referred to as Industrial Internet of Things (IIoT) [2] and Industry 4.0 [3], which both refer to the Internet of Things (IoT) paradigm [4] for industrial environments and processes. The IoT paradigm can introduce several advantages in industrial automation [5] in terms of costs and energy consumptions along with improvement in time-to-market, productivity, and traceability. Moreover, fault tolerance and security can be increased.

To reach the aforementioned targets, conventional machines must be converted into "smart objects" able to monitor own components along with the surrounding environment, to communicate to operators and to each other, and to take decisions too. To this end, Wireless Sensors and Actuators Networks (WSANs) and Machine-to-Machine (M2M) protocols are the most important key enablers. WSAN refers to a set of sensors and actuators nodes, also known as motes, which are able to sense, process and send information through a wireless channel [6]. WSANs can assume a crucial role in industrial applications, for

A. Puliafito et al. (Eds.): ADHOC-NOW 2017, LNCS 10517, pp. 231–238, 2017.
DOI: 10.1007/978-3-319-67910-5_19

instance, in smart metering networks, as well as in HVAC and lighting monitor and control systems, in order to achieve the desired trade-off between production, comfort, costs and energy consumptions. WSANs combined to M2M protocols [7] allow more efficient control with respect to Supervisory Control And Data Acquisition (SCADA) systems. Indeed, in SCADA systems all data must be collected by only one supervisory node with negative impact on network latency and scalability. Open-source and Internet-oriented protocols (i.e. IPv6) are further requirements for flexible IIoT platforms. However, integration of IPv6, M2M, and all the other networking techniques needed for high performance IIoT is not an easy task due to restrained memory resources on commercial motes. Besides, many research works address specific problems (e.g. security, latency, etc.) but neglect WSAN integration into industry as a whole. Conversely, commercial platforms often use proprietary solutions with limitations in interoperability and expandability.

In this paper a new platform for IIoT is proposed, which is able to cope with the above challenges. The new platform, which will be referred to as WEVA (Wireless EVolution for Automation), entirely relies on open-source software and protocols. Moreover, it employs new sensor boards suited for IIoT and encompasses reliability and discovery mechanisms. Furthermore, WEVA provides a graphic interface for management of distributed control networks.

The rest of the paper is organized as follows: in Sect. 2 WEVA architecture is detailed; in Sect. 3 WEVA features are discussed with respect to IIoT requirements; Finally, in Sect. 4 conclusions and future works are drawn.

2 WEVA Architecture

The main WEVA components are listed below and shown in Fig. 1:

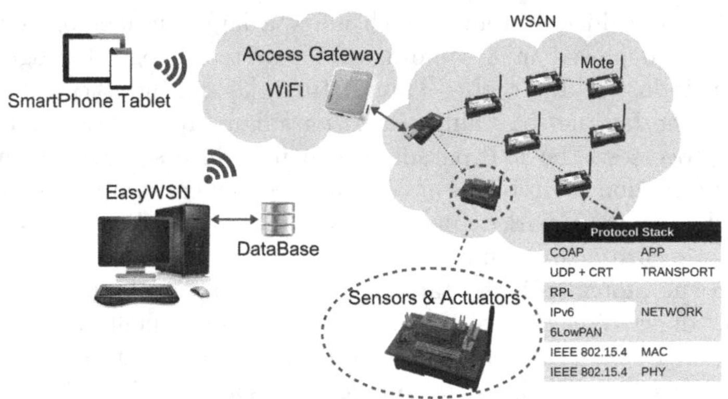

Fig. 1. WEVA architecture.

Sensor and Actuator Boards: WEVA relies on some electronic boards specifically designed for industrial environments (see Subsect. 2.1);

Motes and Operating System: WEVA employs low-cost commercial motes and relies on open-source operating system and development frameworks;

Protocols: the communication protocol stack is entirely based on standard protocols, i.e. IEEE802.15.4 at the physical layer, the IPv6 at the network layer and the CoAP protocol at the application layer, as shown in Fig. 1. Nevertheless, a new mechanism has been introduced to improve reliability (see Subsect. 2.3).

Access Gateway: an Access Gateway (AG) acts as a firewall to control and log accesses; the AG is based on a TelosB mote and a low-cost wireless router, i.e. TP-Link MR3020 [8], where OpenWrt [9] has been installed and configured. Thus, AG allows communication between WSAN motes and WiFi enabled devices that do not support the 802.15.4 protocol.

Services and Applications: WEVA provides a graphic management tool that simplifies setup, monitoring and maintenance of WSANs (see Subsect. 2.4).

2.1 Sensor and Actuator Boards

Three electronic boards have been specifically designed for WEVA:

(1) WEVA-SB (Sensor Board): it provides temperature, humidity and light intensity sensors for environmental monitoring;
(2) WEVA-SM (Smart Meter): it provides smart metering functions to monitor energy consumptions of machines and motors in industrial plants.
(3) WEVA-AB (Actuator Board): it performs all control actions (ON/OFF, PWM and PID) which are needed in industrial environment. Specifically, an on-board 12-bit DAC can be driven by a PID algorithm whose parameters can be remotely reprogrammed. WEVA-AB prototype is shown in Fig. 1.

2.2 Motes and Operating System

Costs, power consumption, computing resources, and coverage range have been the main targets taken into account to select mote platform for WEVA. Among 802.15.4-based motes that have been analysed, IRIS platform [10] has been chosen due to its low-power consumption and its high communication range. To the other hand, a TelosB [11] mote has been preferred for the Access Gateway implementation due to its USB interface and larger RAM respect to IRIS.

An Operating System (OS) is essential for an efficient management of more processing and sensing resources. Today, TinyOS [12] and Contiki [13] are the most used OSs for WSAN applications. However, TinyOS has been preferred to achieve low power consumptions [14].

2.3 Communication Protocols

Figure 1 shows the WEVA protocol stack. As can be observed, all protocols are open and well documented with the exception of the transport layer protocol where a mechanism has been introduced to improve network reliability. For sake of space, only Application layer (i.e. the CoAP protocol) and the Transport layer (i.e. CRT) will be described in details.

CoAP: The Constrained Application Protocol (CoAP) was standardized by IETF as RFC7252 [15] with the main aim to extend the REST architecture, used by web services, to M2M communications. Like HTTP protocol, CoAP uses a request-response mechanism based on four types of messages (GET, PUT, POST and DELETE). Moreover, CoAP defines a *discovery* mechanism that can be used to discover node resources by exploiting a conventional HTML browser (see Fig. 2). Response to the discovery request is the list of available resources (i.e. URI addresses: *dac, k, ...*) that can be also accessed by GET and PUT messages. Thus, CoAP allows to interact with motes and their resources just like with web pages. In particular, PUT messages can be used to configure all parameters (e.g.: duty-cycling period *lpl*, encryption key *k*, etc.). All motes in WEVA support the CoAP protocol and thus the CoAP *observation* mechanism [16], i.e. a resource can be marked as observable so that nodes will be informed when it changes.

Fig. 2. CoAP: discovery request

CRT: To improve network reliability, a new forwarding technique has been integrated in WEVA, which relies on the Chinese Remainder Theorem (CRT) [17]. CRT can be applied in WSANs to split packets into smaller components (i.e. CRT components) and to send them through different paths. Especially, CRT parameters can be chosen so that original packets can be reconstructed even though only a few CRT components are received [18]; this way, CRT can be used as a Forwarding Error Correction (FEC) technique to improve reliability.

Table 1. ARQ vs. CRT

Transmission technique	Parameters	Energy consumptions (J)	PLR
ARQ	$N_{r,max} = 1$	0.0157	7.60%
CRT	$N_{CRT} = 5, f = 2$	0.0158	0.63%
ARQ	$N_{r,max} = 2$	0.0195	4.40%
CRT	$N_{CRT} = 5, f = 3$	0.0203	0.03%

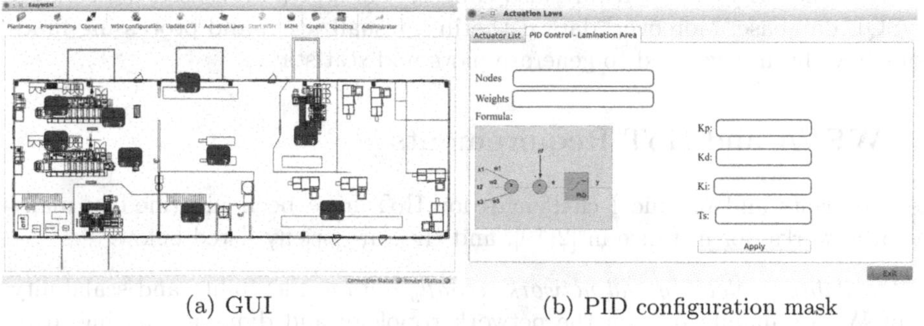

(a) GUI (b) PID configuration mask

Fig. 3. EasyWSN

Experimental results on energy consumptions per node and Packet Loss Ratio (PLR), i.e. percentage of lost packets, for both CRT and ARQ techniques are reported in Table 1. Tests have been carried out for a 3-hop WSAN with 12 motes (one source node, one destination node, and two 5–node clusters) where 5000 packets have been sent. In ARQ, the maximum number of retransmissions $N_{r,max}$ has been changed and a perfect load balancing has been realized. In CRT, source node packets have been split in $N_{CRT} = 5$ components and CRT parameters have been set to reconstruct packets even though at most $f \in \{1, 2, 3\}$ components have been lost. As observed in Table 1, CRT allows to reduce PLR by up to two orders of magnitude.

2.4 Services and Applications

EasyWSN is the graphic management tool of WEVA that allows to simplify setup, configuration, monitoring and maintenance operations. As illustrated in Fig. 3(a), EasyWSN relies on a Graphic User Interface (GUI) that is able to show nodes in actual industrial environment; moreover, GUI allows to configure node parameters (duty-cycling period, sampling time, etc.) and to retrieve information about node state (battery level, link quality, etc.) or available resources. Moreover, EasyWSN allows to specify actuation laws. For instance, as shown in

Fig. 3(b), a configuration mask allows to set PID parameters (K_i, K_d and K_p) to generate control signals according to:

$$y[n] = K_p \cdot e[n] + K_i \cdot \sum_{k=0}^{n} e[k] + K_d \cdot (e[n] - e[n-1]) \tag{1}$$

Moreover, a list of nodes and a set of related weights can be specified by means of the same interface. In this case, the error signal $e[n]$ in Eq. (1) is obtained as a weighted sum of the received data, thus distributed controls are possible in WEVA. The EasyWSN software also collects and stores all sensed data into a MySQL database. Moreover, data gathering, visualization and processing functions have been developed to generate plots and statistics.

3 WEVA and IIoT Requirements

Requirements and technical challenges for IIoT have been identified in many research works, for instance in [2,19], and they are briefly listed below:

- *Flexibility, scalability and network reconfiguration:* Flexibility and scalability of WSAN mainly rely on the network topology and dynamic routing algorithms. Among the most common WSAN topologies, mesh topology should be preferred because it offers multi-path communications which are useful to increase network reliability. Nevertheless, mesh networks need that all devices are able to act as a router. All nodes in WEVA employ RPL [20] protocol thus they can act as routers, providing energy-efficient mesh networks with a standard Internet protocol.
- *Security:* WEVA provides security mechanisms at Data-Link Layer, Transport layer and Application layer. Especially, a 128-bit AES-like encryption algorithm has been implemented for IRIS motes; encryption can be enabled on the per-node basis, and keys can be managed by the EasyWSN software. Moreover, the WEVA Access Gateway acts as a firewall so that accesses can be controlled and logged. Finally, CRT can also reduce the risk of eavesdropping attacks in that CRT parameters must be known in advance and a sufficient number of components must be also collected.
- *Reliability:* Several IoT wireless solutions (ZigBee, WirelessHART and ISA100) use ARQ mechanisms to improve reliability. However, an increasing number of retransmissions has as counter effect on the network latency and this could be a serious issue in real-time control applications. Vice versa, WEVA is based on CRT, thus it is able to decrease the PLR also of several orders of magnitude without impact on network latency.
- *Energy consumptions:* Low power motes have been specifically chosen in WEVA with the aim to minimize power consumptions. Moreover, further reduction in energy consumptions can be obtained by combining CRT with duty-cycling techniques [21] and distributed compression algorithms [22].
- *Availability of Management tools:* WEVA is enhanced by a comprehensive graphic management tool (i.e. EasyWSN) suitable for distributed controls.

– *Discovery mechanisms:* Unlike other wireless solutions, WEVA provides a discovery mechanism based entirely on a standard protocol, i.e. CoAP. As shown in Subsect. 2.4, not only nodes but also their resources can be easily discovered and accessed by means of a web browser.

4 Conclusions

In this paper a new complete solution for Industrial Internet of Things (IIoT), has been presented. WEVA is based on open source software and communication protocols; it encompasses reliability, a discovery mechanism and a graphic tool for WSANs set-up, monitoring and maintenance. As future works a more detailed comparison between WEVA and other wireless solutions will be carried out.

References

1. Beudert, R., Juergensen, L., Weiland, J.: Understanding Smart Machines: How They Will Shape the Future. Schneider-Electric, white paper 2015
2. Xu, L.D., He, W., Li, S.: Internet of Things in industries: a survey. IEEE Trans. Ind. Inf. **10**(4), 2233–2243 (2014)
3. Wollschlaeger, M., Sauter, T., Jasperneite, J.: The future of industrial communication: automation networks in the era of the Internet of Things and industry 4.0. IEEE Ind. Electron. Mag. **11**(1), 17–27 (2017)
4. Hou, L., Bergmann, N.W.: System requirements for industrial wireless sensor networks. In: 2010 IEEE 15th Conference on Emerging Technologies Factory Automation (ETFA 2010), pp. 1–8, September 2010
5. Sasajima, H., Ishikuma, T., Hayashi, H.: Future IIOT in process automation - latest trends of standardization in industrial automation, IEC/TC65. In: 54th Annual Conference of the Society of Instrument and Control Engineers of Japan (SICE). Hangzhou, pp. 963–967 (2015)
6. Akyildiz, I.F., Kasimoglu, I.H.: Wireless sensor and actor networks: research challenges. Ad Hoc Netw. **2**(4), 351–367 (2004)
7. Vogli, E., Alaya, M.B., Monteil, T., Grieco, L.A., Drira, K.: An efficient resource naming for enabling constrained devices in smartM2M architecture. In: 2015 IEEE International Conference on Industrial Technology (ICIT), pp. 1832–1837, March 2015
8. TL-MR3020: DataSheet. http://static.tp-link.com/TL-MR3020(EU)_V3_Datasheet.pdf
9. OpenWrt. https://openwrt.org/
10. Iris: DataSheet. http://www.memsic.com/userfiles/files/Datasheets/WSN/IRIS_Datasheet.pdf
11. TelosB: DataSheet. http://www.memsic.com/userfiles/files/Datasheets/WSN/telosb_datasheet.pdf
12. TinyOS. http://tinyos.stanford.edu/tinyos-wiki
13. Dunkels, A., Gronvall, B., Voigt, T.: Contiki - a Lightweight and Flexible Operating System for Tiny Networked Sensors. In: Proceedings of the First IEEE Workshop on Embedded Networked Sensors, Tampa, Florida, USA (2004)
14. Reusing, T.: Comparison of operating systems TinyOS and Contiki. Sens. Nodes-Oper. Netw. Appl. (SN) **7**, 7–13 (2012)

15. RFC7252: The Constrained Application Protocol (CoAP). https://tools.ietf.org/html/rfc7252
16. Bormann, C., et al.: CoAP: an application protocol for billions of tiny internet nodes. IEEE Internet Comput. **16**(2), 62–67 (2012)
17. Campobello, G., et al.: On the use of Chinese remainder theorem for energy saving in wireless sensor networks. In: Proceedings of IEEE International Conference on Communications, ICC, 2008, Beijing, China, 19–23 May 2008, pp. 2723–2727 (2008)
18. Campobello, G., Segreto, A., Serrano, S.: Data gathering techniques for wireless sensor networks: a comparison. Int. J. Distrib. Sens. Netw. **12**(3), 1–17 (2016)
19. Rotondi, D., Galipó, A., et al.: Iot@work d1.3 - final requirements and reference architecture. Technical report (2013)
20. RFC6550: IPv6 Routing Protocol for Low-Power and Lossy Networks. https://tools.ietf.org/html/rfc6550
21. Campobello, G., et al.: Trade-offs between energy saving and reliability in low duty cycle wireless sensor networks using a packet splitting forwarding technique. EURASIP J. Wirel. Commun. Netw. **2010**, 1–11 (2010)
22. Campobello, G., et al.: Applying the Chinese remainder theorem to data aggregation in wireless sensor networks. IEEE Commun. Lett. **17**(5), 1000–1003 (2013)

Validating Contact Times Extracted from Mobility Traces

Liu Sang[1], Vishnupriya Kuppusamy[2], Anna Förster[2], Asanga Udugama[2], and Ju Liu[1](✉)

[1] School of Information Science and Engineering, Shandong University, Jinan, China
liusang66304@gmail.com, juliu@sdu.edu.cn

[2] Sustainable Communications Networks Group, University of Bremen, Bremen, Germany
{vp,afoerster,adu}@comnets.uni-bremen.de

Abstract. Use of mobility models to model user movement in mobile networks is a key aspect when developing and evaluating networking protocols in simulators. A trace obtained from an actual user movement is considered as being more realistic than using synthetic mobility models in simulators. Though realistic, usually, these traces lack information about the actual wireless contact durations between users. Most simulators use Unit Disk Graph (UDG) model to determine contact durations. However, due to the nature of radio propagations, a simplistic connectivity model (with UDG) may result in simulating unrealistic connectivity patterns. In this work, we have used an Android Smartphone application to collect GPS traces of moving users and their corresponding Bluetooth Low Energy (BLE) contact times to compare the viability of using UDG to determine contact durations. The results show that trace based model with UDG based wireless connectivity is an effective method to determine contact durations.

Keywords: Network simulation · Mobility model · UDG · Contact times · Wireless range · UDG connectivity model

1 Introduction

The significant increase in the number of Internet of Things (IoT) devices has attracted a lot of attention from researchers in the past few years. A class of IoT devices that constantly move forms highly dynamic networks like opportunistic networks (OppNets), where users generally perform peer-to-peer wireless information exchanges when they are in the communication range of each other. Network simulators are widely used to evaluate such dynamic networks, as they provide the support to model real networks and flexibility to configure the scenario specific details as required. A contact time trace is a log of actual contact durations of users over a time period, which is used in simulations to imitate the contacts from real-world in dynamic networks. Unfortunately, there are only

A. Puliafito et al. (Eds.): ADHOC-NOW 2017, LNCS 10517, pp. 239–252, 2017.
DOI: 10.1007/978-3-319-67910-5_20

a limited number of contact traces available for use in public repositories such as CRAWDAD [1]. An alternative to using contact traces is a combination of suitable mobility models and connectivity models in simulations. Mobility models are used to simulate user movement and connectivity models determine if the users are in contact range of others.

When considering mobility in simulations, the use of actual mobility traces to move users in a network is usually considered more realistic than using of synthetic mobility models [2]. Consequentially, many of the simulators, especially in the area of OppNets, use publicly available Global Positioning System (GPS) based mobility traces to build contact times [3,4]. Furthermore, these simulators use the Unit Disk Graph (UDG) model to decide connectivity between users instead of the sophisticated propagation models such as Rician, Rayleigh, Nakagami, etc. However, is the UDG model sufficiently realistic enough to determine the wireless connectivity between the users?

As for OppNets, UDG seems to be practical and a reasonable fit due to the network characteristics exhibited by OppNets. These networks use short range communications through technologies such as Bluetooth where the connectivity is likely to be less than 100 m. The application specific scenarios could involve contact times of a few minutes during which the channel changes that occur may not adversely affect the communications. Due to these reasons, the scenarios involving severe path-loss and fading are uncommon and the necessity of using such complex connectivity models for OppNets becomes questionable.

Further, using UDG for OppNets simulations facilitates a huge advantage in terms of scalability for IoT based scenarios. As IoT scenarios involve massive numbers of mobile devices, use of a simple connectively model like UDG will generate far lesser number of simulation events compared to the events generated by a comprehensive propagation model and thus, will decrease the time required to complete simulations. Hence, it becomes essential to address the question whether the UDG model is suited for determining connectivity and contact times in the OppNets simulations.

Therefore, the major contribution of this work is to verify the applicability of UDG model in determining contact times in simulations when using GPS-based traces. The work presented here can be summarized as follows:

- We deploy an Android application on the Smartphones of a set of users to collect GPS-based mobility traces and the corresponding Bluetooth Low Energy (BLE) contact times over a period of time, and perform field measurements for Bluetooth range and GPS accuracy in the real-world.
- A trace based mobility model and a UDG based connectively model are implemented in OMNeT++. The collected GPS-based mobility traces are used in the simulation models to determine the contact times.
- These contact times are then compared with the actual BLE contact times. The results from the simulations are compared with BLE traces to verify the usefulness of the contact times extracted from mobility traces.

The rest of this paper is organized as follows: Sect. 2 discusses about the related work. The experiment setup, data collection and processing are explained in Sect. 3. The implementation-specific details are described in Sect. 4. Section 5 presents our findings from the simulations and the result analysis. Section 6 concludes this paper.

2 Related Work

Of the various available models to describe wireless connectivity, UDG is the most simplest model. It assumes that two users can contact each other, if the distance between them is less than a threshold. The UDG model does not account for signal attenuation or any other physical layer modeling. Hence, it is believed to be too simplistic and expected to produce unrealistic results. In ad-hoc networks, UDG is not seen as a good fit to simulate the connectivity of the users as the authors in [5, 6] claim that it produces sub-optimal results. Design principles for realistic simulations of ad-hoc networks are suggested in [7], which suggests to use better connectivity models than UDG.

Although ad-hoc networks and OppNets have common characteristics, the delay bounds and connectivity in OppNets are relaxed compared to ad-hoc networks. Thus, in simulation of opportunistic networks, we find that UDG is widely used as the connectivity model for analysis of user contacts derived from GPS traces. For example, the authors in [8] study the human mobility from GPS traces. They used the UDG model with a wireless range of 25 m to obtain contact times. In [9], authors assumed a scenario using UDG model with GPS enabled devices to design a localization scheme. The maximum wireless range of UDG model was set to 10 m. However, the reason for using UDG or the specific wireless range is not given.

Contact traces are obtained either by using Bluetooth or by Wi-fi indoor positioning systems. Plenty of research has been done to identify if Bluetooth based connections and contact times are efficient enough to be used in OppNets. For example, the limited contact traces available in CRAWDAD are based on Bluetooth [10]. The experiments on identifying social groups and the analysis of human mobility [11] use Bluetooth for getting contact times and the authors have been quite successful in arriving at their goal. Further, the reliability of Bluetooth contact times and the use of this technology itself for analyzing social contacts has been studied in [12]. Since a large part of the world population carries Smartphone with it everywhere, Bluetooth has emerged as one of the wireless technologies for opportunistic data exchanges.

Wi-fi based connectivity traces are used to study the association between different users in [13]. Any two users under the same Wi-fi access point are assumed to be in the same wireless range and able to communicate with each other. Although this may not be necessarily true due to the devices not exhibiting good connectivity conditions, the method of checking wireless connectivity is similar to the UDG model. Further, UDG is also used in simulators such as Adyton [4] and ONE [14], which are primarily opportunistic network simulators. Despite the wide usage of UDG as a connectivity model in this field, its usage has neither been justified with theoretical reasons nor been validated with simulation results.

In this work, we probe one strand of this thread as our main goal and check the validity of using UDG in simulations for determining accurate contact times.

3 Data Collection and Management

In order to evaluate the contact times, we collect mobility traces of mobile users and the corresponding contact time traces by converting them into an appropriate format in this section.

3.1 BluetoothContacts Application

An Android application named BluetoothContacts is used for our data collection. The application is developed by Sustainable Communication Networks (ComNets) of University of Bremen, Germany which is available in Google Play Store[1]. BluetoothContacts regularly logs all BLE devices in a user's proximity as well as the user's own GPS coordinates. The BLE module in user's phone is used through an API to record received beacons from nearby BLE devices and the GPS coordinates are updated every 10 min by the GPS module. The user sends the BLE and GPS data to ComNets via email.

The Bluetooth beacons are formatted to identify device ID for advertisement of BLE devices. The BLE transmitter broadcasts these beacons all the time and BLE scanner scans the beacons within its range at the same time. It is to be noted that the scanning interval of beacons is not preset in order to save the battery lifetime.

3.2 Experiment Setup

A number of volunteers (22 in total) were invited to install BluetoothContacts application on their personal Smartphones and requested to send an email to ComNets every few days. The data collection lasted for a duration of six weeks. Due to non-availability of BLE support from some mobile phones, these users only send GPS data to ComNets. Meanwhile, some users hardly met others during these six weeks. Their data are not useful for our research as it lacks valid contact times.

At the end, there were still 5 users left and one week of their data was selected for analysis. All of these users are colleagues, working in the same building. During this week, users mostly went to office every weekday and had lunch together for approximately two hours. This data is useful for our research analysis because the users' mobility and behavior can be regarded as a small opportunistic network.

[1] BluetoothContacts: https://play.google.com/store/apps/details?id=de.uni_bremen.comnets.BluetoothContacts, developed by Jens Dede and Sarmad Ghafoor.

3.3 Data Conversion

GPS Data. It is well known that GPS coordinates are composed of latitude and longitude and, 3-dimensional coordinates even have altitude. They all belong to the spherical coordinate system, while the simulation uses Cartesian coordinate system. Hence, the GPS coordinates obtained from users' data are converted into Cartesian coordinates. Gauss-Krüger projection [15] is used for this conversion, which is one of the transverse Mercator projections and widely used in national and international mapping systems around the world.

Besides the conversion of GPS coordinates to Cartesian coordinates, all the single user GPS data files are put together in a single file with fields titled time, user ID, X coordinate and Y coordinate. At the same time, by searching the maximum value of X and Y coordinates, the first line of final file is marked to confirm the simulation range.

BLE Data. As the BLE module only records beacons received time and device ID but not the real contact times, a parser is used to calculate contact times between different neighbours from the received time of the beacons. Since the scanning interval is not fixed and beacons can be lost in a complex environment, the contact threshold is set to 90 s, i.e., if the next beacon received time is 90 s more than current beacon time, the contact of the two users had been momentarily interrupted.

Similarly, all the single user BLE data are collected together to form a large file with the fields titled contact start time, contact end time, user ID, neighbour ID and contact continuing time.

4 Implementation

This section describes the simulation models implemented in OMNeT++ to simulate our opportunistic network. OMNeT++ [16] is an open-source modular network simulator. It is bundled with a model framework called INET to perform simulations using networking protocols such as the Internet Protocol (IP). In this section, we extend the functionality of the INET framework to implement a trace based mobility model and a simple UDG model as described below.

4.1 Trace-Based Mobility Model

Trace based mobility model is implemented by updating users' positions at different time as per the trace file to simulate their movement. Position of every user is initialized according to their first location coordinates from the trace file. The subsequent movements for a user are implemented as follows: The user obtains next position and arrival time from trace file, firstly. The average speed of the user at any time slot is the distance between his current and next position divided by the duration of time slot between the two location points. Then the user moves along the direction of his next position at calculated average speed until arriving at the next position. When the user reaches to the new position, his next position and arrival time are read from trace file again and this approach is carried forth for all the data points. If there is no more position left in trace file, the user stays at present position waiting.

The user movement type described above is similar to the line segment mobility model which is readily available in INET framework. Hence, the existing Line-Segment mobility model is inherited as the movement method and we develop our own trace based mobility model to get user positions by inputing the trace file.

4.2 UDG Connectivity Model

As the wireless connectivity of users in UDG model only depending on wireless communication range, neighbours of a user and their corresponding contact times can be changed by adjusting the wireless range.

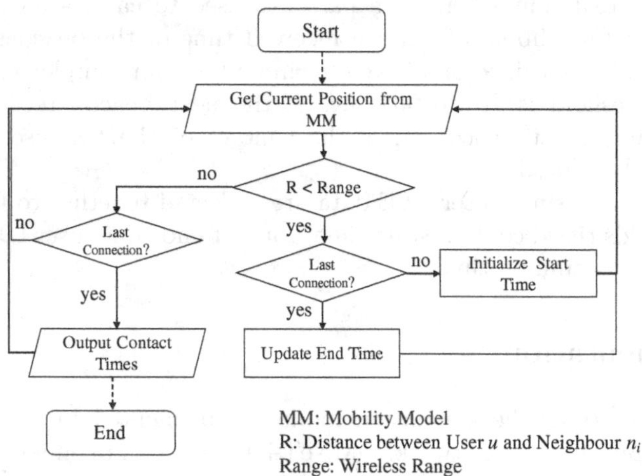

Fig. 1. UDG model flow chart

Figure 1 shows the flow chart of the UDG model, for each user, where the mobility model (MM) is called firstly to get the current position and calculate the distance between the user u and his neighbours $n_i (i = 1, 2, ..., n)$. For instance, when the distance between u and n_i is within the wireless range, the last connection situation of u with n_i is checked. If the last connection situation is false, then this is a new contact of u with n_i. Hence, the contact start time of u with n_i is initialized to current time. Otherwise, the contact end time of u with n_i is updated to current time. Similarly, when the distance between u and n_i is larger than the wireless range, the last contact situation of u with n_i also needs to be checked. If their last connection situation is true, it implies that n_i is going out of range of u now. We store these contact times in an output file, formatted similar to the output file of Android BLE data file. On the other hand, if the last connection between u and n_i is false, it means the n_i has left the wireless range for a while, then we wait until BLE of u starts its next scan for neighbours.

In selecting the parameters of UDG, the neighbour scanning interval is set to 90 s to maintain consistency with the BLE data. The last contact situation of all the neighbours for any user is initialized to be false because there is no contact at the start of the simulation.

The network consists of users which are built up by existing node model that is configured with the mobility and connectivity models. All the parameters are identified in a configuration file.

5 Simulation Results and Analysis

After the implementation of models in the simulator, a set of field measurements were taken to further understand our experimental environment. Based on the knowledge gained, the simulation scenario is set up as close as possible to the real environment. The related simulation results obtained and their analysis are discussed in this section.

5.1 Field Measurements

Real Wireless Range. Due to the presence of walls and obstructions in the indoor environment, the broadcast of BLE beacons are hindered and as a result, the wireless range is smaller compared to an outdoor environment. For outdoor measurement, we chose an open space to measure outdoor BLE wireless range. The results show that the maximum value of wireless range to be at most 100 m, while beacons may get lost if the distance is beyond 80 m. The indoor measurements were done in separate rooms and corridors which had various obstructions. In this environment, wireless range is limited to between 15 m and 25 m.

GPS Accuracy. To obtain an understanding of the mobility traces, the measurements on GPS position accuracy and distribution of users' distances are carried out. We select 7 different positions to measure the distance from a given point (origin). Figures 2 and 3 show the GPS accuracy indoors and outdoors, respectively. Though the GPS traces follow the same pattern as manual measurements, some deviation exists between them and the GPS data from outdoor environment has higher location accuracy than the GPS data from the indoor environment.

Distribution of Mobility Traces. Figure 4 shows the histogram of distance between mobile users (from the data collected through the Smartphone application). The frequencies of 30 m, 50 m and 70 m are significantly higher than others, which implies that the users stay in positions for long time that are separated

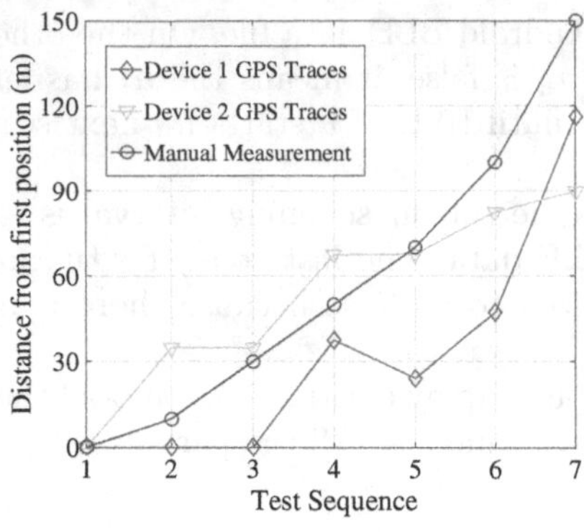

Fig. 2. Indoor GPS Accuracy.

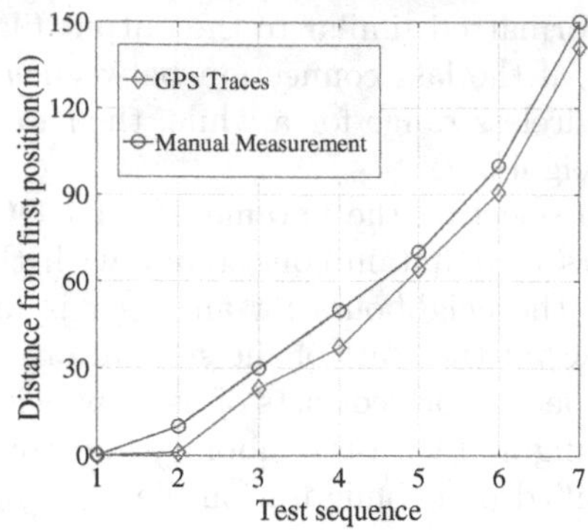

Fig. 3. Outdoor GPS Accuracy.

from each other by 30 m, 50 m or 70 m. Except these points, the frequency of the other distances is almost uniform, meaning that users most probably just pass away when they get to these distances rather than staying somewhere at a high frequency for a long time. It is worthwhile noting that the minimum distance can be less than 5 m so users can be situated very closed to each other.

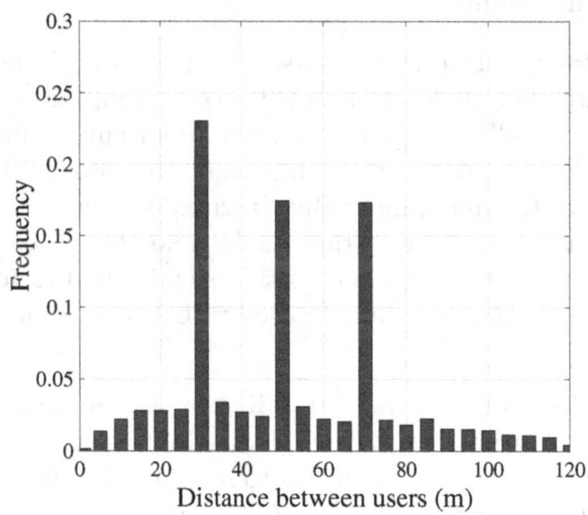

Fig. 4. Histogram of distance between mobile users.

5.2 Simulation Scenario Setup

Since the valid collected data were from 5 users, the number of users in the simulation model is also set to 5. For the mobility of the nodes, we use the trace based mobility model that we developed in this work. The model is given the trace file that we generated from the collected data as the input.

An existing node model of OMNeT++ is configured to use the UDG based network interface model developed for this work. There are two parameters to be configured. One parameter is *wirelessRange* which affect the contact times directly in the simulation. We set *wirelessRange* from 1 m to 120 m in different scenarios to compare the contact times between BLE data and simulation results. The other parameter is *neighbourScanInterval*. As we discussed above, for the robustness of contact times and to maintain consistency with formatted BLE data, 90 s is set as *neighbourScanInterval*. Table 1 shows the important parameters of the models.

Table 1. Network Parameters

Parameters	Purpose	Value
numHosts	The number of users in network	5
wirelessRange	Maximum wireless range for obtaining contact times	selected between 1 and 120 m
neighbourScanInterval	Time interval used to check the update of neighbours	90 s
nodeId	The ID of user, -1 represents automatically obtaining user ID	-1
mobilityType	The type of mobility model	traceBasedMobility
is3D	The coordinates are 3 dimension	false
traceFile	The name of trace file	CartesianTraces.txt

5.3 Results and Analysis

Prior to analyzing contact times, we check the behavior of neighbour availability between two users to determine the usefulness of UDG model. Figure 5 shows four separate neighbour availability graphs. *True* refers to the situation when a user has contact to a neighbour. BLE1 and BLE2 are named to distinguish different users using BLE data to determine neighbour availability for each other while Sim represents these two users' neighbour availability in simulations. The graphs show the duration of contact and the different possibilities of the contact time shift. Due to the update interval for GPS coordinates, the neighbour contact from the simulation are delayed. The shift possibilities are where, (a) simulation is earlier, (b) later, (c) small-size shift movement or (d) almost the same compared with BLE data. Ideally, neighbour availability should be the same between two users with BLE. But in Fig. 5, the two BLE users have different neighbour contact times and durations. A separate investigation found that BLE interfaces perform differently in different Smartphones and hence, the asymmetry of contact times and durations is existent. So, for the analysis, we consider three cases of BLE data according to contact conditions.

Because of the lack of uniformity of BLE data as we mentioned in Fig. 5, the BLE devices of two users at the same time usually have different contact times.

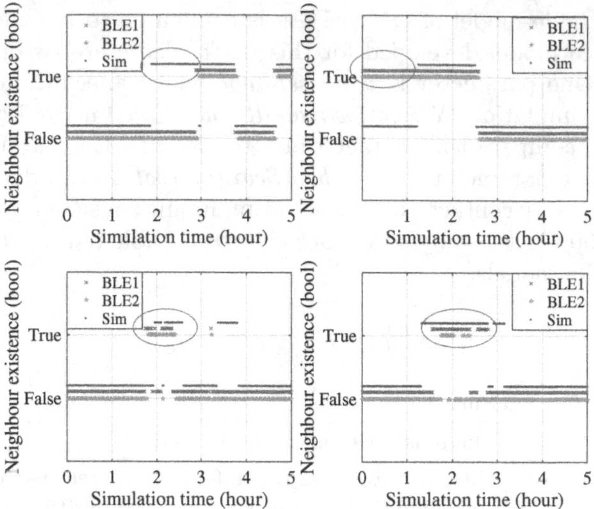

Fig. 5. Four Different Neighbour Availability Situations between Two Users - based on BLE and Simulations (with GPS Traces)

For a pair of users, the longer contact times are selected as BLE contact time *better case* while the shorter contact times are named *worse case*. *Average case* is acquired by getting the mean value of two cases.

Along with comparing the results, we use statistic measures to understand the usability of the data. A metric named Absolute Difference of contact times (AbsD) is defined to validate contact times extracted from mobility traces. In short, AbsD is the mean value of deviation between contact times from simulation and contact times from BLE data for all the users. There are N users in the network and the whole simulation time for each user is divided into K parts. As the Eq. 1 shows, $T_{S_{ij}}$ and $T_{B_{ij}}$ represent the contact times of the j_{th} part of i_{th} user in the simulation network and BLE data, respectively. T_{ij} means the simulation time for i_{th} user's j_{th} part. $|a|$ is the absolute value of a. All the i and j are traversed to obtain the absolute difference of contact times for the whole network's simulation duration. Once the absolute differences for each user pair is computed, these multiple absolute differences are averaged.

$$AbsD = \frac{1}{NK} \sum_{i=1}^{N} \sum_{j=1}^{K} \frac{|T_{S_{ij}} - T_{B_{ij}}|}{T_{ij}} \tag{1}$$

Figure 6 shows three AbsD curves, obtained as per *better case* device, *worse case* and *average case*. All the curves have a slight decline at the beginning because the expansion of wireless range in simulation allows mobile users to maintain contact prolonging their simulation contact times. In doing so, the contact times from simulation get closer to real BLE contact times for wireless ranges of 20 m to 30 m, thus achieving the lower bound. It means that wireless

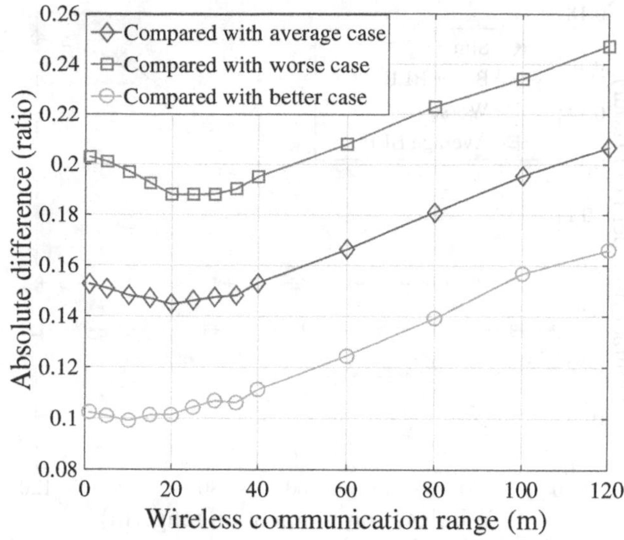

Fig. 6. Absolute Difference of Contact Times against Wireless Range.

ranges between 20 m–30 m are more suitable for reducing the absolute difference so as to get closer to real contact times. When the wireless range is larger than 30 m, AbsD has a significant growth with the increase in wireless range. The reason is that, too long wireless range translates to users being in contact in simulation even though they may not be having contact in reality. Further, even with the worse case, the absolute difference is less than 0.2 which is a quite good result to validate contact times extracted from mobility traces.

Figure 7 shows average contact times per hour for different wireless ranges. We sum the total contact times of available user pairs and it is divided by the sum of total simulation time of available user pairs. The curve represents the changes of average contact times in simulation results compared with the constant BLE lines. Similarly, three constant lines are used to represent *better case*, *worse case* and *average case* of BLE contact times per hour, respectively. These straight lines don't change with the increase in wireless range because all of them are users' real contact times and they aren't affected by simulated range. As these constant values come from real-world, we assume the values of *better BLE* and *worse BLE* are the upper and lower bounds of real contact time. Therefore, the suitable wireless range should be selected from 30 m to 40 m to get similar average contact times per hour from the simulation and the BLE.

In summary, Figs. 6 and 7 evaluates the experiment from different perspectives, but both of them are sensible. Absolute difference of contact times tells us the instantaneous deviation between simulation and BLE at every contact moment for each user while average contact times represents the mean value of the whole network duration of contact times. Because of the GPS accuracy, the contact delay and other factors, deviation of results always exists in our simulation but can be

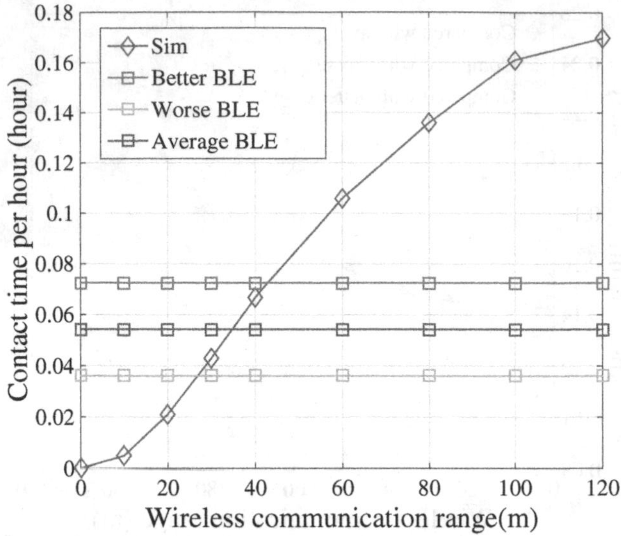

Fig. 7. Average Contact Times per Hour against Wireless Range.

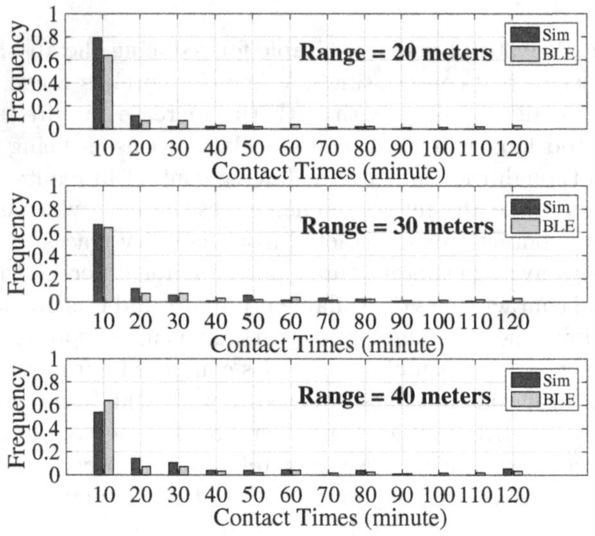

Fig. 8. Histogram of Contact Times. (Range = 20 m, 30 m and 40 m)

decreased by adjusting the wireless range. It is worth noting that in either view, the contact times are realistic compared with BLE data.

Further, Fig. 8 shows the contact times frequency histogram for determining a suitable wireless range for our scenario. We set the wireless range as 20 m, 30 m and 40 m because they are suitable values from the views of absolute difference of contact times and average contact times. Under 30 m wireless range setting, the frequency of contact times in simulation almost has the same distribution as BLE's. In other words, it validates that the trace based mobility model with UDG connectivity model is realistic in a new view using the distribution of contact times. Mostly, BLE contact times are less than 10 min as shown by Fig. 8. We are certain of the lost beacons bringing about this situation according to the BLE data recorded.

6 Conclusions and Future Work

In this work, we have verified whether the UDG is a suitable connectivity model for determining contact times using GPS traces in simulators. In other words, we have validated whether the contact times extracted from GPS traces using UDG model are realistic to be used in simulations.

Through an Android Smartphone application, we collected GPS and BLE data to compare the contact times measured by Bluetooth and contact times obtained from simulations. A trace-based mobility model and UDG model are implemented in OMNeT++ to calculate contact times extracted from mobility traces.

From our simulation results, it is found that the contact times from simulation based on UDG model follow the same pattern as contact times from Bluetooth traces. Hence, it is verified that the UDG is effective as a connectivity model and thus, is suitable to extract contact times from GPS traces in simulations. Under such experiment scenario, an appropriate value of wireless range is selected in the simulation making it as realistic as possible.

Since the UDG model is found to be an effective connectivity model, it can be further evaluated in the context of large scale networks such as with certain scenarios of the IoT. Such an environment is influenced by a myriad of parameters and movement traces that ultimately are plus points for using UDG based model.

Acknowledgements. This work was partly supported by the National Natural Science Foundation of China (61371188). Ju Liu is the contact author of this paper. Liu Sang was supported by the China Scholarship Council for a year of study at the Sustainable Communication Networks, University of Bremen, Germany.

References

1. CRAWDAD: A community resource for archiving wireless data at dartmouth. http://crawdad.org/. Accessed 4 Apr. 2010
2. Kim, M., Kotz, D., Kim, S.: Extracting a mobility model from real user traces. In: Proceedings of INFOCOM 2006, 25th IEEE International Conference on Computer Communications, pp. 1–13. IEEE (2006)
3. Herrera-Tapia, J., Hernández-Orallo, E., Tomás, A., Calafate, C.T., Cano, J.C., Zennaro, M., Manzoni, P.: Evaluating the use of sub-gigahertz wireless technologies to improve message delivery in opportunistic networks (2017)

4. Papanikos, N., Akestoridis, D.G., Papapetrou, E.: Adyton: a network simulator for opportunistic networks (2015). https://github.com/npapanik/Adyton
5. Khadar, F., Simplot-Ryl, D.: Connectivity and topology control in wireless ad hoc networks with realistic physical layer. In: Third International Conference on Wireless and Mobile Communications, ICWMC 2007, p. 49. IEEE (2007)
6. Stojmenovic, I.: Simulations in wireless sensor and ad hoc networks: matching and advancing models, metrics, and solutions. IEEE Commun. Mag. **46**(12), 102–107 (2008)
7. Stojmenovic, I., Nayak, A., Kuruvila, J.: Design guidelines for routing protocols in ad hoc and sensor networks with a realistic physical layer. IEEE Commun. Mag. **43**(3), 101–106 (2005)
8. Zignani, M., Gaito, S., Rossi, G.: Extracting human mobility and social behavior from location-aware traces. Wirel. Commun. Mobile Comput. **13**(3), 313–327 (2013)
9. Yadav, V., Mishra, M.K., Sngh, A., Gore, M.: Localization scheme for three dimensional wireless sensor networks using gps enabled mobile sensor nodes. Int. J. Next-Gener. Netw. (IJNGN) **1**(1), 60–72 (2009)
10. Akestoridis, D.G.: CRAWDAD dataset uoi/haggle (v. 2016–08-28): derived from cambridge/haggle (v. 2009–05-29). http://crawdad.org/uoi/haggle/20160828
11. Chaintreau, A., Hui, P., Crowcroft, J., Diot, C., Gass, R., Scott, J.: Impact of human mobility on opportunistic forwarding algorithms. IEEE Trans. Mobile Comput. **6**(6), 606–620 (2007)
12. Cabero, J.M., Urteaga, I., Molina, V., Liberal, F., Martín, J.L.: Reliability of bluetooth-based connectivity traces for the characterization of human interaction. Ad Hoc Netw. **24**, 135–146 (2015)
13. Chery, P., Li, J., Burge III, L.L.: Characterizing the association between mobile users using wireless network traces. In: The Fifth Richard Tapia Celebration of Diversity in Computing Conference: Intellect, Initiatives, Insight, and Innovations, pp. 70–74. ACM (2009)
14. Keränen, A., Ott, J., Kärkkäinen, T.: The ONE simulator for DTN protocol evaluation. In: Proceedings of the 2nd International Conference on Simulation Tools and Techniques, ICST (Institute for Computer Sciences, Social-Informatics and Telecommunications Engineering), p. 55 (2009)
15. Deakin, R., Hunter, M., Karney, C.: The gauss-krüger projection. In: Proceedings of the 23rd Victorian regional survey conference (2010)
16. Varga, A.: Omnet++. Modeling and Tools for Network Simulation, pp. 35–59 (2010)

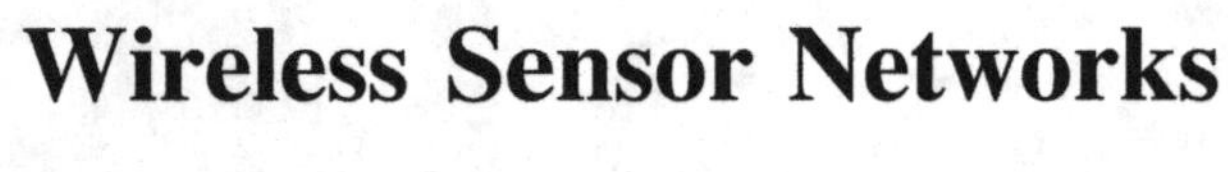

Wireless Sensor Networks

Optimising Wireless Sensor Network Link Quality Through Power Control with Non-convex Utilities Using Game Theory

Evangelos D. Spyrou(✉) and Dimitrios K. Mitrakos

School of Electrical and Computer Engineering, Aristotle University of Thessaloniki, Egnatia Odos, University Campus, 54124 Thessaloniki, Greece
{evang_spyrou,mitrakos}@eng.auth.gr

Abstract. The asymmetric and dynamic properties of the wireless channel required the derivation of a plethora of link quality metrics that assisted in the enhancement of wireless link reliability. Expected Transmission Count is a robust link quality metric used in most of the state-of-the-art works, due to its consideration of bidirectional links. Transmission power plays a key role to mitigate interference, which affects link quality. Power control problems, often exhibit non-convex behaviour and can not be solved using traditional methods. In this paper, we optimise transmission power taking onboard the Expected Transmission Count metric. We construct our game-theoretic model with pricing and show that it can reach Pareto-dominant equilibrium. Finally, we create a learning algorithm using logit dynamics and show that it guarantees probabilistic convergence of the joint action to the potential function maximisers.

Keywords: Non-convex utility function · Logit learning · Power control · Potential game · ETX

1 Introduction

Due to the asymmetric and dynamic properties of the wireless channel, a number of link quality metrics have been proposed [2,3], in order to enhance wireless link reliability. Transmission power plays a key role to mitigate interference; thus, maximising the quality of service and link quality. Non-convexity of utility functions under the interference model is a well-known problem, since the coupling between nodes' interference is rather complicated [4,10–13].

The literature provides us with distributed power control approaches, which accomplish local optimality; however, global optimal conditions cannot be guaranteed except in some special cases, such as log-concave utility functions [5,6]. Other works suggested game-theoretic approaches to find the optimal power allocation, where nodes act as selfish players when they wish to transmit [7,8]. Game theory studies mathematical models of conflict and cooperation between wireless devices. The rationality of a node is satisfied if it pursuits the satisfaction of its preferences through the selection of appropriate strategies.

A. Puliafito et al. (Eds.): ADHOC-NOW 2017, LNCS 10517, pp. 255–261, 2017.
DOI: 10.1007/978-3-319-67910-5_21

In this paper, we investigate the optimisation of link quality Expected Transmission Count (ETX) [2] through power control with a potential game approach [9]. Potential games are games where the incentive of players to change their strategy can be expressed in a single global function, the potential function. We prove that there is an equilibrium point and that we can reach the Pareto dominant one. We design an algorithm based on logit learning that guarantees probabilistic convergence of the joint action to the potential function maximisers.

The paper is structured as follows: Sect. 2 formally describes the system model and the potential game approach proposed, Sect. 3 presents the power control learning algorithm, Sect. 4 provides the results from the simulations and Sect. 5 presents the conclusions.

2 Utility Function of Game with Pricing

We consider a set of nodes $N = 1, ..., i$ and denote by their power allocations by $P = (p_1, ...p_i)$. We assume that each node i can adjust its transmission power p_i within a range and the P is a compact set.

$$0 \leq p_i^{min} \leq p_i \leq p_i^{max} \tag{1}$$

We design our utility function using the ETX link quality metric, whose relationship with the transmission power is given in [1]. ETX is widely known as the inverse of the probability of Packet Success Delivery given as

$$ETX_l = \frac{1}{PRR_{frwd} PRR_{bkwd}} \tag{2}$$

In this work, we employ pricing, which creates revenue for the network and it will promote utilisation of resources more efficiently by the nodes.

We define the utility function of the game and we normalise it take values between $[0, 1]$ to result in the optimisation problem

$$\max U_i = -ETX_{ij} + p_i \text{ s.t } 0 < p_i^{min} \leq p_i \leq p_i^{max}, p_i \in A \tag{3}$$

Note that this utility function is non-concave in p_i.

Lemma 1. *The game $\Gamma =< N, A, u >$ with the individual utility function (3) is an exact potential game with the potential function given by*

$$V(\mathbf{p}) = -\sum_m ETX_m + \sum_m p_m, p_m \in N \tag{4}$$

Proof. For a game to be a potential game it is straightforward to show that $\frac{\partial V(\mathbf{p})}{\partial p_m} = \frac{\partial u_m(\mathbf{p})}{\partial p_m}, m \in N$. □

Proposition 1. *The potential game with pricing exhibit at least one Nash equilibrium. Moreover, in the case that we have more than one Nash equilibrium, the most efficient one is accomplished for* p_i^* *when*

$$u^* = -\frac{1}{PRR(p_i^*)_{frwd} * PRR(p_j)_{bkwd}} + p_i^* \tag{5}$$

where p_j *for* $PRR(p_j)_{bkwd}$ *is the node's transmission power to send the acknowledgments.*

Proof. If we set $p_i = p_i^*$, the resulting ETX will be ETX^* of the utility $u^* = -\frac{1}{PRR(p_i^*)_{frwd}*PRR(p_j)_{bkwd}} + p_i^*$, meaning that each user uses its best-response strategy. Therefore, p_i^* is a Nash equilibrium.

To generalise the above, let $p_i = \tilde{p}_i \geq p_i^*$, then $\tilde{p}_i$ is a Nash equilibrium where

$$u^* = -\frac{1}{PRR(\tilde{p}_i)_{frwd} * PRR(p_j)_{bkwd}} + \tilde{p}_i \tag{6}$$

Therefore, when the smallest transmission power $\tilde{p}_i$ that is a Nash equilibrium is employed, the larger the utility we accomplish. This is due to the fact that a higher transmission power that achieves ETX^* causes more interference for other nodes, which raise their transmission power as well; hence, their utilities will be decreased, since the pricing will be higher. Thus, the Nash equilibrium with $p_i = p_i^*$ is the most efficient Nash equilibrium. □

Corollary 1. *If* E *is the set of Nash equilibria, for the potential game with pricing,* $p_i^* \in E$ *is the Pareto-dominant equilibrium,* $u_i(\mathbf{p}_i^*) \leq u_i(\mathbf{p})$ *for all* $i, \mathbf{p} \in E$.

Proof. Initially, we show that if we have two Nash equilibria in the potential game with pricing, $\mathbf{x}, \mathbf{y} \in E$ where $\mathbf{x} \geq \mathbf{y}$, then $u_i(\mathbf{x}) \leq u_i(\mathbf{y}), \forall i$. Indeed, for a p_i, the utility decreases with the increase of $\mathbf{p}_{-j}$. This is the case, since interference decreases at both PRR_{frwd} and PRR_{bkwd}. Note that for PRR_{bkwd} interference increases less since $p_k \in \mathbf{p}_{-j}$ is on its numerator. Thus, since $\mathbf{x}_{-j} \geq \mathbf{y}_{-j}$, we have

$$u_i(x_i, \mathbf{x}_{-j}) \leq u_i(y_i, \mathbf{y}_{-j}) \rightarrow u_i(\mathbf{x}) \leq u_i(\mathbf{y}) \tag{7}$$

It follows from the above and Proposition 1 that the Nash equilibrium with the smallest transmission powers employed by the nodes is the Pareto-dominant one. □

3 Continuous Learning Power Control Potential Game

We employ the logit dynamics [14], which operate in the following manner to construct a learning algorithm. We separate time in time steps t. At each time step one player may update her state in a random fashion. The selection of the strategy is undertaken from the strategy set P, according to the Boltzman distribution with parameter $\beta \geq 0$. Specifically, the strategy $p_i \in P$ is chosen from the subset $B_i \in \sigma(P)$ with the probability, which we denote with ρ as

$$\rho(B_i)(t) = \frac{\int_{B_i} \exp\{\beta u_i(x^i, p^{-i}(t-1))dx^i)\}}{\int_P \exp\{\beta u_i(x^i, p^{-i}(t-1))\}dx^i} \tag{8}$$

Essentially, for a fixed parameter $\beta \geq 0$, the logit dynamics is a homogeneous Markov chain $M_\beta = (\{X_t\}_{t\in\mathbb{N}}, P, \rho)$ on the strategy set P. Since in our problem we have continuous strategy sets, the transition probability kernel is

$$\rho(p, B) = \frac{1}{N}\sum_{i=1}^{N} \rho(B_i) \tag{9}$$

where $B \in \sigma(P)$, $B_i = l_i(p) \cap B$ is an intersection of the line of the ith strategy passing through p with B, $p \in P$ and $\rho(B_i)$ is given in (8), with $p^{-i}(t-1) = p^i$.

From [14,15] we have the following theorem for potential games.

Theorem 1. *For the potential game with pricing $\Gamma = (N, \{P_i\}, \{u_i\})$ with the potential function V defined in (4), the Markov chain defined in (9) has the stationary distribution $\Pi(\cdot)$ with density given by*

$$\Pi(\boldsymbol{x}) = \frac{\exp\{\beta V(\boldsymbol{x})\}}{\int_{\boldsymbol{P}} \exp\{\beta V(\boldsymbol{y})\}d\boldsymbol{y}} \tag{10}$$

With this approach the Boltzmann-Gibbs distribution is used to update a node's strategy. Furthermore, the convergence to stationary distributions may be reduced from exponential to linear times with respect to the number of players. In the case of exact potential games, the Boltzmann-Gibbs distribution is provided by:

$$\Pi(\mathbf{x}) = \frac{\exp\{\beta u_i(\mathbf{x})\}}{\int_{\mathbf{P}} \exp\{\beta u_i(\mathbf{y})\}d\mathbf{y}} = \frac{\exp\{\beta V(\mathbf{x})\}}{\int_{\mathbf{P}} \exp\{\beta V(\mathbf{y})\}d\mathbf{y}} \tag{11}$$

As we can see from (10) as the parameter $\beta \to \infty$, the weight of the stationary distribution Π is on the strategies that are the maximisers of the potential function V. We do not claim, however, that the optimal states may be reached in a probabilistic manner. Every fixed value of β, results in the non-optimal outcomes in limiting Π with some positive probabilities. Hence, it is necessary to examine the setting of β as a function of time that will end up converging to optimal states. From [16] we can guarantee convergence of the nodes' strategies to the maximum values of the potential function as

$$U_i = \frac{-ETX_{ij} + p_i + M}{2M} \in [0, 1] \tag{12}$$

The potential function of this game is $V(\mathbf{p}) = \frac{-\sum_m ETX_m + \sum_m p_m}{2M}$. Our primary effort resides with the setting of parameter β as a function of time, where that the Markov chain that is created is inhomogeneous [17,18]. Therefore, the parameter β must be set in such a way, such that the Markov chain is strongly ergodic [19,20]. Thus, we utilise [16] to show that the Markov chain is weakly ergodic and the logit learning algorithm of our potential game reaches the global optimal condition. Thus, we have

Theorem 2. *For the potential game with pricing defined in Lemma 1, there exists a finite integer c, such that the logit learning algorithm with* $\beta(t) = \frac{ln(t+1)}{c}$ *applied to the game guarantees probabilistic convergence of the joint action* $(p_i p_{-i})$ *to the set of potential function maximisers* $P* = \{p* \in P | V(p*) = \max_p V(p)\}$, *namely* $\lim_{t\to\infty} Pr\{p(t) \in P*\} = 1$.

Finally, the determination of the constant c is provided in the same paper as the minimal integer c such that $(1 - \frac{1}{N^N})^{c/N} \leq \frac{1}{2+\epsilon}$ by considering Markov chains with continuous states [21]. Here, we consider only time-homogeneous logit dynamics with $\beta = 0$ and attempt to locate the smallest constant c such that $\alpha(P^c) > 0$, where $\alpha(\cdot)$ is the ergodicity coefficient defined in [17,22] as

$$\alpha(\rho) = \inf_{x',x''} \inf X_i \sum_{t=1}^{l} \min(\rho(x', X_i), P(x'', X_i)) \tag{13}$$

where we take the first infimum over $x', x'' \in X$ and we take the second infimum over all partitions of X into pairs of the non-intersecting and non-empty subsets $X_i, i = 1, ..., l$ Putting this value to the $\beta(t)$ function, the logit dynamics accomplish convergence to the potential function maximisers.

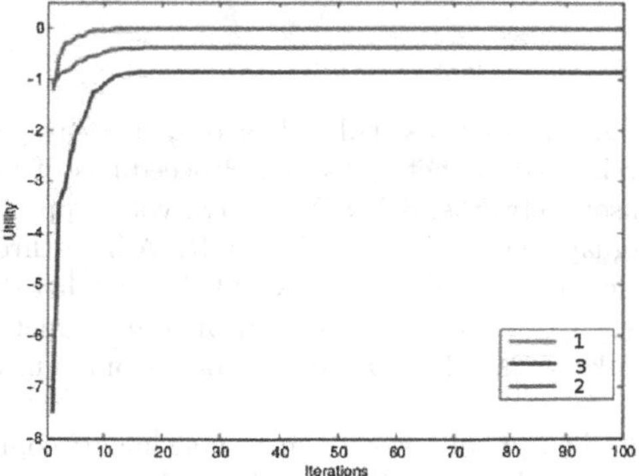

Fig. 1. Utilities of nodes 1, 2 and 3

4 Results

We decided to put our approach to the test by using a network of 7 nodes. The ETX value is being obtained every 10 packets and each node transmits at a rate of 4 packets per second. This means that the channel gain of every link remains constant. Each node starts transmitting with a random transmission power level.

In order to show the performance of our approach we selected three links from the network. The first link has one interferer, the second link is suffering from interference from two nodes and the third link is susceptible to interference from three nodes. We measured the utilities of the three links under the aforementioned conditions as we can see in Fig. 1. It is clear that link 1 achieves a higher utility, converging to approximately −0.0265, which is very close to the highest utility with value 0. Link 2, which suffers from more interference shows a utility of approximately −0.3882. Lastly, link 3 converges to a utility with value −0.8525.

5 Conclusions

In this work, we addressed the problem of optimising link quality with power control using a game theoretic approach. We created a utility function that encapsulates ETX with transmission power as pricing. We established a game and proved that it is an exact potential game. We provided evidence that we able to locate the Nash equilibrium that is Pareto dominant. Thereafter, we presented a learning algorithm that is based on logit dynamics and showed that it reaches the potential function maximisers. We simulated a network of 7 nodes and selected three links to show the performance of our algorithm in terms of utilities. For future work, we will attempt to put our approach to a wireless sensor network simulator such as TOSSIM.

References

1. Spyrou, E.D., Yang, S., Mitrakos, D.K.: Discrete strategy game-theoretic topology control in wireless sensor networks. In: Proceedings of the 6th International Conference on Sensor Networks, SENSORNETS, vol. 1, pp. 27–38 (2017)
2. De Couto, D.S., Aguayo, D., Bicket, J., Morris, R.: A high-throughput path metric for multi-hop wireless routing. Wirel. Netw. **11**(4), 419–434 (2005)
3. Draves, R., Padhye, J., Zill, B.: Comparison of routing metrics for static multi-hop wireless networks. ACM SIGCOMM Comput. Commun. Rev. **34**(4), 133–144 (2004)
4. Luo, Z.-Q., Zhang, S.: Dynamic spectrum management: complexity and duality. IEEE J. Sel. Top. Signal Process. **2**(1), 57–73 (2008)
5. Huang, J., Berry, R.A., Honig, M.L.: Distributed interference compensation for wireless networks. IEEE J. Sel. Areas Commun. **24**(5), 1074–1084 (2006)
6. Hande, P., Rangan, S., Chiang, M., Wu, X.: Distributed uplink power control for optimal sir assignment in cellular data networks. Netw. IEEE/ACM Trans. **16**(6), 1420–1433 (2008)

7. Alpcan, T., Başar, T., Srikant, R., Altman, E.: CDMA uplink power control as a noncooperative game. Wirel. Netw. **8**(6), 659–670 (2002)
8. Spyrou, E.D., Mitrakos, D.K.: Approximating nash equilibrium uniqueness of power control in practical WSNs, arXiv preprint arXiv:1512.05141 (2015)
9. Monderer, D., Shapley, L.S.: Potential games. Games Econ. Behav. **14**(1), 124–143 (1996)
10. Qian, L.P., Zhang, Y.J., Chiang, M.: Globally optimal distributed power control for nonconcave utility maximization. In: Global Telecommunications Conference (GLOBECOM 2010), pp. 1–6. IEEE (2010)
11. Yang, L., Sagduyu, Y.E., Zhang, J., Li, J.H.: Distributed power control for adhoc communications via stochastic nonconvex utility optimization. In: 2011 IEEE International Conference on Communications (ICC), pp. 1–5. IEEE (2011)
12. Zhou, S., Wu, X., Ying, L.: Distributed power control and coding-modulation adaptation in wireless networks using annealed gibbs sampling. In: INFOCOM, 2012 Proceedings IEEE, pp. 3016–3020. IEEE (2012)
13. Ru, G., Li, H., Tran, T., Lin, W., Liu, L., Wu, H.: Distributed optimal power control for multicarrier cognitive systems. In: Global Communications Conference (GLOBECOM), pp. 1132–1137. IEEE (2012)
14. Blume, L.E.: The statistical mechanics of strategic interaction. Games Econ. Behav. **5**(3), 387–424 (1993)
15. Lasaulce, S., Tembine, H.: Game Theory and Learning for Wireless Networks: Fundamentals and Applications. Academic Press, Waltham (2011)
16. Tatarenko, T.: Log-linear learning: convergence in discrete and continuous strategy potential games. In: 2014 IEEE 53rd Annual Conference on Decision and Control (CDC), pp. 426–432. IEEE (2014)
17. Dobrushin, R.L.: Central limit theorem for nonstationary markov chains. i. Theory Probab. Appl. **1**(1), 65–80 (1956)
18. Isaacson, D.L., Madsen, R.W.: Markov Chains. Theory and Applications. Wiley, New York (1976)
19. Dorea, C.C., Cruz, J.A.R.: Approximation results for non-homogeneous Markov chains and some applications. Sankhyā: Indian J. Stat. 243–252 (2004)
20. Hajnal, J., Bartlett, M.: Weak ergodicity in non-homogeneous Markov chains. In: Mathematical Proceedings of the Cambridge Philosophical Society, vol. 54, no. 02, pp. 233–246. Cambridge Univ Press (1958)
21. Roberts, G.O., Rosenthal, J.S., et al.: General state space Markov chains and mcmc algorithms. Probab. Surv. **1**, 20–71 (2004)
22. Dobrushin, R.: Central limit theorem for nonstationary Markov chains. ii. Theory Probab. Appl. **1**(4), 329–383 (1956)

Routing Protocol Enhancement for Mobility Support in Wireless Sensor Networks

Jinpeng Wang(✉), Gérard Chalhoub, Hamadoun Tall, and Michel Misson

Université Clermont Auvergne/LIMOS CNRS, Aubiére, France
{jinpeng.wang,gerard.chalhoub,hamadoun.tall,michel.misson}@uca.fr

Abstract. Wireless sensor networks (WSNs) technology is attracting more attention with the emergence of the Internet of Things (IoT). Wireless sensor networks have been intensively studied for optimizing their life time by enhancing their low power operations. Few studies have been developed in order to deal with mobility. In some cases, nodes move around a specific area and the network topology should adapt fast. Routing protocols should react to topology changes and build new routes. Traditionally, control traffic is used to update routing metrics. Each node will then choose a next-hop among its neighbor nodes based on this routing metric. In this paper, we propose an enhancement based on signal strength monitoring and depth updating in order to improve the routing protocol performance in mobility scenarios where all nodes are moving. This enhancement can be integrated into most data collection routing protocols. It helps them to better cope with mobility and to make faster decisions on updating their next-hop neighbors. We also implemented this enhancement in the Routing Protocol for Low-Power and Lossy Networks (RPL). We show that through simulations using our technique allows better performance for different routing metrics.

Keywords: Wireless sensor networks · Mobility · Routing · RPL · Ad hoc

1 Introduction

Wireless sensor networks (WSNs) technology is one of the building blocks of the Internet of Things. Sensor nodes produce data which is usually collected using wireless networking in a multihop manner to reach a gateway that leads to a monitoring station. Indeed, sensor nodes have limited communication coverage and need to send data in a hop by hop manner to reach the destination. This operation is managed by the routing protocol that based on a certain routing metric. It will choose a next-hop between its available neighbor nodes.

Traditionally a WSN is formed by static sensor nodes that rarely change positions [1]. Nevertheless, mobility is one of the requirement for several emerging applications, including health-care monitoring, intelligent transportation system, smart cities, farming equipment monitoring, etc. [2]. In some applications it is required that all nodes need to be mobile, such as health-care monitoring and

A. Puliafito et al. (Eds.): ADHOC-NOW 2017, LNCS 10517, pp. 262–275, 2017.
DOI: 10.1007/978-3-319-67910-5_22

intelligent transportation system. For instance, in a health-care monitoring system all the patients are free to move inside a hospital and need to be equipped with sensor nodes to collect data in a continuous manner [3]. Whereas in some applications only some of the nodes need to be mobile, such as farming equipment monitoring and smart cities [4]. In mobile scenarios nodes are free to move and organize themselves into a connected network. Hence, the topology is continuously changing due to the movement of nodes and radio links instability. This is a challenging factor for routing protocols. Every time the topology changes, the routing protocol needs to update the path to reach the destination. Therefore, routing protocols should be able to adapt rapidly to the topological changes and reconstruct routes in a timely manner.

In this paper, we concentrate on mobile data collection applications, where all nodes are mobile and all traffic is destined to a single sink node. For these applications, we propose an enhancement mechanism based on a combination of Received Signal Strength Indicator (RSSI) monitoring and level updating that we called RL (RSSI and Level) mechanism. This mechanism makes faster decisions for updating next-hop neighbors. RL is based on level knowledge, link existence monitoring and movement direction estimation. When applied to traditional routing protocols, RL enhances the speed of coping with mobility scenarios, and thus, enhances the overall performance of the network.

The rest of this paper is organized as follows. In Sect. 2 we present the related work concerning mobility in routing protocols. Section 3 describes our enhancement mechanism, RL. In Sect. 4, we analyse the results obtained by including RL to data collection routing protocols. Finally, we conclude the paper along with future investigations in Sect. 5.

2 Related Work

In [5], authors proposed a fuzzy logic system (FLS) for MANETs to convert all the available network metrics of the routes into a single metric, which is called fuzzy cost (FC). The routes, which have minimum fuzzy cost, will be considered as optimal paths. Three metrics Delay, Bandwidth and Residual Energy are used in FLS and for each metric three linguistic variables Low, Medium and High are applied to quantify the metric value. However, the source node needs to go through all the nodes in the topology before sending data packets, this increases the network delay and overhead.

In [6], authors presented a Tree-Based Routing Protocol (TBRP) for mobile sensor networks. TBRP contains three phases: tree formation phase, data collection using Time Division Multiple Access (TDMA), and transmission phase, and purification phase. TBRP is a dynamic routing protocol and it can efficiently update the topology of network. However, even in static networks synchronization is a challenge when using TDMA schedule; with mobility the network cannot guarantee a precise synchronization.

RPL is a Routing Protocol designed for Low-Power and Lossy Networks (LLNs) [7]. In [8], authors proposed Co-RPL, which is an extension of RPL based on the Corona mechanism in order to support mobility. They modified the trickle timer to allow the DAG root to broadcast DIO messages periodically. Mobile routers do not broadcast DIO messages until they receive DIO messages from the DAG root or neighbor nodes. Although a modified timer can help nodes update routing metrics timely, authors did not consider rank problem in mobility. Rank calculated according to the different metrics received from parent nodes is an important mechanism in RPL. With rank mechanism nodes only send data packets to nodes that have lower ranks. RPL has a movement limitation which causes slow rank update and loops will occur [7].

In [9], Cluster Based Routing (CBR), an adaptive TDMA scheduling and cluster head protocol in which cluster head is randomly chosen in each round is proposed. This protocol is an enhancement of LEACH-Mobile protocol that reduces the packet loss by 25% compared to LEACH-Mobile according to an evaluation using MATLAB. In this protocol, cluster head receives data not only from its members during TDMA allocated time slot but also from other sensor nodes that are moving from one cluster to another. According to CBR, a mobile node could be a cluster head, but nothing prevents packets loss when cluster head has a weak link.

MoMoRo, a mobility support layer that can be easily applied to existing data collection protocols, was introduced in [10]. Authors used Expected Transmission Count (ETX), average RSSI and Symbol Error Rate (SER) as routing metrics and have proposed a fuzzy estimator to supervise the quality of the link. They set a threshold for the metrics in the fuzzy estimator. When the receiving metrics are higher or lower than the threshold, other links with better metrics will be suggested. However in MoMoRo control packets are only triggered when routing metric is over threshold. This will lead to a slow metrics update in mobility.

In [11], authors proposed an RSSI based hop count calculation method that can reflect the distance between adjacent nodes. They established a new hop count calculation method by introducing a traditional path loss model. Therefore according to the relationship between the hop count and received power in the path loss model, a new and minimum hop count will be obtained. But this method has a premise that there is a linear relationship between the hop count and distance of two communication nodes. In real condition path loss will be affected by many factors and this premise will cause a deviation.

Authors of [12] proposed CoLBA, Collaborative Load Balancing Algorithm, routing protocol that achieves load balancing in order to avoid queue overflow in nodes located near the sink in many to one traffic scenarios. CoLBA uses the time spent by data packets in the queues as a routing metric. Each node computes the difference between the queuing time and the dequeuing time of a packet as the queuing delay and then calculates an average value over the ten last dequeued packets. CoLBA offers a quite stable algorithm for selecting parent nodes in static networks, but in a mobile topology average queueing delay does not give an up-to-date image of the topology.

In [13], authors presented an opportunistic mechanism that supervises mobile nodes and receives critical periodical information transmitted by mobile nodes using ZigBee hierarchical routing protocol. Mobile nodes select a neighbor to be its parent according to LQI (Link Quality Indicator) or RSSI (Received Signal Strength Indicator) values [15]. Authors proposed a mechanism that ensures delivery of critical data in networks where every packet is a critical packet. However, movement of a node is often unpredictable and the process of retrieving packets from old parent is time consuming.

In [14], authors proposed a handoff mechanism called smart-HOP. In smart-HOP mechanism, mobile nodes monitor the link quality by receiving reply packets from the serving Access Point (AP) during data transmission phase. The reply packets contain the average RSSI, or signal to noise ratio (SNR), of the n packets received by the AP. However, smart-HOP is designed for cellular networks and not ad-hoc networks. Indeed, in cellular networks, all APs are fixed and mobile nodes are only one hop away from them.

In mobile wireless network three methods are often used to help routing protocol make an enhancement. In [6], sending packets from lower level to higher level is used to avoid loops in mobility. Periodically sending control messages helps re-construct network fast when topology changes frequently in mobility [8]. In [10,11,13], ETX, RSSI and SER are applied to monitor link quality between two nodes and help give a better routing. In this paper, we aim to combine these three methods for the selection of the next-hop neighbor. We will explain our protocol in the sections below.

3 RSSI and Level Monitoring Enhancement

In many-to-one multi-hop wireless sensor networks applications, data is transmitted by sensor nodes towards the sink node which is the only destination. Before sending data, nodes need to select a next-hop based on one or more routing metrics. In mobile scenarios, changes in the topology of the network increase the risk of link failures. In addition, if the routing metrics cannot be updated timely, it is likely to cause loops. In what follows, we present our proposal which is an enhancement to routing protocols that copes with topology changes, prevents routing loops, and allows nodes to have more than one potential next-hop options. This enhancement is based on RSSI values and positions of the nodes relative to the sink.

3.1 Level Calculation

During the routing process, a level metric is used in addition to the routing metric. The construction of the initial topology of the network starts at the sink node that begins to broadcast control messages which contain routing metrics. Each node that receives these messages builds a list of potential next-hops that we call parents set in what follows. The level of a node is a value that defines the position of the node relative to the sink in terms of number of hops, with sink node being at level 0. Based on the hop count obtained from control messages, a node will know the level of sender node. If the level of the receiver node is higher, it means this control message is from a potential next-hop. The level of the receiver node will be equal to the smallest received level plus one. Note that the receiver node adds all senders with lower levels to its parents set. Our level mechanism is similar with the rank mechanism of RPL using the hop-count metric.

3.2 Metric Diffusion

With RL mechanism, control messages have three functions: (i) broadcasting metrics, (ii) monitoring link existence, (iii) and monitoring movement direction. Control messages are sent periodically in order to update routing metrics timely. At the beginning of each period the sink is the first to broadcast a control message. When a node receives a control message from a node in its parents set, it updates the routing metrics and then broadcasts its level and metrics. Nodes only broadcast control messages once they receive them from nodes with lower level. This insures broadcast process to have a downwards direction from the sink to the leave nodes.

3.3 Link Existence and Movement Direction Monitoring

Every node in the network needs to make sure that the link between itself and nodes in the parents set exists. We use the RSSI values to allow nodes to monitor links [16]. Nodes obtain RSSI values from control messages. A node stores two RSSI values for each parent, Old RSSI value and New RSSI value. We use them to monitor movement direction. Old RSSI value is obtained from the previous control message and New RSSI value is obtained from the newest control message. We consider that the node is getting closer to its parent if the New RSSI value is bigger than the Old RSSI value. In case the New RSSI value is smaller than the Old RSSI, we consider that the node is getting further away from its parent.

In addition, we set a RSSI threshold within which the node has a good link quality with its parent node. When the RSSI value is smaller than the threshold, the node will consider whether to stop using the link or not. If the node is getting near to its parent, we consider that this parent node can still be used as a next-hop; otherwise this parent will be deleted from the parents set.

Each time a node adds a new node into the parents set, a timer is set for this parent. We call this timer the lifetime of a parent. Nodes are kept in the parents set for the duration of their lifetime. Before the lifetime of a parent expires, if a new control message is received from this parent, the timer will be reset. According to RL mechanism, we set two sorts of lifetimes: long lifetime and short lifetime.

Long lifetime is given to nodes that have a RSSI value that is higher than the threshold, and short lifetime is given to nodes that have a RSSI value that is lower than the threshold. Indeed, when the parent node is in a zone where the radio link is about to fail, a short lifetime will help avoid using this parent node for a long period in case we do not receive its control messages.

3.4 Loop Avoidance and Level Update

A routing loop is a common problem in mobility scenarios. Due to the changes in the topology, a current next-hop may become a descendant, and loops will occur. In order to avoid this, we set the following three rules. (i) The level of a node must be greater than the level of all nodes in its potential parents set. (ii) Nodes cannot forward data packets to nodes with higher or equal levels. (iii) Nodes must ignore control messages which are received from higher level nodes.

Monitoring link existence and movement direction allows nodes to update their levels. Algorithm 1 depicts the level update process.

$Level_r$ stands for the level of the receiver of the control message and $Level_s$ stands for the level of the sender of the control message. $Lifetime_l$ stands for the long lifetime and $Lifetime_s$ represents the short lifetime.

```
while true do
    if Level_r > Level_s then
        if NewRSSI ≤ THRESHOLD then
            if NewRSSI ≤ OldRSSI then
                Level_r = Level_r + 1;
            else
                Level_r = Level_s + 1;
                Setting Lifetime_s for this parent node;
            end
        else
            Level_r = Level_s + 1;
            Setting Lifetime_l for this parent node;
        end
    else
        Ignore
    end
end
```

Algorithm 1. Level Update

In case the RSSI value is bigger than the threshold, we set a long lifetime for this parent and we update the level of the receiver $Level_r = Level_s + 1$.

If the level of receiver node ($Level_r$) is bigger than the level of the sender node ($Level_s$), nodes need to check the RSSI value of the control message. If this RSSI value is smaller than the threshold, then nodes compare it to the Old RSSI value. If the New RSSI value is bigger than the old one, node C gives a short lifetime for node A and updates its level $Level_r = Level_s + 1$. Otherwise, if New RSSI value is smaller than the old one, this means that the node is getting away from its parent. In this case, the node increments its own level $Level_r = Level_r + 1$. We do this because the node does not know if it is the one that is moving or if it is the sender node that is moving. This way, if the sender node is moving, the node will receive new control messages from other neighbors and update its level according to these control messages. And in case it is the receiver node that is moving away, it will help avoid this node being selected as a parent by its children nodes. Indeed, this part of the algorithm avoids creating loops.

Finally, in case the $Level_r$ is smaller than the $Level_s$ we ignore the control message.

4 Simulation Environment and Performance Evaluation

In this section, we evaluate the performance of RL mechanism by doing simulations with Cooja simulator. Cooja is a flexible Java-based simulator that is designed for simulation networks of sensors running the Contiki operating system [17]. We applied RL mechanism to 4 different routing metrics.

4.1 Routing Metrics on Which We Applied the RL Mechanism

When a node receives control messages from other nodes, it builds a neighbors list which we call neighbors set. Note that the parents set of RL mechanism contains nodes from the neighbors set that have the smallest level. Based on different routing metrics the node selects a best next-hop in the neighbors set. Different metrics can be used in combination with RL mechanism. When RL mechanism is applied, nodes can only select the next-hop from the parents set.

In our simulations, we considered data collection scenarios only. All nodes in the network send their traffic to the sink. We chose to show the efficiency of our RL mechanism using three of the most used metrics number of hops (HopCount), latency, and ETX. HopCount is a metric that is based on the number of hops that separates source node from destination (sink node). Nodes select the node that has the smallest hop count in the neighbors set or parents set as their next-hop. For the latency metric, we calculate the queuing delay, which is the time spent by a packet in the packet queue of the MAC layer. The node computes the difference between the queuing time and dequeuing time of a packet in the queue. We make an average calculation of the last ten packets as it is done in [12]. If the number of dequeued packets is less than ten, the average calculation is only computed over the number of dequeued packets. Nodes choose the next-hop based on the smallest path delay, which represents the estimated time needed to reach the sink based on the queueing delay of intermediate nodes. For the ETX, we use an average value of last ten packets as ETX metric and select the next-hop based on the smallest path ETX in the neighbors set or parents set in a similar way to what is done in [18].

We applied HopCount, latency and ETX with RL mechanism and we call these new combinations as RL_HopCount, RL_Latency and RL_ETX. Unlike the latency and ETX, HopCount is a quantized metric. Normally in the neighbors set or parents set many nodes have the same smallest number of hops to the sink. Using HopCount metric, we always select the first node in the neighbors set or parents set with the smallest number of hops. In addition to these 3 metrics, we also introduce a random variant method which consists on randomly choosing the next-hop from the nodes that have the same smallest hop count. We call this metric Random and its implementation with RL mechanism RL_Random. We also applied RL mechanism to RPL. Since the native RPL does not introduce Random as its metric, we only included RL to RPL_Hopcount, RPL_ETX and RPL_Latency, and we called them RL_RPL_Hopcount, RL_RPL_ETX and RL_RPL_Latency.

4.2 Simulations Parameters

Cooja offers four propagation models [19]. We used Multi-path Ray-tracer Medium (MRM), which takes reflections and refractions into account to simulate real environment. We added a Gaussian random variable in the path loss formula in order to simulate instability in the radio links. Random variable makes transmission range randomly fluctuate between 30 m to 50 m. We used CSMA/CA to access the medium. Each simulation lasts 5 min. We emulated Z1 platform. Data packet size is 30 Bytes. Nodes are randomly deployed in a 200 m * 200 m area.

4.3 Mobility Model

The mobility model we used is Random Waypoint Model that is usually used to evaluate MANET routing protocols [20]. At the beginning, all nodes, except the sink, are randomly deployed in a 200 m $*$ 200 m area. The sink is located at the center of the area. When the simulation starts, each mobile node randomly selects one location in the simulation field as the destination. Then, it moves towards this destination with velocity which is randomly chosen from [1 m/s, 3 m/s] every 5 s. We add rest time for each node. Each time a node reaches its location, the node will rest in new location for 5 s and then select a new destination for the next movement. All nodes, except the sink, repeat these steps until the simulation stops.

4.4 Results Analysis

We used five performance metrics to evaluate the enhancement of RL: (i) packet delivery ratio, (ii) number of sent packets, (iii) number of retransmissions, (iv) number of dropped packets and (v) average end-to-end delay.

Packet Delivery Ratio. Packet delivery ratio is defined as the ratio of received data packets at the sink to those generated by the source nodes.

Figure 1 shows that Random, HopCount and ETX outperform Latency, which is mainly due to the fact that Latency has more loops and collisions during the transmission. All four metrics have better packet delivery ratio when used with RL mechanism. RL_Random performs better than the other variants mainly due to the fact that the random choice in parents set balances the traffic load. Indeed, sending packets to a different parent node at each transmission increases the probability of success of packet delivery and decreases the probability of sending packets to a node that is moving away. Having more nodes in the network increases the traffic load. This may cause network collisions, however, this also helps the parents set contains more potential next-hop options to reduce congestions. Therefore, RL_Random has almost the same packet delivery ratio with 40 nodes and 60 nodes.

Figure 2 shows the packet delivery ratio results for RPL and RL_RPL. With RPL, delivery ratio of Latency metric does not decrease much with the increase of number of nodes compared to the routing based on latency presented in Fig. 1. This is mainly due to the rank mechanism in RPL that can help nodes avoid some loops compared with the routing protocols that do not have rank mechanism. Note that all the three variants with RL outperform native RPL.

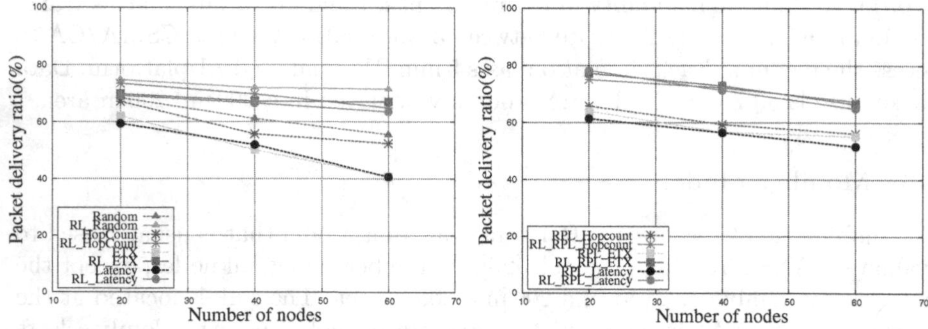

Fig. 1. Packet delivery ratio

Fig. 2. Packet delivery ratio of RPL

Number of Sent Packets. Total number of sent packets refers to the number of packets that are sent over the medium during the simulation. Nodes in our simulations generate the same number of packets but each node may send a different number of frames depending on the retransmission attempts due to collisions and packet loss. This performance metric gives us the total number of packets that each metric sent for the same number of generated packets. An efficient metric should have fewer sent packets for the same number of generated and received packets.

Figure 3 shows that Random and HopCount send less packets than ETX and Latency. This is mainly due to the fact that ETX and Latency select the next-hop with best ETX and Latency values, which might lead in some cases to choosing a longer path. The number of sent packets depends on the length of the path that packets followed to reach the sink. Results show that all four metrics send less packets when used with the RL mechanism, because the level setting in RL mechanism helps avoid loops occurrence which in turn helps reduce the number of transmission hops travelled by packets. Random and HopCount have almost the same number of sent packets with and without RL mechanism with 20 nodes and 40 nodes scenarios, this is essentially due to the fact that they use a metric based on the shortest path and inherently avoid loops. Whereas we can see that RL mechanism enhances ETX and Latency because it helps them avoid loops and update the next-hop in a timely manner. In four variants, RL_Random outperforms other variants. This is mainly thanks to the random choices in parents set as explained in the packet delivery ratio results.

Figure 4 shows that with 20 and 40 nodes the performances are almost the same, but when the number of nodes increases to 60 RL variants in RPL outperform native RPL. This is mainly due to the fact that in small network the probability of loops is smaller and the rank mechanism of RPL can still help to avoid them. Whereas with 60 nodes the probability of loops with mobility increases and the rank mechanism is not able to cope with it. Furthermore, Fig. 2 shows a 10% increase in packet delivery ratio. With 60 nodes RPL_Latency sends more packets compared with RPL_ETX and RPL_Hopcount, which is mainly due to the fact that the metric of latency will dramatically increase, which easily causes a longer path selection, when there are many collisions in the network. Also note that RL helps reduce the number of sent packets for all metrics in Figs. 3 and 4. This shows the efficiency of RL in avoiding loops and in finding shorter paths.

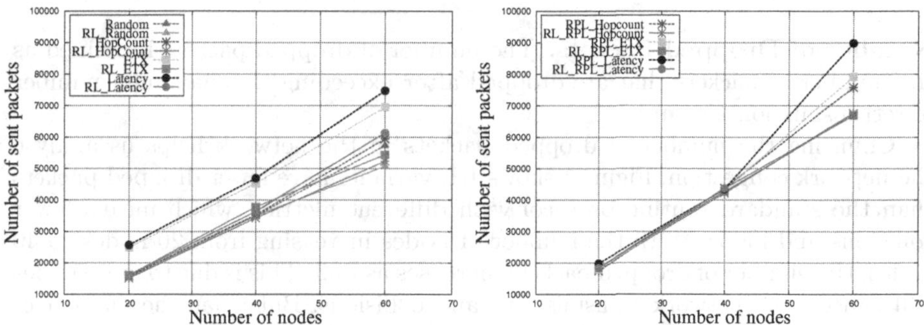

Fig. 3. Number of sent packets

Fig. 4. Number of sent packets of RPL

Number of Retransmissions. According to CSMA/CA, nodes retransmit frames until the number of retransmissions reaches a fixed maximum value after which the frame is dropped. This gives an idea about the congestion of the network and how efficient each metric is in avoiding to overload the network with retransmissions.

Figure 5 shows RL variants have better performance when it comes to the number of retransmissions. The Latency metric sends much more packets than other metrics. This is mainly due to the time needed for Latency to update its metric value. A node will wait for new latency values to update its own latency value by doing an average of the past values. This takes too long before and become outdated very quickly because the topology keeps changing.

Figure 6 shows that all the RPL metrics and their variants have almost the number of retransmissions with 20 and 40 nodes. However, when the number of nodes increases to 60, all the RL variants in RPL outperform the native RPL metrics. Although the native RPL metrics ETX and Latency have almost the same performance in packet delivery ratio with 60 nodes, Latency needs more transmissions to achieve it due to its slow convergence perspective.

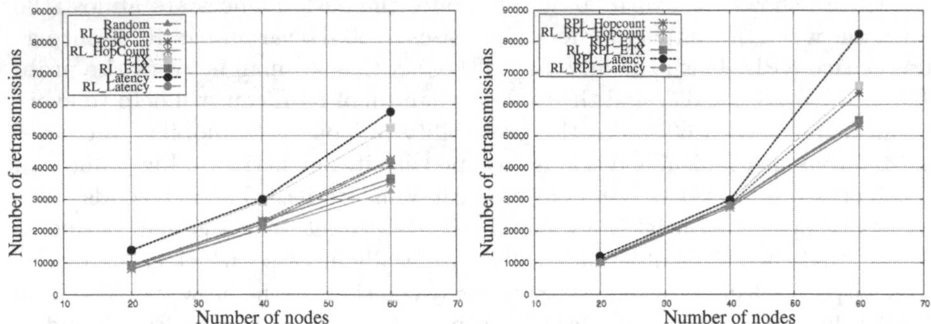

Fig. 5. Number of retransmissions

Fig. 6. Nb. of retransmissions of RPL

Number of Dropped Packets. The number of dropped packets is defined as the number of packets that are dropped after exceeding the maximum number of retransmission attempts.

Counting the number of dropped packets in the network helps us analyse the network congestion. Figure 7 shows RL variants have fewer dropped packets than the standard routing protocol with different metrics, which means fewer collisions and loops. With the number of nodes increasing from 20 nodes to 40 nodes, the number of dropped packets increases as well. This is due to more nodes will cause more network transmissions and collisions. But when the number of nodes increases to 60 nodes, all routing metrics and RL variants do not have a huge increase in the number of dropped packets. This is mainly because more number of nodes gives every node more choices to select next hop and this can efficiently reduce collisions in the network. Latency has a much higher number of dropped packets compared to other metrics, which is mainly due to the fact that the metric of Latency is very depend on the stability of the network. With the movement of the nodes, the value of metric will fluctuate greatly, which makes Latency easily to select a faraway node to be the next-hop.

Figure 8 shows that all the RL variants in RPL outperform the native RPL. This is due to the fact that RL mechanism helps node select optimal path to the sink, which has less loops and less collisions.

Average End-to-End Delay. The end-to-end delay refers to the duration taken by a packet from a node to the sink. To compute the average end-to-end delay we measure the delay taken by total received packets and divide it by the number of successful received packets at the sink. The packet average end-to-end delay can be computed using the following two equations.

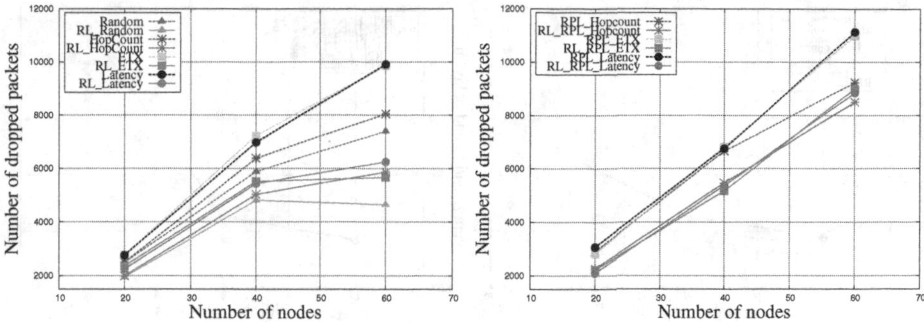

Fig. 7. Nb. of dropped packets

Fig. 8. Nb. of dropped packets of RPL

$$Total\ Delay = \sum_{i=1}^{n} (RecvTime(i) - SentTime(i)) \quad (1)$$

$$Average\ end\ to\ end\ Delay = Total\ Delay/n \quad (2)$$

Where n stands for the number of successfully received packets.

Figure 9 shows that except Latency, the other metrics with and without RL mechanism have very similar average delay with 20 nodes and 40 nodes. When the number of nodes increases to 60, RL variants show their enhancement in network delay performance. This is mainly due to the fact that with 20 and 40 nodes there are not so many collisions and retransmissions. When the number of nodes is 60 and the topology of network becomes more complex, the performance of delay will begin to make difference. Latency shows much higher average end to end delay for all scenarios mainly because the routing metric of Latency is not being updated fast enough to cope with the topology changes as explained earlier. In addition, Latency suffers from a high retransmission attempts using CSMA/CA. Indeed, after each unsuccessful transmission attempt a packet will wait for a backoff time to be sent again, which will cause huge delay for a packet. Note that RL_Random outperforms other RL variants due to the random choice for the parent with small number of retransmission attempts.

Figure 10 shows that all the metrics with and without RL mechanism have very similar average delay with 20 and 40 nodes. RL variants only show their improvement when the number of nodes is 60. This is mainly due to the fact that rank mechanism activates its function with small sized network in mobility. Note that there is dramatically increase of average end-to-end delay when the number of nodes changes from 40 to 60. This is essentially due to the result of the increase in the number of sent packets, number of retransmissions, and the number of dropped packets. All these factors contribute to the increase in the end-to-end delay.

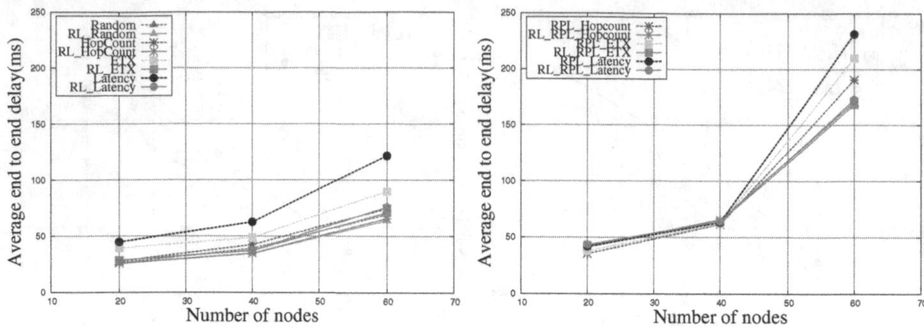

Fig. 9. End-to-end delay

Fig. 10. End-to-end delay of RPL

5 Conclusion

In this paper, we proposed a generic mechanism, called RL mechanism, that enhances the performance of routing protocols in convergecast mobile data collection scenarios. RL takes into account RSSI values and level knowledge in order to avoid loops and make faster updating on next-hop nodes list that we called parents set.

RL monitors the direction of moving nodes based on the variation of RSSI values. When a node detects that one of its potential next-hops in the parents set is moving away, it will anticipate a link failure and try to use another node from the parents set. This helps nodes update their next-hop choice in a timely manner in mobile scenarios.

Simulation results show that our RL mechanism enhances the network performance on many levels (successful delivery packets, number of packet loss, number of transmissions and retransmissions, and end-to-end delay). We have proven that load balancing has a good effect on the network performance because it avoids sending packets to the same node even though it is moving away. We have also proven that using slow convergence routing metric such as Latency metric is not suitable for mobile scenarios. Furthermore, we also successfully applied RL mechanism in RPL and proved that RL mechanism enhances RPL performance especially in networks of 60 nodes.

In our future work we plan on modifying the CSMA/CA parameters based on the RSSI values monitoring mechanism. In addition, RL mechanism can be included in the native control traffic of routing protocols, such as DIO and DIS of RPL. This will help reduce the overhead of RL.

Acknowledgment. This research was conducted with the support of the European Regional Development Fund (FEDER) program of 2014–2020 and the region council of Auvergne.

References

1. El Korbi, I., Brahim, M.B., Adjih, C., Saidane, L.A.: Mobility enhanced RPL for wireless sensor networks. In: IEEE NOF, pp. 1–8 (2012)
2. Bouaziz, M., Rachedi, A.: A survey on mobility management protocols in wireless sensor networks based on 6LoWPAN technology. Comput. Commun. **74**, 3–15 (2016)
3. Aminian, M., Naji, H.R.: A hospital healthcare monitoring system using wireless sensor networks. J. Health Med. Inform. **4**, 121 (2013)
4. Rawat, P., Singh, K.D., Chaouchi, H., Bonnin, J.M.: Wireless sensor networks: a survey on recent developments and potential synergies. J. Supercomput. **68**(1), 1–48 (2014)
5. Yadav, A.K., Das, S.K., Tripathi, S.: EFMMRP: design of efficient fuzzy based multi-constraint multicast routing protocol for wireless ad-hoc network. Comput. Netw. **118**, 15–23 (2017)
6. Singh, M., Sethi, M., Lal, N., Poonia, S.: A tree based routing protocol for Mobile Sensor Networks (MSNs). Int. J. Comput. Sci. Eng. **2**(1S), 55–60 (2010)
7. Winter, T.: RPL: IPv6 routing protocol for low-power and lossy networks (2012)
8. Gaddour, O., Koubâa, A., Rangarajan, R., Cheikhrouhou, O., Tovar, E., Abid, M.: Co-RPL: RPL routing for mobile low power wireless sensor networks using Corona mechanism. In: IEEE SIES, pp. 200–209 (2014)
9. Awwad, S.A., Ng, C.K., Noordin, N.K., Rasid, M.F.A.: Cluster based routing protocol for mobile nodes in wireless sensor network. Wireless Pers. Commun. **61**(2), 251–281 (2011)
10. Ko, J., Chang, M.: MoMoRo: providing mobility support for low-power wireless applications. IEEE Syst. J. **9**(2), 585–594 (2015)
11. Ding, J., Zhang, L., Cheng, G., Ling, Z., Zhang, Z., Lei, Y.: Study on DV-Hop algorithm based on modifying Hop count for wireless sensor networks. IJCSET **2**, 1452–1456 (2012)
12. Tall, H., Chalhoub, B., Misson, M.: CoLBA: a collaborative load balancing algorithm to avoid queue overflow in WSNs. In: IEEE DSDIS, pp. 682–687 (2015)
13. Mouawad, A., Chalhoub, G., Misson, M.: Data management in a wireless sensor network with mobile nodes: a case study. In: IEEE ICWCUCA, pp. 1–6 (2012)
14. Fotouhi, H., Zuniga, M., Alves, M., Koubaa, A., Marrón, P.: Smart-HOP: a reliable handoff mechanism for mobile wireless sensor networks. In: Picco, G.P., Heinzelman, W. (eds.) EWSN 2012. LNCS, vol. 7158, pp. 131–146. Springer, Heidelberg (2012). doi:10.1007/978-3-642-28169-3_9
15. Hu, S.C., Lin, C.K., Tseng, Y.C., Chen, W.T.: Distributed address assignment with address borrowing for ZigBee networks. In: IEEE ICC, pp. 454–459 (2014)
16. Adewumi, O.G., Djouani, K., Kurien, A.M.: RSSI based indoor and outdoor distance estimation for localization in WSN. In: IEEE ICIT, pp. 1534–1539 (2013)
17. Osterlind, F., Dunkels, A., Eriksson, J., Finne, N., Voigt, T.: Cross-level sensor network simulation with COOJA. In: IEEE LCN, pp. 641–648 (2006)
18. Ancillotti, E., Bruno, R., Conti, M.: Reliable data delivery with the IETF routing protocol for low-power and lossy networks. IEEE Trans. Ind. Inf. **10**(3), 1864–1877 (2014)
19. Stehlk, M.: Comparison of simulators for wireless sensor networks. Faculty of Informatics (2011)
20. Bai, F., Helmy, A.: A survey of mobility models. In: Wireless Adhoc Networks, vol. 206, p. 147. University of Southern California, USA (2004)

Correlation-Free MultiPath Routing for Multimedia Traffic in Wireless Sensor Networks

Dhouha Ghrab[1(✉)], Imen Jemili[1], Abdelfettah Belghith[2], and Mohamed Mosbah[3]

[1] HANAlab, University of Manouba, Manouba, Tunisia
ghrab.dhoha@gmail.com, jemili.imen.ensi@gmail.com
[2] College of Computer and Information Sciences, King Saud University, Riyadh, Saudi Arabia
abelghith@ksu.edu.sa
[3] LaBRI, Bordeaux INP, Universiy of Bordeaux, CNRS, Bordeaux, France
mohamed.mosbah@labri.fr

Abstract. Delivering multimedia content, requiring high throughput and reduced end-to-end delay, is a hard task in Wireless Sensor Networks, due to the limited bandwidth and energy. Multipath routing is an attractive alternative to remedy to these constraints by allowing bandwidth aggregation to achieve high throughput and load balancing to reduce latency and energy consumption. However, inter-path correlation can throttle concurrent data transfer through different paths. Besides, operating under duty-cycle mode, adopted by many long-term applications, incurs additional challenges due to the intermittent unavailability of nodes. In this context, we propose CFMPR-M, a Correlation-Free MultiPath Routing protocol for multimedia traffic in duty-cycled networks, providing simple mechanisms to construct non-correlated routes towards the sink. It relies on a cross-layer interaction between MAC and routing levels, useful to avoid the drawbacks inherent to duty-cycle mode. Simulation results show the good performance of our protocol in terms of energy consumption, establishment time and control overhead.

Keywords: Multipath routing · Multimedia traffic · Correlation · Wireless Sensor Network · Duty-cycle

1 Introduction

The technological advances in Micro-Electro-Mechanical Systems (MEMS) and miniaturization technologies allowed the availability of low cost multimedia sensors endowed with appropriate hardware such as CMOS[1] cameras. These advances fostered the proliferation of potential multimedia applications in various domains, such as smart monitoring, video surveillance, smart Home, etc. In fact, multimedia content, like images, audio or video, can describe better

[1] CMOS: Complementary Metal Oxide Semiconductor.

A. Puliafito et al. (Eds.): ADHOC-NOW 2017, LNCS 10517, pp. 276–289, 2017.
DOI: 10.1007/978-3-319-67910-5_23

the occurred events which can help to better understand the situation and so, react appropriately. However, delivering multimedia data through a WSN while ensuring QoS guarantees is a hard task [3,10] in such constrained environment, characterized by limited resources in energy, bandwidth and memory capacity. Actually, multimedia applications are generally resource-hungry since multimedia streams induce large size data requiring large buffer size at intermediate nodes and high bandwidth. Besides, delivering large data lasts more transmission time and consequently consumes more energy amount, causing fast energy drain. This problem is more stressed when operating under duty-cycled networks, where nodes switch between active and inactive states to reduce energy consumption. In fact, the intermittent availability of nodes leads to an extra delay for data transfer and reduces throughput. Thus, operating under duty-cycled environments makes the transfer of multimedia traffic, while meeting all the mentioned requirements, more challenging. Providing sufficient bandwidth to transfer data remains the most important problematic for multimedia routing. Hence, it is necessary to conceive an appropriate routing protocol that takes into account the specificity of large size multimedia traffic and able to ensure the QoS requirements.

In this context, forwarding multimedia traffic through multiple paths is an efficient solution, since it enables bandwidth aggregation, traffic load balance and reduces end-to-end delay. Moreover, splitting the transmission load among multiple paths distributes the energy consumption more equitably. In this regard, we introduce CFMPR-M: a Correlation Free Multipath Routing Protocol for multimedia traffic in duty-cycled WSNs. The salient feature of our work is to build non correlated paths from one source towards the sink and adapt the duty-cycling scheduling of involved nodes in a way to be able to exploit the different routes simultaneously and equally distribute energy consumption among different nodes in the paths. CFMPR-M relies on simple methods to build required routes without resort to localization or high computing algorithms.

The rest of the paper is structured as follows. Section 2 outlines related work. The description of our protocol is introduced in Sect. 3. Then, we propose to evaluate the performance of our design is Sect. 4. Finally, Sect. 5 summarizes our contribution and discusses future directions.

2 Related Work

Multipath routing [8] is a promising approach in WSNs that copes with the deficiencies of single path routing. In fact, single-path routing can't satisfy the performance demands of diverse applications, due to the resource constraints of sensor nodes in terms of limited power, storage and processing capacities in addition to the unreliability and limited bandwidth of wireless links. Indeed, multipath routing enables the establishment of multiple routes from the source towards the sink which allows to achieve the following benefits: Reliability, fault-tolerance, load balancing and QoS improvements. Load balancing allowing bandwidth aggregation through splitting traffic load over several routes is suitable for multimedia traffic characterized by high generation rate and large data size. Consequently, congestion will be reduced and energy consumption will be equally distributed among multiple nodes. To enable simultaneous data transfer through multiple paths, these paths should be non correlated to avoid inter-paths interference. In this context, authors, in [13], present EECA[2] a multipath routing protocol handling correlation issue. EECA exploits node position information to build two collision-free paths where their distance from each other exceeds the interference range R. Initially, the source node classifies its neighbors into two distinct groups, where nodes in each group lay at one side of the source-destination line, opposite to the side of the other group, and should be distanced more than $R/2$ from the source-destination line. Only these nodes will participate in Route-request (RREQ) flooding initiated by the source node. Using backoff-timers according to the node's distance towards the sink, only one node wins to broadcast its received RREQ packet at each stage of the RREQ flooding. Upon receiving a RREQ message, the sink sends a Route-reply (RREP) packet in the reverse path towards the source. The source will use the two established routes simultaneously and adjust the traffic load according to the energy situation of each route. To save energy, data transmission power is adjusted to the level just sufficient to reach the next hop in the route. I2MR[3][11] aims to increase throughput by discovering zone-disjoint paths and distribute the traffic over the discovered paths. I2MR starts by discovering the shortest path towards the sink. During the RREP phase of the first path discovery, I2MR marks one and two-hop neighboring nodes of all the intermediate nodes along this path as the interference zone of the primary path. Based on location information, I2MR selects a secondary and a backup destinations that are beyond the interference range of the primary destination and have the shortest euclidean distances from the source. Then, it constructs the shortest secondary and backup paths, towards respectively the secondary and backup destinations, through nodes outside the interference zone of the primary path. For load balancing, the source uses the primary and secondary paths concurrently and keeps the third one as the backup path. EECA and I2MR relies on location information requiring special hardware support and extra control overhead which can be costly for resource-constrained WSNs. IAMDV[4], proposed in [7], attempts to find two node-disjoint paths with minimal interference from a source towards a destination in two rounds, without needing any hardware support for location information. In the first round, IAMVD discovers the shortest path and on the way back to the source, it tries to block the neighbors of this established first route which are not supposed to participate in the routing process, using a sleeping mechanism. In the second round, the two last shortest paths, which are far away from the first one by 2R, are constructed. IAMVD splits the video stream to I-frames, P-frames and B-frames and assigns the shortest path to the most important frames (I-frames) and the alternate paths to the less important frames P and B.

[2] EECA: Energy Efficient and Collision Aware multipath routing.
[3] I2MR: Interference-Minimized Multipath Routing.
[4] IAMDV: Interference-Aware Multipath Routing for Video Delivery.

MR2[5] [6] relies on an adaptive incremental technique to construct minimum-interfering routes to satisfy the bandwidth requirements of multimedia applications. MR2 starts by building the first shortest path towards the sink through RREQ and RREP flooding. When nodes along this path receives data packets, they should notify their neighboring nodes through sending *bepassive* messages, to act as the passive nodes in order to prevent them from participating in any route discovery process. Additional paths are constructed, similarly, whenever the active paths are unable to provide the required bandwidth of the available network traffic. MR2 suffers from extra control overhead caused by sending *bepassive* messages. Similarly, LIEMRO [9][6] resorts to an adaptive iterative approach to built a sufficient number of interference-minimized node disjoint paths towards the sink. Each iteration is composed of a discovery and an establishment phases, building a new path to be used for data transfer in case of throughput enhancement. Otherwise, it will be disabled and the route construction process will be terminated. To build interference-minimized paths, LIEMRO uses a cost function, which considers the interference level that a node has experienced. For load balancing, LIEMRO splits data traffic over multiple established routes, according to the relative quality of each path. Contrarily to previous protocols, FMRP [5][7] attempts to construct multiple node-disjoint and non correlated paths within only two steps. The discovery phase initiated by the sink through flooding an interest message throughout the network. Each node includes in its interest message the list of its known predecessors and neighbors within the same level. In order to collect information about the predecessors neighborhood, each node differs its interest transmission for a predefined time sufficient enough to receive all interest messages from these neighbors. Based on hop count information included in the interest message, each node stores information about its predecessors and neighbors within the same level into two different tables serving as a local view for the second step. The second step is the route construction phase triggered by a source node through sending RREP packets. The source node selects from its predecessors nodes which are not immediate neighbors, as next hop nodes. If no such combination exists, it resorts to the neighbors belonging to the same level. The source node, as well as intermediate nodes, include information about common neighbors of selected next hop nodes. These nodes are to avoid, whenever possible, by the subsequent nodes when selecting their next hop nodes in order to have node-disjoint paths. FMRP resorts to REFUSE messages to correct or adjust built paths.

Most of existing multipath routing protocols are designed for always-on networks without considering duty-cycled environments. NDRECT[8], proposed in [2], is a multipath routing protocol operating under duty-cycled network. However, it focuses only on building node-disjoint paths to handle urgent traffic and ignores correlation problem. In the best of our knowledge, our design is the

[5] MR2: Maximally Radio-Disjoint Multipath Routing.
[6] LIEMRO: Low-Interference Energy efficient Multipath ROuting.
[7] FMRP: Fast Multipath Routing Protocol.
[8] NDRECT: Node-Disjoint Routes Establishment for Critical Traffic.

first multipath routing protocol dealing with correlation issue in a duty-cycled environment.

3 Protocol Overview

The objective of CFMPR-M is to satisfy the requirements of multimedia applications, mainly related to high throughput and acceptable end-to-end delay while prolonging the network lifetime. Thus, we propose to construct non correlated node-disjoint paths towards the sink in order to smoothly deliver multimedia traffic without being disturbed by neighboring transmissions, without the need of a global view of the network topology or location awareness of the different nodes. Nodes disjointedness enables load balancing by splitting the traffic load among different paths which allows an efficient resources use. However, it is not sufficient to provide high throughput and reduced delay since we can't avoid inter-paths correlation preventing the use of different paths simultaneously. To build required paths, only two steps are needed. The first step is the discovery phase, initiated by the source through flooding RREQ packets. The basic idea consists in eliminating nodes, able to be involved in correlated or braided paths, from participation in routing process and creating local view to help each node during the next hop selection. Besides, the objective of this phase is to gather information about eventual routes towards the sink, able to satisfy application requirements. The second step is route establishment phase, which is triggered by the sink node after receiving all RREQ sent by the source. In this step, the sink, as well as eligible intermediate nodes, will exploit local information acquired during the first step to decide next hop node in each path.

For energy efficiency, our routing protocol is performed under a duty-cycled network, based on E-ECAB[9][1,4] as an underlying MAC protocol. E-ECAB supports efficiently broadcast operations, useful for disseminating queries during routes discovery and route establishment phases. In E-ECAB, nodes can adjust their scheduling according to multiple contexts which are: neighborhood knowledge, traffic load, medium awareness and history awareness. Thanks to neighborhood knowledge acquired by means of beacons exchange and to the traffic load observation, nodes can adapt their functional behavior and adjust their scheduling according to the actual network state. Depending on current circumstances, nodes will be either active for a maximal period (PAmax) and inactive for a minimal period (PImin) or active for a minimal period (PAmin) and inactive for a maximal inactive period (PImax). Upon starting the discovery phase, each node switches to maximal period activity, in order to ensure the well reception and overhearing of subsequent routing queries, until being involved in a route or the expiration of a predefined timer. The flexibility of E-ECAB scheme makes it suitable also for unicast transmission. In order to avoid the negative impact of duty-cycling on network performances, we resort to the cross-layer coordination between MAC and routing layers to adapt nodes scheduling according to network

[9] E-ECAB: Enhanced Efficient Context-Aware multihop Broadcast over duty-cycled WSNs.

layer needs. In fact, selected nodes will adjust their scheduling according to the source's transmission rate, in addition to the other contexts used by E-ECAB, in order to allow fluent data transfer in a pipeline fashion.

Duty-cycle adjusting process is out the scope of this paper.

In the following, we provide details of the discovery and establishment phases of our protocol then we describe the maintenance process.

3.1 Discovery Phase

The objective of the discovery phase is to acquire sufficient information allowing to build a local view for each node. These collected information will guide the sink node as well as intermediate nodes to make local decisions when selecting the best next hop for each path. This phase is triggered by the source node through flooding a RREQ packet. To be able to construct two non correlated and non braided paths, we need to select as First Hop, nodes which are non immediate neighbors in order to reduce the risk of crossed or neighboring paths, which suffer from inter paths correlation. Then, we discard nodes able to create correlated or crossed paths from the routing process.

Thanks to neighborhood knowledge acquired by means of beacons exchange when running E-ECAB protocol at the MAC level, the source node includes in an added field a list of First Hop pairs. Each pair is composed of the IDs of 2 successor nodes that are non immediate neighbors and each first hop can be included only in one pair. This list specifies the list of First Hop nodes that are allowed to forward the received RREQ packet, sent by the source node. This list is the key that will guide to construct non correlated and non braided node-disjoint paths.

First Hop pairs selection and RREQ handling: The source node selects a first neighboring node A from the neighboring list. Then, it picks out another node from the list that is a non immediate neighbor with node A. The source then repeats the same instructions with the remaining neighbors. Selected nodes to form a pair must not figure in already formed pairs. The selection process ends after parsing all the source neighbors.

Upon receiving a RREQ from the source, each selected First Hop node will broadcast it to its neighbors. However, if all the first hop nodes contend at the same time for channel access, the risk of collisions increases. To overcome such problem, we should differ the transmission of RREQ packets by the different First Hop pairs. Thus, the source node assigns for each First Hop pair pi different boundaries $[Bi1, Bi2]$ to select a backoff timer.

Before forwarding the RREQ packet, each First Hop node adds the list of its neighbors. This information will be used by the sink during the establishment phase to avoid selecting routes with neighboring First Hop nodes. The objective of the discovery phase is to gather information in order to create a local view for each node useful during the second step for routes selection. Thus, each node must wait the reception of all RREQ packets sent by its predecessors. Thanks to the included hop count information, only RREQ packets, received from the predecessors providing a minimal hop count and from a novel First Hop node, are considered, the others are discarded. If a node receives 2 RREQ packets with the same hop count from two different First Hop belonging to the same First Hop pair, it must keep silent and cancel the RREQ forwarding for that pair. The aim is to exclude some nodes, eligible candidates in many braided paths, from the participation in the forwarding of some RREQ packets during the discovery phase. To this end, we introduce a selective RREQ forwarding algorithm operating as follows: A node receiving 2 RREQ packets, originated from 2 different First Hop nodes with the same hop count, checks if these nodes constitute a pair figuring in the list of First Hop pairs. In such situation, the node keeps silent and discards the according RREQ packets. However, if the hop counts are different, the node will consider only the packet with less hop count and discards the other one. Otherwise, the node will consider the two packets and proceeds then to forwarding process.

To optimize the functional behavior of our protocol, we exploit the RREQ packets aggregation which reduces control overhead.

Algorithm 1. RREQ selection and forwarding algorithm: without aggregation

```
n: number of received RREQ with minimal hop count;
RcvRREQList={RREQ[i]/i ≤ n}: List of all received RREQ with minimal hop count;
validRREQList=∅;RejectedFH=∅;
for (i = 0;i < n i++) do
   for (k=0;k<n;k++) do
      if RREQ[i].FH=RREQ[k].FH then
         RejectedFH=RejectedFH ∪ {RREQ[i].FH; RREQ[k].FH}
      end if
   end for
end for
for (j=0;j<n i++) do
   if RREQ[j].FH ∉ RejectedFH then
      validRREQList=validRREQList ∪ {RREQ[j]}
   end if
end for                        ▷ validRREQList, the list of valid RREQ packets to be forwarded
for pkt in validRREQList do
   pkt.src=node ID
   Forward pkt
end for
```

Aggregation process: In order to alleviate control packets overhead incurred during RREQ flooding, we can have recourse to aggregation. We propose three types of forwarding: Without aggregation, partial aggregation and full aggregation.

- Without aggregation: As explained in Algorithm 1, each node should wait the reception of all RREQ packets. After filtering eligible packets, nodes forward each packet, while considering the fixed boundaries for backoff selection of each First Hop pair, in order to avoid eventual collisions. Redundant transmissions may reduce the risk of RREQ packets losses, which improves route discovery. However, such alternative may induce extra delays spent in transmitting successive multiple packets and increase control overhead.

- Partial aggregation: As detailed in Algorithm 2, partial aggregation allows each node to send an aggregated packet of RREQ packets originated from the same First Hop node. For example in Fig. 1a, the node L receives 2 RREQ packets from nodes E and F. These two packets are issued from the same First Hop node A. Thus, node L will send only one RREQ packet where it specifies node A as the First Hop.
- Full aggregation: In this variant, described by Algorithm 3, all considered packets will be aggregated in one packet while specifying in the First Hop field, the list of all First Hop traversed by the considered RREQ packets. As an example, when node F in Fig. 1a receives two RREQ packets from the First Hop A and the First Hop B, which are not associated to the same pair, it forwards only one aggregated packet containing in the First Hop field the IDs of A and C.

Algorithm 2. RREQ selection and forwarding algorithm: partial aggregation

```
n: number of received RREQ with minimal hop count;
RcvRREQList=RREQ[i]/i ≤ n: List of all received RREQ with minimal hop count;
validFH=∅; RejectedFH=∅;
for (i=0;i<n i++) do
   for (k=0;k<n;k++) do
      if RREQ[i].FH=RREQ[k].FH then
         RejectedFH=RejectedFH ∪ {RREQ[i].FH; RREQ[k].FH}
      end if
   end for
end for
validFH=validFH \ RejectedFH
i=0;
for fh in validFH do
   i++;
   aggRREQ[i].FH=fh
   aggRREQ.hopcount=minHopcount(fh)+1
   aggRREQ.originalSrc=RREQ[0].originalSrc
   aggRREQ.src= node ID
   Forward aggRREQ
end for
```

For each RREQ receipt, each node creates a new entry in RREQ table where it stores information about the original source and the sender of the RREQ packet, the First Hop node and the number of hops towards the source. This table will help the node to select the appropriate next hop during the second phase.

Figure 1a summarizes RREQ propagation during the discovery phase. Selected First Hop pairs are (A, C) and (B, D). The RREQ packets coming from A and C are propagated through green links while RREQ packets coming from B and D are propagated through pink links. Nodes discarded from forwarding RREQ packets associated to the pair (A, C) are G, M and R. Those eliminated from participating in forwarding RREQ packets associated to the pair (B, D) are nodes H, N and S.

Algorithm 3. RREQ selection and forwarding algorithm: full aggregation

```
n: number of received RREQ with minimal hop count;
RcvRREQList=RREQ[i]/i ≤ n: List of all received RREQ with minimal hop count;
validRREQList=∅; aggRREQ=∅;RejectedFH=∅;aggFH=∅;nAgg=0;
for (i=0;i<n i++) do
   aggFH=aggFH ⋃RREQ[i].FHlist
   for (k=0;k<n;k++) do
      for fh inRREQ[i].FHlist do
         if ∪fh ∈ RREQ[k].FHlist then
            RejectedFH=RejectedFH⋃{fh; ∪fh}
         end if
      end for
   end for
end for
aggFH=aggFH \ RejectedFH
aggRREQ.FHlist=aggFH
aggRREQ.hopcount=minHopcount+1
aggRREQ.originalSrc=RREQ[0].originalSrc
aggRREQ.src= node ID
Forward aggRREQ
```

3.2 Path Establishment Phase

The second phase is initiated by the sink node. Upon receiving all RREQ from its predecessors, the sink node starts the establishment phase by associating to each selected predecessor a distinct First Hop node, based on its local view. In order to avoid correlated paths, the selected predecessors as well as their corresponding First Hop nodes must be non immediate neighboring nodes. To ensure such condition, the sink node will favor the selection of two First Hop nodes forming a pair whenever it is possible. Otherwise, it can check if two First Hop nodes are immediate neighbor thanks to the neighbor list associated to each First Hop node included in received RREQ packets. After, selecting appropriate pairs (predecessor, corresponded First Hop), the sink sends to each selected neighbor a RREP packet containing all necessary information to guide intermediate nodes to select their next hop to reach the source while traversing the selected First Hop. The RREP packet includes the following fields: The selected pair (predecessor, corresponded First Hop), the sender ID, the selected next hop ID, the original source ID, hop count and the sequence number. RREP packets must be acknowledged to ensure their well reception by the concerned nodes. Each node upon receiving a RREP packet will choose as a next hop towards the source, a predecessor from which it has received a RREQ, with same original source and through the First Hop node specified in the RREP packet. Thanks to the elimination of some nodes from the forwarding of RREQ packets reached from two First Hop forming a pair, a node can be ensured that any eligible predecessor to be selected in a route can't be involved in the second route.

The establishment phase ends when the source node receives the 2 RREP packets associated to the 2 selected paths. Upon RREP receipt, each node stores, in its routing table, information about the next hop ID related to each source. During data transfer, each node will refer to the routing table to forward data to the corresponding next hop related to the original source of received data.

The example shown in Fig. 1 illustrates the path establishment phase. At the end of the discovery phase, the sink node receives RREQ packets from its neighbors Q, R, S and T coming from distinct First Hop nodes A, B, C and D, where (A, C) and (B, D) form the 2 considered First Hop pairs. The sink node starts by trying to associate each First Hop pair to two predecessors which are non immediate neighbors, such as Q and S associated to the pair (A, C). Assuming the sink will select this last combination. Hence, it will associate to node Q the First Hop A and to node S the First Hop C, as illustrated in Fig. 1b. The first RREP sent to node Q will construct the first route, for example the route (Q, K, F, A, SRC). And, the second RREP sent to node S will construct the route (S, N, I, C, SRC), as shown in Fig. 1b.

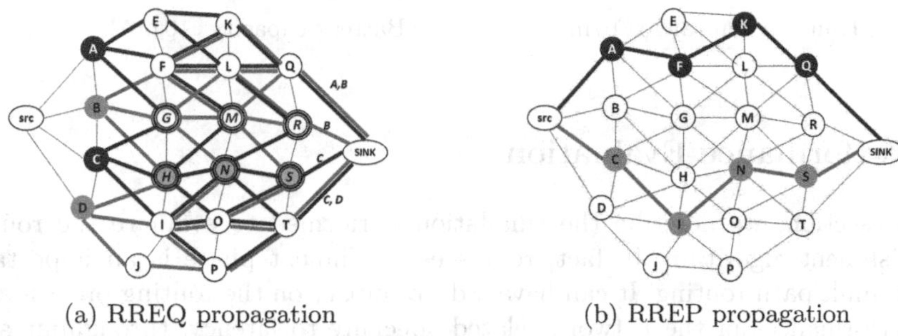

Fig. 1. Example of routes selection and RREP propagation

3.3 Maintenance Phase

The validity of selected routes is constrained by nodes remaining energy. Hence, in order to anticipate routes recalculation before a path failure, we propose to estimate the lifetime of current routes. The lifetime of a route R, $RouteLife(R)$, is equal to the minimal lifetime of nodes belonging to that route. To estimate the route lifetime, we should estimate how long nodes in the route could keep alive. Thus, being aware of the traffic load and generation rate, each node N computes its energy consumption $EC_k(N)$ each period k, a period should be enough large allowing a node to receive and forward a data packet. Then, the node Ni can compute an estimation of its lifetime, $lifetime(Ni)$, as follows:

$$lifetime(Ni) = \frac{RE(Ni)}{Moy(EC_j(Ni)\ ;\ 0 < j \leq k)} \tag{1}$$

where $RE(Ni)$ is the residual energy of node Ni and $Moy(EC_j(Ni)\ ;\ 0 < j \leq k)$ represents the mean of energy consumption during a period, computed as the sum of energy consumed during all past periods divided by the number of these periods.

When the node's lifetime exceeds a predefined threshold $T > 1$, it notifies the source node through transmitting an *ALERT* frame. In fact, when $lifetime(Ni)$ is below 1, it means a node failure. Hence the threshold value should allow the node to send an ALERT message before its energy depletion. Upon receiving the ALERT frame, the source will trigger a novel discovery phase.

Table 1. Simulation parameters

Parameter	Value	Parameter	Value
Network size	50; 100; 150; 200	PA_{max}	0.00445 s
Average degree	7	PI_{min}	0.0017 s
Bandwidth	1 Mbits/s	PA_{min}	0.002225 s
Transmission range	10 m	Battery capacity	10 mAh

4 Performance Evaluation

In this section, we describe the simulation performed to evaluate the routes establishment algorithm. In fact, routes establishment phase is an important step in multipath routing. It can have a deep impact on the routing process and the performances of the network related specially to latency, throughput and energy, since route maintenance may be costly for one session routing approach.

4.1 Evaluation Metrics and Simulation Parameters

We focus on evaluating the performances related to the control overhead and the time needed for paths construction. Besides, we have to quantify the amount of energy consumed during this phase to show the benefits of operating in duty-cycle mode. Hence, we consider the following metrics:

- Control overhead: it refers to the sum of all control packets sent during the routes construction phase, like Interest, ACK, REJECT and RREP packets for FMRP and RREQ, ACK and RREP messages for our protocol.
- The total time spent for paths establishment: It refers, for CFMPR-M, to the time spent from sending the first RREQ packet until receiving the RREP packets by the source. For FMRP, it represents the time spent from the interest flooding by the sink node until the reception of the RREP packet by the sink.
- The average amount of energy consumed during paths establishment: It is equal to the sum of energy consumed by all the nodes during the simulation time divided by the number of nodes in the network.

We test the different protocols in multiple network topologies with various network sizes, with one sink and a unique source. We put the source node at the left side while the sink is placed at the right side of the simulation area, in order to allow a multihop communication. We consider dense networks where each node has at least 4 neighbors with an average degree equal to 7, in order to maximize the number of routes from the source towards the sink. Simulation parameters are summarized in Table 1. For CFMPR-M protocol, we suggest to compare 3 variants where each variant adopts a different RREQ aggregation type:

- CFMPR-M (WA): RREQ packets are forwarded without aggregation.
- CFMPR-M (PA): RREQ packets are forwarded using partial aggregation.
- CFMPR-M (FA): RREQ packets are forwarded using full aggregation.

The objective of CFMPR-M is to build two correlation-free paths in order to enable a concurrent transfer of data allowing load balancing and bandwidth improvement. Hence, we choose to compare our protocol against FMRP which similarly considers correlation issue to build paths but it is not designed for duty-cycled environment. So, we can assess the gain as well as penalties when operating with duty cycled mode.

For our evaluation, we conduct several simulations under the OMNeT++ simulator [12], where we implemented all proposed variants and protocols.

4.2 Results

Figure 2a shows the time duration needed to build the required routes for CFMPR-M variants and FMRP. Related results are very close mainly in small topologies but FMRP and CFMPR-M (FA) achieve less delays for route construction while the variant CFMPR-M (WA) shows significantly higher delay. We can remark clearly that the delay increases when the aggregation rate reduces. The extra delay of each variant is caused by the RREQ forwarding strategy. For CFMPR-M (FA), only one RREQ packet will be sent. However, for the other variants, each node may send subsequently more than one packet; As the number of packets to be forwarded increases, the forwarding time increases leading to more extra delays. For CFMPR-M (FA), each node will forward at most x packets, where x is the number of First Hop nodes. However, in CFMPR-M (WA), the maximal number of forwarded packets is multiplied by the number of predecessors. In FMRP, nodes also have to wait for a time before forwarding an Interest message; This additional time is introduced at each node allowing it to receive all the interests sent from its predecessors which are the eventual successors on the reverse path towards the sink. In all variants of CFMPR-M, nodes have also to wait the reception of all RREQ packets from all their predecessors. The difference between FMRP and our protocol is that, in FMRP, only one Interest message is propagated while in our protocol each RREQ coming from a distinct First Hop node is considered as a distinct RREQ packet. As nodes of each First Hop pair send their RREQ packets in differed time, the time spent waiting the reception of these distinct RREQ increases. The waiting time is reduced in CFMPR-M (FA) thanks to full aggregation which makes the forwarding process similar to FMRP. The extra delay introduced in CFMPR-M (FA) is due to the duty-cycle mode. Results related to construction time are consolidated by measurements related to the control overhead, illustrated in Fig. 2b.

We can notice the huge differences between sent control packets between all variants of CFMPR-M. The partial aggregation in CFMPR-M (PA) and mainly the full aggregation in CFMPR-M (FA) allow a significant decrease in control overhead. CFMPR-M (FA) shows better results than FMRP thanks to, first the full aggregation process and secondly to the preventive RREQ forwarding methods that discards some nodes from forwarding some RREQ packets when received from two First Hop nodes forming a First Hop pair. Besides, it allows to prevent the sink node as well as intermediate nodes from selecting as next hop, nodes that can be involved in correlated paths. In this way, our protocol doesn't need extra control packets such us REJECT and REFUSE messages useful to remedy to paths correlation caused by selecting wrong nodes as next hop.

The impact of control overhead on energy can't be clearly shown, since the size and the number of exchanged messages are not important, as we can remark in Fig. 2c when comparing results of CFMPR-M variants. Nevertheless, we can observe noticeably the difference between energy consumption in FMRP and CFMPR-M. CFMPR-M consumes less energy amount than FMRP, thanks to operating with a duty-cycle mode which allows to reduce energy dissipation in idle listening and unnecessary overhearing.

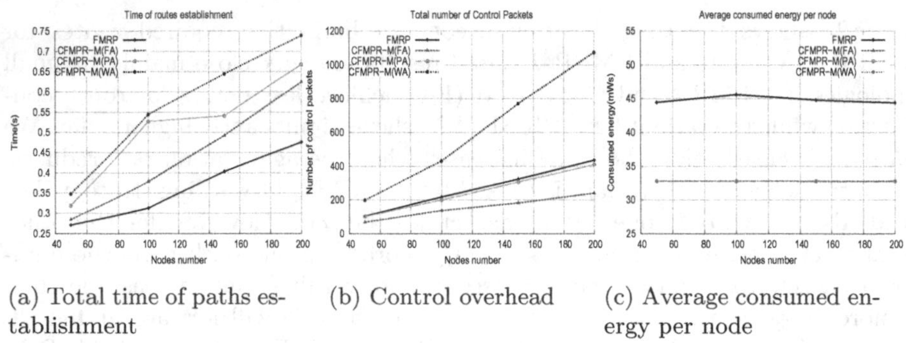

(a) Total time of paths establishment

(b) Control overhead

(c) Average consumed energy per node

Fig. 2. Simulation results

5 Conclusion

In this paper, we have presented CFMPR-M, a correlation-free multipath routing protocol to support multimedia traffic in duty-cycled environment. The objective is to satisfy QoS requirements in terms of high throughput and reduced latency while prolonging network lifetime. CFMPR-M allows to build two non correlated paths from a source towards the sink in order to enable simultaneous data transfer through the different paths. Simulation results show the effectiveness of aggregation process of CFMPR-M to unburden the network from additional control overhead. Besides, thanks to the preventive RREQ forwarding method, nodes can safely select appropriate next hop nodes relying on their local view created during RREQ propagation. Such mechanisms allowed to decrease the impact of duty-cycling on time establishment and achieve a trade off between time establishment and energy efficiency. As a future work, we intend to test data transfer through established paths to highlight more the benefits of the cross layer coordination between MAC and routing levels to ensure multimedia QoS needs along with energy saving guaranteed by operating in duty-cycle mode.

Acknowledgments. This work has been funded by PHC-Utique project No. 17G 1417.

References

1. Ghrab, D., Jemili, I., Belghith, A., Mosbah, M.: ECAB: an Efficient Context-Aware multi-hop Broadcasting protocol for wireless sensor networks. In: 2015 International Wireless Communications and Mobile Computing Conference (IWCMC), pp. 1023–1029. IEEE (2015)
2. Ghrab, D., Jemili, I., Belghith, A., Mosbah, M.: NDRECT: Node-Disjoint Routes Establishment for Critical Traffic in WSNs. In: 2016 International Wireless Communications and Mobile Computing Conference (IWCMC), pp. 702–707. IEEE (2016)
3. Han, G., Jiang, J., Guizani, M., Rodrigues, J.J.C.: Green routing protocols for wireless multimedia sensor networks. IEEE Wirel. Commun. (2016)
4. Jemili, I., Ghrab, D., Belghith, A., Mosbah, M.: Context-aware broadcast in duty-cycled wireless sensor networks. Int. J. Semant. Web. Inf. Syst. (IJSWIS) **13**(3) (2017)
5. Jemili, I., Tekaya, G., Belghith, A.: A fast multipath routing protocol for wireless sensor networks. In: 2014 IEEE/ACS 11th International Conference on Computer Systems and Applications (AICCSA), pp. 747–754. IEEE (2014)
6. Maimour, M.: Maximally radio-disjoint multipath routing for wireless multimedia sensor networks. In: Proceedings of the 4th ACM Workshop on Wireless Multimedia Networking and Performance Modeling, pp. 26–31. ACM (2008)
7. Nikseresht, I., Yousefi, H., Movaghar, A., Khansari, M.: Interference-aware multipath routing for video delivery in wireless multimedia sensor networks. In: 2012 32nd International Conference on Distributed Computing Systems Workshops (ICDCSW), pp. 216–221. IEEE (2012)
8. Radi, M., Dezfouli, B., Bakar, K.A., Lee, M.: Multipath routing in wireless sensor networks: survey and research challenges. Sensors **12**(1), 650–685 (2012)
9. Radi, M., Dezfouli, B., Bakar, K.A., Razak, S.A., Nematbakhsh, M.A.: Interference-aware multipath routing protocol for QoS improvement in event-driven wireless sensor networks. Tsinghua Sci. Technol. **16**(5), 475–490 (2011)
10. Shen, H., Bai, G.: Routing in wireless multimedia sensor networks: a survey and challenges ahead. J. Netw. Comput. Appl. **71**, 30–49 (2016)
11. Teo, J.Y., Ha, Y., Tham, C.K.: Interference-minimized multipath routing with congestion control in wireless sensor network for high-rate streaming. IEEE Trans. Mob. Comput. **7**(9), 1124–1137 (2008)
12. Varga, A.: OMNeT++ User Manual, Version 4.1 (2010). www.omnetpp.org
13. Wang, Z., Bulut, E., Szymanski, B.K.: Energy efficient collision aware multipath routing for wireless sensor networks. In: IEEE International Conference on Communications, 2009, ICC 2009, pp. 1–5. IEEE (2009)

Impact of Simulation Environment in Performance Evaluation of Protocols for WSNs

Affoua Therese Aby[1(✉)], Marie-Françoise Servajean[1], Nadir Hakem[2], and Michel Misson[1]

[1] UCA (Université Clermont-Auvergne), LIMOS, CNRS, UMR 6158, BP 10448, 63175 Aubière, France
{affoua_therese.aby,m-francoise.servajean,michel.misson}@uca.fr

[2] UQAT (Université du Québec en Abitibi-Temiscamingue), LRTCS, Val d'Or, Qc, Canada
nadir.hakem@uqat.ca

Abstract. Wireless sensor networks are increasingly used in many emerging applications. This type of network is composed of hundreds of low-cost sensor nodes, but with a limited budget batteries, low communication range, limited throughput, reduced computing power, low memory and low storage capacity. Communication protocols are proposed in the literature to deal with technical challenges coming from low intrinsic resources of sensor nodes. In most of these studies, simulations comparing a proposed protocol with other existing protocols are performed to show that the proposed protocol provides overall better performance. However, the environmental specification that made these comparisons is very often neglected or non-existent. In this study we show that it is essential to have the simulation environment very well defined before considering whether a protocol provides better performance than others. To do this we use two duty cycle MAC protocols, the standard IEEE 802.15.4 and SlackMAC (a protocol that we proposed). The aim of the paper is not to make an exhaustive comparison of protocols. We intuitively know that SlackMAC provides a better overall performance than the standard. What we are trying to show is the gap between performances according to simulation conditions. We will mainly focus on the topologies used and the capture effect. The results draw attention to the fact that it is essential to clearly define the simulation environment and also to reconcile the chosen conditions with the results when comparing the performances of two protocols.

Keywords: WSNs · QoS · Performance evaluation · Simulation conditions · Topologies impact · Capture effect impact

1 Introduction

Wireless Sensor Networks (WSNs) have attracted great interest in the last decade. This explains the rich and active research about communication protocols for WSNs in order to deal with technical constraints related to sensor

A. Puliafito et al. (Eds.): ADHOC-NOW 2017, LNCS 10517, pp. 290–304, 2017.
DOI: 10.1007/978-3-319-67910-5_24

nodes. In most of these studies, in the absence of real conditions to validate the proposed protocols, many simulations are carried out to evaluate their performances. Mostly, these simulations are done with performance comparisons of proposed protocols compared to existing one to show that the new protocol provides better performance. However, sometimes the level of detail on the results surpasses the description of the simulation conditions used to obtain them.

In this paper we raise the curtain on this issue and we draw attention regarding the quantitative results of simulation when communication protocols are compared. Let us take the case of works that conduct to establish efficient mechanisms to ensure a long lifetime to the network. We take as examples energy-efficient MAC protocols based on sequences of active periods (during which the radio module is on) and sleep periods (during which it is off) called duty cycle. The duty cycle represents the proportion of the active period over the total duration of the cycle (active period + sleep period). The main energy-saving MAC protocol based on this mechanism is the standard IEEE 802.15.4 [1] in beacon-enabled mode. In this type of protocols nodes agree on a common calendar for their periods of activity and sleep. This category is called synchronous duty-cycle MAC protocols. Other categories based on an asynchronous mechanism are also proposed in the literature. In these second categories, nodes do not have a common calendar for their period of activity and sleep. Most of these protocols are proposed as improvements to some existing protocols. However, there is a lack of important information on simulation conditions such as: radio frequency, antenna, propagation model, transmission power, topology information, data traffic information, capture effect, etc.

These omission do not allow to judge the relevance of the results and also does not allow reproducibility. There are studies that show the impact of different propagation models and topologies on simulation results as in [2,3]. However, to the best of our knowledge, no studies have shown these impacts with such a high level of detail as that in this paper. We use two energy-efficient MAC protocols, the standard IEEE 802.15.4 [1] and SlackMAC [4,5] (a protocol that we have proposed). The aim of the paper is not to make an exhaustive comparison of protocols. We know intuitively that SlackMAC generally provides better performance than the standard. What we are trying to demonstrate through intensive simulations is the diversity in the performance gaps according to simulation conditions. Mainly, we will focus on the topologies used and the capture effect.

This paper is organized as follows. In Sect. 2, we make a summary of duty cycle MAC protocols and take note of the levels of detail on the missing simulation conditions. In Sect. 3, we analytically compare the two protocols that serve tests in our study and give the technical details of our study. In Sect. 4, we show simulation results from this comparative study. Finally, we conclude our work in Sect. 5.

2 State-of-the-art

The MAC protocols of the literature based on the duty cycle mechanism can be classified into two main categories such as synchronous duty cycle MAC protocols and asynchronous duty cycle MAC protocols.

2.1 Synchronous Duty Cycle MAC Protocol

In synchronous duty cycle MAC protocols synchronization can be global or local.

In the case of global synchronization, all nodes share a common schedule for their periods of activity and inactivity. The main MAC protocol based on this mechanism is the standard IEEE 802.15.4 [1] in beacon enabled mode. This is also the case of protocol in [6] and LO-MAC (*Low Overhead* MAC) [7].

In the case of local synchronization, the nodes are synchronized by neighborhood, very often according to a tree topology as a function of their position relative to the sink. One of the first protocols based on this mechanism is S-MAC (*sensor-MAC*) [8] then D-MAC (*Data-gathering MAC*) [9] and TreeMAC [10]. Improvements of these protocols are proposed in ID-MAC [11], DW-MAC (*Demand Wakeup MAC*) [12], DSF (*Dynamic Switch-based Forwarding*) [13] and iCore [14].

2.2 Asynchronous Duty Cycle MAC Protocol

In asynchronous duty cycle MAC protocols, nodes do not have a common calendar for their activity and sleep periods. A distinction can be made between sender-initiated and receiver-initiated MAC protocols.

In sender-initiated MAC protocols, most of the communication load is supported by sender nodes. The main protocol based on this principle is B-MAC protocol *Berkeley MAC*) [15]. In B-MAC, a sender sends a long preamble before sending data frame. X-MAC [16] is one of the first improvement of B-MAC. Many other protocols such as BoX-MAC [17] and OSX-MAC [18] have been proposed thereafter.

Unlike sender-initiated MAC protocols, in receiver-initiated MAC protocols, most of the communication load is supported by the receiver nodes. In receiver-initiated MAC protocols the receiver initiates the communication by sending a beacon frame to express its ability to receive data frame. RI-MAC [19] is the main protocol based on this mechanism. RIX-MAC (*Receiver-Initiated X-MAC*) [20], ERI-MAC [21], OC-MAC [22] and protocol in [23] are improvement of RI-MAC.

Other hybrid MAC protocols that are both sender-initiated and receiver-initiated have been proposed to balance the communication load on the sender and receiver nodes. This is the case of protocol in [24] and SlackMAC [4,5].

2.3 Overview of Simulation Conditions

In most of the protocols cited above comparisons by simulation on NS2 [25] are performed to show that the proposed protocols provide better overall performance than those of the existing. Table 1 gives a summary of simulation conditions description used in these comparisons. These simulation conditions essentially concern radio frequency, antenna (type, gain, height), propagation model, transmission power (noted TX power in Table 1). It is also important to provide information about topology used (type, size, transmission range that provides connectivity, number of nodes, and sink position). Parameters such as receive power threshold and carrier sense threshold (noted RXThresh_ and CSThresh_), and capture

Table 1. Overview on specification of NS2 simulation conditions

Simulation conditions / Protocols	DW-MAC [12]	LO-MAC [7]	ID-MAC [11]	DSF [13]	iCore [14]	OSX-MAC [18]	OC-MAC [22]	RIX-MAC [20]	[23]	ERI-MAC [21]	[24]	SlackMAC [5,4]
Radio Frequency	✓	✓	✗	✓	✗	✗	✗	✗	✗	✓	✗	✗
Antenna: -type -gain -height	1/3	1/3	✗	✓	✗	✗	✗	1/3	✗	✗	✗	✗
Propagation model	✓	✓	✗	✗	✓	✗	✗	✓	✓	✗	✓	✓
TX power	✓	✓	✓	✗	✗	✓	✗	✓	✗	✓	✓	✓
Topology: -type -size -TX range -number of nodes -sink position	✓	✓	✓	3/5	✓	1/5	✓	3/5	✓	✓	✓	✓
-RXThresh_ -CSThresh_	✗	✓	✗	✗	✗	✓	✗	✓	✗	✗	✗	✗
Traffic: -type -size -period	✓	2/3	2/3	1/3	1/3	1/3	2/3	✓	2/3	2/3	2/3	2/3
Capture Threshold	✗	✗	✗	✗	✗	✗	✗	✗	✗	✗	✗	✗

threshold (noted CPThresh_) are all equally important. Traffic information such as type, size in byte and generation period of data packets are also required.

It can see from Table 1 that these important elements of simulation conditions are often not fully defined and sometimes non-existent.

3 Study Framework

In this section we first describe the operating mechanism of the reference synchronous MAC protocol described in the standard IEEE 802.15.4 [1] in beacon enabled mode and the asynchronous MAC protocol SlackMAC [4,5]. Then we give the technical details of the implementation of our comparative study.

3.1 Description of the Operating Mechanism of IEEE 802.15.4 and SlackMAC

In the beacon enabled mode of the standard IEEE 802.15.4, the medium is accessed using the slotted CSMA/CA (Carrier-Sense Multiple Access with Collision Avoidance) algorithm. All nodes wake up periodically together and share a common activity throughout the SD (Superframe Duration) period and change to sleep mode the rest of the BI (Beacon Interval) period.

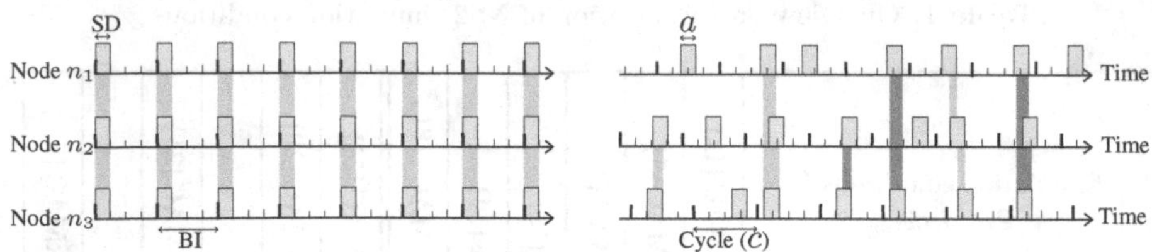

Fig. 1. Example of activity of three nodes n_1, n_2 and n_3 in range, with a duty cycle of 25% for respectively the standard IEEE 802.15.4 (left) and SlackMAC (right).

The left side of Fig. 1 shows an example of the activity and sleep mechanism in the standard IEEE 802.15.4 with a duty cycle of 25% ($\frac{SD}{BI} \times 100$) for three nodes (n_1, n_2 and n_3) in range. It can be seen that all the activity of the nodes is concentrated on the period SD. This reduces the communication time given to each node and increases the risk of collision depending on the number of nodes in competition to access the medium.

In SlackMAC, the medium is also accessed using the slotted CSMA/CA algorithm. The right side of Fig. 1 shows an example of activity cycle for three nodes (n_1, n_2 and n_3) in range with a duty cycle of 25% ($\frac{a}{c} \times 100$) in SlackMAC. Unlike the standard in SlackMAC, initially all nodes choose their activation times uniformly at random in the cycle. When a node chooses a time that yields to successful communications (reception or transmission of a frame), it memorizes it and the probability to choose this time increases, as can be seen in darker towards the right of Fig. 1. It is noted that the activity of the nodes can be distributed over the whole cycle unlike the standard. Indeed, on average there are few active nodes at the same time, which reduces the risk of collision and allows a higher communication time per nodes than the standard for the same duty cycle as shown in [4].

According to such conditions, it is intuitively known that SlackMAC will ensure overall better performance than the standard. However, what we are trying to show in what follows is the diversity in the performance gaps according to simulation conditions.

3.2 Technical Details

In this part we detail the technical environment used for our simulations. We want to highlight the fact that in a deployment study and a priori evaluation of its performance by simulation of WSNs, it is essential to be attentive and precise on the description of simulation conditions. In a simulation process we know how the choice of the propagation model is important for the relevance of the results. In this paper we will consider that this choice has been made at best and we focus on the impact of the chosen type of topology and on the effects of the given value of capture threshold parameter, often neglected in literature. The topology (or distance between nodes), the propagation model, and the capture threshold are three elements concerned to distinguish a collision of a successful reception. In a first step we propose a small synthesis of the topologies usually exploited but not often justified and/or not sufficiently detailed. On the second hand we provide some clarification on why it is important to specify the capture effect.

Topology Production Strategy. In addition to specifying the propagation model and its associated parameters, it is also important to give the rules that allowed to generate the topology (or topologies) used for the simulation. Generally, in the used WSNs topologies for communication protocol comparison tests, nodes are positioned according to a grid or randomly distributed over a given area.

Figures 2 and 3 illustrate respectively the topologies in square and pseudo-linear, with in each case, one hundred nodes are positioned both in grid and randomly.

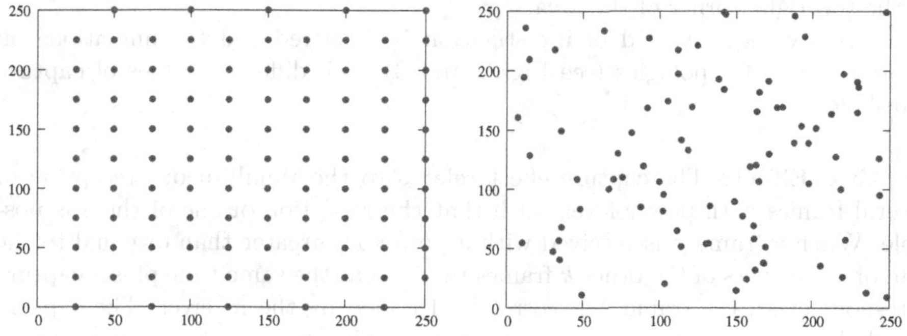

Fig. 2. Topology of 100 nodes on square area of 250 m × 250 m with respectively a grid positioning (left) and random positioning (right).

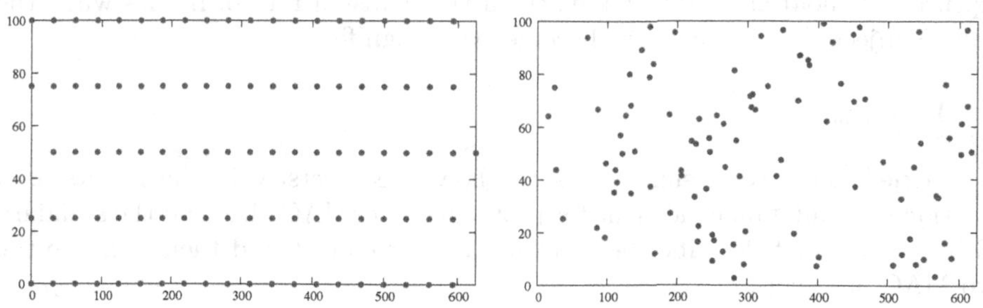

Fig. 3. Topology of 100 nodes on square area of 625 m × 100 m with respectively a grid positioning (left) and random positioning (right).

The random positioning is more closer to a realistic positioning. However, the concept of random position can change depending on how the placement is carried out. Indeed, to randomly place N nodes on a given area by hand, the person will try to cover the whole area and avoid nodes overlapping. For the same surface S and for the same number of nodes, the topologies obtained by a generator of random positions (x, y) lead to much greater diversity of solution but are not necessarily more representative for WSNs field. It can have less covered places than others, overlays and nodes that are too distant from others. To avoid the latter two situations, once the positions are randomly drawn, we use two filters in the control of network connectivity by Prim's algorithm (developed by Robert C. Prim (1957)). We set a minimum value of 1 m and a maximum of 50 m between any neighbors nodes. The surface of the coverage area (62500 m^2), the propagation model, the data packets production, the total number of nodes and the percentage of source nodes are identical in each case. We consider two forms of topologies (square of size 250 m × 250 m and pseudo-linear of size 625 m × 100 m) and for each type, nodes are positioned both in grid and randomly. In any case, the sink is located at the top right corner of the area.

In this work, our field of investigation is thus reduced to simulations on these 4 types of topologies (see Figs. 2 and 3) with different values of capture threshold.

Capture Effect. The capture effect relates to the simultaneous reception of several frames with power levels such that the reception of one of them is possible. When a frame f is received with a power P_f greater than or equal to the sum of the powers of the other k frames received at the same time plus a capture threshold fixed, the frame f is correctly decoded by the receiver. The capture threshold is the minimum power ratio in dB which enables the receiver in case of simultaneous reception to decode the strongest signal correctly. Therefore, if the difference between the power of a received signal and the sum of power of the other signals is greater than this threshold, reception is considered as a capture and the dominant signal is decoded correctly. Otherwise, the simultaneous reception will cause a collision and the signals can not be interpreted. This capture threshold has a direct impact on the collision rate of frames when the traffic subjected to the network becomes very significant.

4 Results

We carried out several simulations to show the diversity in the results as a function of simulation conditions for protocols SlackMAC [4,5] and the standard IEEE [1]. We first describe the simulation environment and then compare the two MAC protocols.

4.1 Simulation Parameters

Our simulations are performed using the network simulator NS-2 [25]. Global simulation parameters are given in Table 2. We use 10 of each type of topology of 100 nodes described in Sect. 3 (see Figs. 2 and 3). We generate a convergecast communication (from the nodes to the sink), for 30 source nodes located randomly in the network. These source nodes perform periodic measurements and route them via other nodes to the sink. Nodes have a duty-cycle of 1% and the global cycle is 5 s (that is, nodes are active during $A = 50$ ms every $C = 5$ s). The same gradient-based routing protocol is used to route packets hop by hop towards the sink for both MAC protocols. All presented results, in each case are averaged over 10 repetitions per topology for the 10 topologies. We compute transmission power (in dBm) using outage probability method (defined in [26, 27]). For transmission range of 50 m and shadowing parameters defined in Table 2, the equivalent transmission power which can ensure stability of the radio links for 95% of the reception between two nodes is equal to 10.5342501084 dBm. This value is used as a transmission power in all simulations.

4.2 Simulation Results

Results with Square Topologies of Size 250 m × 250 m. In this first scenario, we perform tests using both a grid and random topologies (respectively noted SquLat and SquRnd). We vary the traffic generation period from 5 s to 60 s with capture threshold of 2 dB and 10 dB.

Table 2. Global simulation parameters

Topologies area	250 m × 250 m and 625 m × 100 m
Transmission range	50 m
Number of nodes	100
Number of source nodes	30
Radio frequency	2.4 GHz
Receive threshold (RXThresh)	−85 dBm
Carrier-sense threshold (CSThresh)	−92 GHz
System loss (L)	1
Antenna type	Omnidirectional
Antenna gain (Gt, Gr)	1
Antenna height (Z)	1.5 m
Propagation model: shadowing	Path loss exponent = 2.5
	Shadowing deviation = 4.0 dB
Data traffic	Constant-bit rate (CBR)
Data frame size	30 bytes
Data traffic period	From 5 s to 60 s
Maximum send queue size	20 frames
Number of topology	10
Number of repetitions per topology	10
Simulation duration	3600 s

Figures 4 and 5 show respectively the delivery ratio and the end-to-end delay of data frames, as a function of the traffic generation period (from 5 s to 60 s) for IEEE 802.15.4 and SlackMAC, when square topologies are into grid and random.

For IEEE 802.15.4:

- when nodes are positioned in a grid, the delivery ratio increases from 23.43% to 66.34% for the capture threshold of 2 dB and from 14.37% to 49.56% for the capture threshold of 10 dB. The average delay decreases from 155.75 s to 10.02 s for the capture threshold of 2 dB and from 148 s to 10.52 s for the capture threshold of 10 dB.
- when nodes are randomly positioned, the delivery ratio increases from 25.18% to 66.07% for the capture threshold of 2 dB and from 15.81% to 50.2% for the capture threshold of 10 dB. The average delay decreases from 144.69 s to 7.22 s for the capture threshold of 2 dB and from 145.28 s to 9.66 s for the capture threshold of 10 dB.

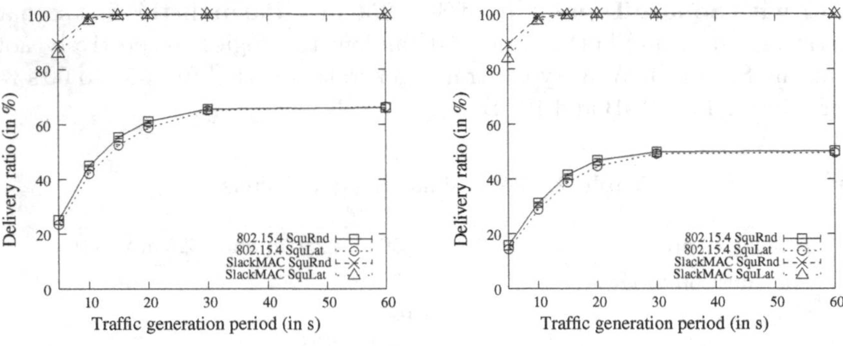

Fig. 4. The packets delivery ratio as a function of the traffic generation period with respectively the capture threshold of 2 dB (left) and 10 dB (right) for IEEE 802.15.4 and SlackMAC when square topologies are into grid and random.

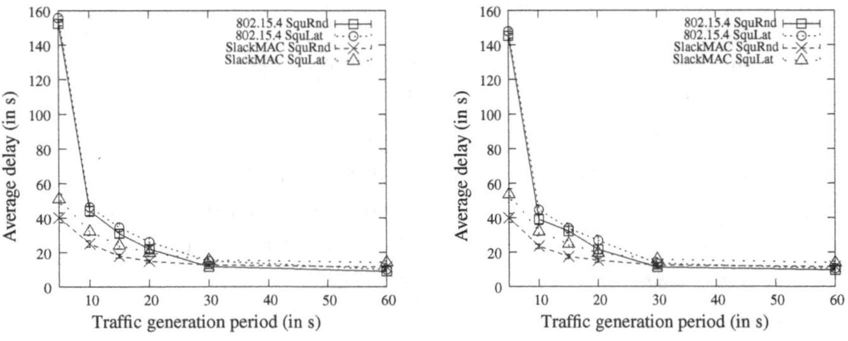

Fig. 5. Average delay as a function of the traffic generation period with respectively the capture threshold of 2 dB (left) and 10 dB (right) for IEEE 802.15.4 and SlackMAC when square topologies are into grid and random.

For SlackMAC:

- when nodes are positioned in a grid, the delivery ratio increases from 85.91% to 99.07% for the capture threshold of 2 dB and from 83.89% to 99.9% for the capture threshold of 10 dB. The average delay decreases from 59.09 s to 14.19 s for the capture threshold of 2 dB and from 53.52 s to 13.90 s for the capture threshold of 10 dB.
- when nodes are randomly positioned, the delivery ratio increases from 89.29% to 99.93% for the capture threshold of 2 dB and from 88.08% to 99.92% for the capture threshold of 10 dB. The average delay decreases from 40.14 s to 11.47 s for the capture threshold of 2 dB and from 39.96 s to 11.55 s for the capture threshold of 10 dB.

The results of this scenario show that for SlackMAC, whether the nodes are positioned in grid or randomly on a square area, the delivery ratio vary very little according to the traffic and the capture threshold. However, the average delay is better in the case of random positioning when the traffic varies regardless of the capture threshold. For the standard, the results also vary very little with the type of positioning. On the other hand, there is a difference of 16% in the delivery ratio between a capture threshold of 2 dB and 10 dB for the two types of positioning. The average delay also varies little, but remains better when the positioning is random.

Results with Pseudo-linear Topologies of Size 625 m × 100 m. In this second scenario, we perform tests using both a grid and random topologies (respectively noted StrLat and StrRnd). We also vary the traffic generation period from 5 s to 60 s with capture threshold of 2 dB and 10 dB.

Figures 6 and 7 show respectively the delivery ratio and the end-to-end delay of data frames, as a function of the traffic generation period (from 5 s to 60 s) for IEEE 802.15.4 and SlackMAC, when pseudo-linear topologies are into grid and random.

For IEEE 802.15.4:

- when nodes are positioned in a grid, the delivery ratio increases from 18.34% to 66.98% for the capture threshold of 2 dB and from 15.42% to 54.65% for the capture threshold of 10 dB. The average delay decreases from 157.94 s to 29.49 s for the capture threshold of 2 dB and from 145.99 s to 19.79 s for the capture threshold of 10 dB.
- when nodes are randomly positioned, the delivery ratio increases from 18.09% to 66.67% for the capture threshold of 2 dB and from 15.28% to 56.48% for the capture threshold of 10 dB. The average delay decreases from 152.20 s to 33.02 s for the capture threshold of 2 dB and from 142.32 s to 22.59 s for the capture threshold of 10 dB.

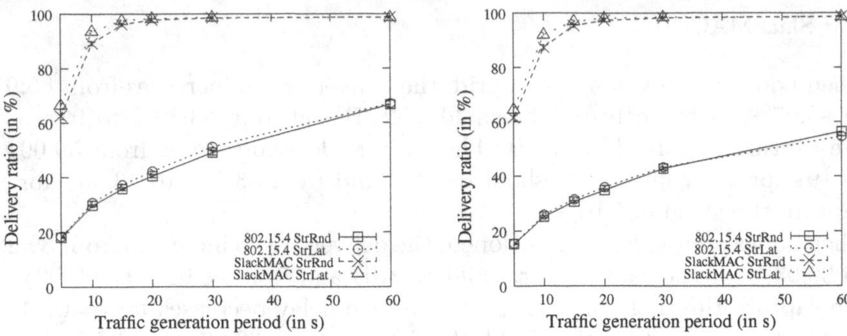

Fig. 6. The packets delivery ratio as a function of the traffic generation period with respectively the capture threshold of 2 dB (left) and 10 dB (right) for IEEE 802.15.4 and SlackMAC when pseudo-linear topologies are into grid and random.

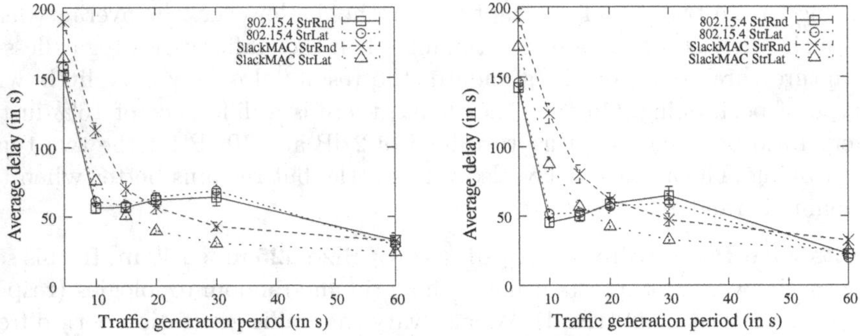

Fig. 7. Average delay as a function of the traffic generation period with respectively the capture threshold of 2 dB (left) and 10 dB (right) for IEEE 802.15.4 and SlackMAC when pseudo-linear topologies are into grid and random.

For SlackMAC:

- when nodes are positioned in a grid, the delivery ratio increases from 66.40% to 98.72% for the capture threshold of 2 dB and from 64.41% to 98.57% for the capture threshold of 10 dB. The average delay decreases from 164.26 s to 24.39 s for the capture threshold of 2 dB and from 171.67 s to 25.10 s for the capture threshold of 10 dB.
- when nodes are randomly positioned, the delivery ratio increases from 62.55% to 98.8% for the capture threshold of 2 dB and from 61.49% to 98.73% for the capture threshold of 10 dB. The average delay decreases from 190.27 s to 32.32 s for the capture threshold of 2 dB and from 192.74 s to 32.43 s for the capture threshold of 10 dB.

The results of this second scenario show that for SlackMAC, whether the nodes are positioned in grid or randomly on a square area, the delivery ratio vary very little according to the traffic and the capture threshold. However, unlike the square zone where the best average delay is obtained with a random positioning, that is not the case with a pseudo-linear zone in which the best average delay is with a grid positioning regardless of the capture threshold. For the standard, the results vary very little with the type of positioning for the delivery ratio and the average delay regardless of the capture threshold.

Effect of Capture Threshold. In this last scenario, we perform tests using the two topologies area such as square and pseudo-linear. We set the traffic generation period to 60 s and vary the capture threshold from 2 dB to 10 dB.

Figures 8 and 9 show respectively the delivery ratio and the end-to-end delay of data frames, as a function of the capture threshold (from 2 dB to 10 dB) for IEEE 802.15.4 and SlackMAC.

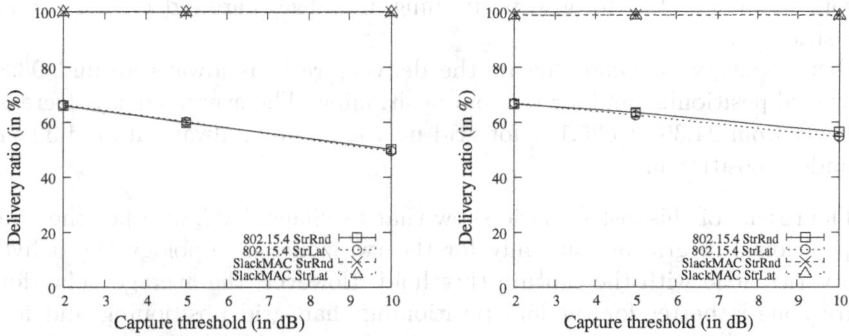

Fig. 8. The packets delivery ratio as a function of the capture threshold with respectively square topology (left) and pseudo-linear topology (right) into grid and random for IEEE 802.15.4 and SlackMAC when the traffic generation period is 60 s.

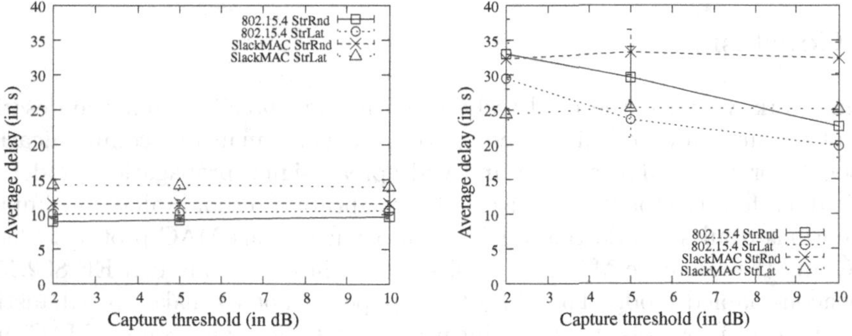

Fig. 9. Average delay as a function of the capture threshold with respectively square topology (left) and pseudo-linear topology (right) into grid and random for IEEE 802.15.4 and SlackMAC when the traffic generation period is 60 s.

For IEEE 802.15.4:

- when topology is square, the delivery ratio decreases from 66.34% to 49.56% for grid positioning and from 66.07% to 50.20% for random positioning. The average delay increases from 10.02 s to 10.52 s for grid positioning and from 8.93 s to 9.66 s for random positioning.
- when topology is pseudo-linear, the delivery ratio decreases from 66.98% to 54.65% for grid positioning and from 66.67% to 56.48% for random positioning. The average delay decreases from 29.49 s to 19.79 s for grid positioning and from 33.02 s to 22.59 s for random positioning.

For SlackMAC:

- when topology is square, the delivery ratio is always around 99.9% for grid positioning and for random positioning. The average delay decreases slowly from 14.19 s to 13.9 s for grid positioning and always around 11 s for random positioning.
- when topology is pseudo-linear, the delivery ratio is always around 98.8% for grid positioning and for random positioning. The average delay decreases slowly from 24.39 s to 25.10 s for grid positioning and always around 32 s for random positioning.

The results of this last scenario show that for SlackMAC, whether the nodes are positioned in grid or randomly for the two zones of topology the delivery ratio varies little with the capture threshold. However, the average delay for a square zone is better for random positioning than grid positioning and for a pseudo-linear zone the better average delay is obtained with a grid positioning. Unlike SlackMAC, the delivery ratio in the standard decreases when the capture threshold increases regardless of the zone and the positioning. The average delay increases with the capture threshold for square zone and decreases with the capture threshold for pseudo-linear zone regardless of positioning.

5 Conclusion

In this work, we show the need to be attentive and precise about the description of the simulation conditions comparing the performance of communication protocols for WSNs. For a given and widespread signal propagation model we pointed out how the topology choice and the capture threshold value may impact simulations results. To do this we used an asynchronous MAC protocol SlackMAC and the reference MAC protocol specified in the standard IEEE 802.15.4 in beacon-enabled mode. The aim of the paper is not to make an exhaustive comparison of protocols. We knew intuitively that the asynchronous MAC protocol would provide better overall performance than the standard. What we have demonstrated with intensive simulations, is the diversity in the results when the simulation conditions change. The conditions we have dealt with are the topology types and the capture threshold used in a WSN simulation process: (i) grid or random topologies for squared or stretched areas and (ii) capture threshold varying from 2 dB to 10 dB. The results showed great diversity such as for example for the same form of topology (square or pseudo-linear), either one of the two protocols gives better performances when the positioning is in grid and the other one rather with a random positioning. These observations are also made for two different forms of topology. These results also showed a very significant impact of the capture threshold on one protocol than the other. These remarkable differences between performances when simulation conditions change, confirm that it is essential to clearly define the simulation environment and also to reconcile the chosen conditions with the results when comparing the performances of protocols.

Acknowledgment. This research was partially supported by the "Digital Trust" Chair from the University of Auvergne Foundation.

References

1. IEEE 802.15: IEEE standard for local and metropolitan area networks - part 15.4: Low-rate wireless personal area networks (LR-WPANs). ANSI/IEEE, Standard 802.15.4 R2011 (2011)
2. Kanthe, A.M., Simunic, D., Prasad, R.: Effects of propagation models on AODV in mobile ad-hoc networks. Wireless Pers. Commun. **79**(1), 389–403 (2014)
3. Alduais, M., Abdulwahab, N.: The performance evaluation of different logical topologies and their respective protocols for wireless sensor networks, Ph.D. dissertation, Universiti Tun Hussein Onn Malaysia (2015)
4. Aby, A.T., Guitton, A., Lafourcade, P., Misson, M.: History-based MAC protocol for low duty-cycle wireless sensor networks: the Slack-MAC protocol. EAI Endorsed Trans. Mob. Commun. Appl. (2016). European Union Digital Library, EUDL
5. Aby, A.T., Guitton, A., Lafourcade, P., Misson, M.: SLACK-MAC: adaptive MAC protocol for low duty-cycle wireless sensor networks. In: Mitton, N., Kantarci, M.E., Gallais, A., Papavassiliou, S. (eds.) ADHOCNETS 2015. LNICSSITE, vol. 155, pp. 69–81. Springer, Cham (2015). doi:10.1007/978-3-319-25067-0_6
6. Chalhoub, G., Guitton, A., Misson, M.: MAC specifications for a WPAN allowing both energy saving and guaranted delay - Part A: MaCARI: a synchronized tree-based MAC protocol. In: IFIP WSAN (2008)
7. Nguyen, K., Ji, Y., Yamada, S.: Low overhead MAC protocol for low data rate wireless sensor networks. Int. J. Distrib. Sens. Netw. (2013)
8. Ye, W., Heidemann, J., Estrin, D.: An energy-efficient MAC protocol for wireless sensor networks. In: Proceedings of the IEEE Twenty-First Annual Joint Conference of the IEEE Computer and Communications Societies, INFOCOM 2002, vol. 3, pp. 1567–1576. IEEE (2002)
9. Lu, G., Krishnamachari, B., Raghavendra, C.: An adaptive energy-efficient and low-latency MAC for tree-based data gathering in sensor networks. Wirel. Commun. Mob. Comput. **7**, 863–875 (2007)
10. Song, W.Z., Huang, R., Shirazi, B., LaHusen, R.: TreeMAC: localized TDMA protocol for real-time high-data-rate sensor networks. Pervasive Mob. Comput. **5**(6), 750–765 (2009)
11. Cunha, F.D., Cunha, I., Wong, H.C., Loureiro, A.A., Oliveira, L.B.: ID-MAC: an identity-based MAC protocol for wireless sensor networks. In: 2013 IEEE Symposium on Computers and Communications (ISCC), pp. 000 975–000 981. IEEE (2013)

12. Sun, Y., Du, S., Gurewitz, O., Johnson, D.B.: DW-MAC: a low latency, energy efficient demand-wakeup MAC protocol for wireless sensor networks. In: Proceedings of the 9th ACM International Symposium on Mobile Ad Hoc Networking and Computing. ACM, pp. 53–62 (2008)
13. Gu, Y., He, T.: Dynamic switching-based data forwarding for low-duty-cycle wireless sensor networks. IEEE Trans. Mob. Comput. **10**(12), 1741–1754 (2011)
14. Cheng, L., Gu, Y., Niu, J., Zhu, T., Liu, C., Zhang, Q., Hel, T.: Taming collisions for delay reduction in low-duty-cycle wireless sensor networks. In: The 35th Annual IEEE International Conference on IEEE, IEEE INFOCOM 2016, pp. 1–9 (2016)
15. Polastre, J., Hill, J., Culler, D.: Versatile low power media access for wireless sensor networks. In: ACM Sensys, November 2004
16. Buettner, M., Yee, G.V., Anderson, E., Han, R.: X-MAC: a short preamble MAC protocol for duty-cycled wireless sensor networks. In: Proceedings of the 4th International Conference on Embedded Networked Sensor Systems, pp. 307–320. ACM (2006)
17. Moss, D., Levis, P.: BoX-MACs: exploiting physical and link layer boundaries in low-power networking. Computer Systems Laboratory, Stanford University, pp. 116–119 (2008)
18. Kim, G., Ahn, J.: On-demand synchronous X-MAC protocol. In: Computer Science and Software Engineering (JCSSE), pp. 1–6. IEEE (2016)
19. Sun, Y., Gurewitz, O., Johnson, D.B.: RI-MAC: a receiver-initiated asynchronous duty cycle MAC protocol for dynamic traffic loads in wireless sensor networks. In: ACM Sensys (2008)
20. Park, I., Lee, H., Kang, S.: RIX-MAC: an energy-efficient receiver-initiated wakeup MAC protocol for WSNs. KSII Trans. Internet Inf. Syst. **8**(5) (2014)
21. Nguyen, K., Nguyen, V.-H., Le, D.-D., Ji, Y., Duong, D.A., Yamada, S.: A receiver-initiated MAC protocol for energy harvesting sensor networks. In: Jeong, Y.-S., Park, Y.-H., Hsu, C.-H., Park, J. (eds.) Ubiquitous Information Technologies and Applications. LNEE, vol. 280, pp. 603–610. Springer, Heidelberg (2014). doi:10.1007/978-3-642-41671-2_77
22. Wang, X., Zhang, X., Chen, G., Zhang, Q.: Opportunistic cooperation in low duty cycle wireless sensor networks. In: 2010 IEEE International Conference on Communications (ICC), pp. 1–5. IEEE (2010)
23. Yoo, H., Shim, M., Kim, D.: Dynamic duty-cycle scheduling schemes for energy-harvesting wireless sensor networks. IEEE Commun. Lett. **16**(2), 202–204 (2012)
24. Aby, A.T., Guitton, A., Misson, M.: Study of blind rendez-vous in low power wireless sensor networks. In: 2014 IEEE 79th Vehicular Technology Conference (VTC Spring), pp. 1–5. IEEE (2014)
25. Network simulator 2 (2002). http://www.isi.edu/nsnam/ns
26. Mezghanni, M.S., Kandil, N., Hakem, N.: Performance study of IEEE 802.15. 4/4g waveforms over the mobile underground mine radio-channel. In: IEEE 84th Vehicular Technology Conference (VTC-Fall), pp. 1–6. IEEE (2016)
27. Simon, M.K., Alouini, M.: Outage performance of multiuser communication systems. In: Digital Communication over Fading Channels, 2nd edn., pp. 638–680 (2005)

Data Management

A Real-Time Query Processing System for WSN

Abderrahmen Belfkih(✉), Claude Duvallet, Bruno Sadeg, and Laurent Amanton

Laboratoire d'Informatique, de Traitement de l'Information et des Systèmes, Normandie Univ., UNIHAVRE, UNIROUEN, INSA Rouen, LITIS, 76600 Le Havre, France

{Abderrahmen.Belfkih,Claude.Duvallet,Bruno.Sadeg, Laurent.Amanton}@univ-lehavre.fr

Abstract. Wireless Sensor Networks (WSN) are considered as a large distributed database, which can be queried using SQL-like language. Several Query Processing Systems (QPS) have been proposed, to extract data from WSN. Many WSN-based applications require fresh data and strict deadlines when they receive responses from the WSN. However, the QPS are not dedicated to ensure the temporal constraints like the temporal validity of data and the queries deadlines. In this paper, we propose a new model for Real-Time Query Processing System (RTQPS) in order to guarantee temporal constraints in WSN. Our system is based on the QPS architecture, including some components of real-time database management system (DBMS). RTQPS ensures efficient query dissemination and data collection. It provides deadline controllers to ensure real-time constraints and real-time schedulers to meet query deadlines. We have carried out many simulations in order to validate our approach.

Keywords: Wireless sensor networks · Query processing system · Sensor databases · Real-time scheduling

1 Introduction

Wireless Sensor Networks (WSN) is composed of small wireless nodes, deployed throughout an area of interest, to monitor physical or environmental phenomena such as humidity, light and temperature. WSN has been used in many applications which require consistent data and a strict delivery time, to be efficient and to take preventive actions if required. However, the limited capacities of sensor nodes (data storage, limited memory, etc.) and the large amount of data exchanged between the base station and the sensor nodes can cause long delays during the data delivery. It can affect the temporal validity of data.

In this paper, we propose a real-time query processing system (RTQPS), designed to ensure temporal constraints in WSN. We have based our system architecture on the basic query processing system architecture, including real-time components, like real-time schedulers and deadline controllers. RTQPS ensures a query dissemination and a data collection using paths selected based

A. Puliafito et al. (Eds.): ADHOC-NOW 2017, LNCS 10517, pp. 307–313, 2017.
DOI: 10.1007/978-3-319-67910-5_25

on the minimum latency between nodes. It includes deadline controllers and real-time schedulers to assign priorities to transactions. We have used different scenarios with different real-time scheduling algorithms, to test the RTQPS performances.

This paper is organized as follows. In Sect. 2, we present some QPS in WSN. In Sect. 3, we present our system architecture. In Sect. 4, we give the simulation results. Finally, in Sect. 5, we conclude and we give some perspectives to this work.

2 Background

Many QPS have been proposed in WSN like Cougar [6], TinyDB [3] and Corona [2]. Cougar presents an infrastructure dedicated to the sensor data management [6]. It provides an SQL-Like query language to facilitate the users query writing. It uses the `<DURATION>` clause to specify the query lifetime and the `<EVERY>` clause to define the query execution cycle. However, it does not include clauses which can specify the query deadlines or the temporal validity of data.

TinyDB is a distributed query processing system for WSN [3]. It provides a declarative SQL-Like query language, named TinySQL. It uses the `<epoch>` clause, that defines the period of time between the start of each sample period, to structure computation in the network and to minimize power consumption. But, it does not have components that can check for the temporal validity of data. Corona is a distributed in-network query processing system [2]. It provides a declarative SQL-Like language to formulate queries. It introduces the notion of freshness into WSN, allowing the user to obtain data from a sensor network with data freshness guarantees. Corona includes the `<EPOCH>` clause which defines the time between two query executions, the `<RUNCOUNT>` clause which defines the number of times of query execution and the `<FRESHNESS>` clause which specifies how much time it is possible to wait since the last sensor acquisition.

3 RTQPS Architecture

3.1 Base Station

It consists of the following elements: (1) a *Cache Memory* is used to keep a historical of measurements. Old data are deleted periodically from the cache. (2) An *arrival query list* is used to receive queries and to put them inside. (3) The *Deadline Controller* is used to check for the query deadlines. (4) The *Transaction Scheduler* is used to ensure that transactions meet their deadlines. The queries are sorted depending on the scheduling method used by the system. (5) The *Data Validity Controller* checks the validity time of data and removes invalid data from the data list.

3.2 Relays Node

It consists of the following elements: (1) an *arrival query list*, which receives queries from the base station. (2) A *Deadline Controller*, which verifies query deadlines. (3) The *Transaction Scheduler*, which assigns priorities to transactions. The first query in the list is injected for child nodes. (4) The *Data Validity Controller*, which removes invalid data from the data list.

3.3 Child Node

It consists of *Deadline Controller* and *query processor*. When it receives the query from relay node, the *Deadline Controller* verifies the query deadline. If deadline is not exceeded, the query is executed by the *query processor*. Then, it builds the response message and sends it to its parent node.

4 Query Processing Plan

4.1 Network Builds and Relays Nodes Selection

This step is dedicated to build the tree based-topology of the network and to select relays nodes. Relays nodes represent relays points in the tree, dedicated to disseminate queries and to collect responses from sensor nodes. The relays nodes are selected by their neighbours, based on an auto-selection program [1]. We have used the RPL[1] protocol [5] operations to perform this step. The relay selection process is based, on the notion of rank. It starts with the construction of the tree topology called DODAG[2]. First, the root node (DAG) sends a DODAG Information Object (DIO) to its neighbours. The DIO message contains: the node identifier, the objective function, the node rank, sending date and a parameter called the "Minimum Rank" to help the receiving node to compute its own rank. The root node has a default rank. Each node receives the DIO messages, selects the best parent node, which represents the sender that has the minimum rank. Then, it computes its own rank and sends a DIO message to its neighbours, including its rank, and so on, until each node has defined its best parent in the network. Finally, we obtain a tree composed of the root node (the base station), parent nodes (the relays nodes) and child nodes (leaf nodes). So, all parent nodes can form paths to the base station. This way, we ensure that the base station can communicate and receives responses from any node in the tree.

4.2 Query Dissemination

The user sends a query to the base station throw a graphical interface. When the base station receives the query, it compares the query with query history. If the query has been already used, the base station uses simply the previous

[1] Routing Protocol for Low Power and Lossy Wireless Networks.

[2] Destination Oriented Directed Acyclic Graph.

measurements from its cache. If a more recent measurement is needed, the query is placed in the active list. The *Deadline Controller* compares the absolute deadline (di) of transaction with current time. If the current time is less than the transaction deadline, the transaction is transmitted to the scheduler. Otherwise, the transaction is aborted because it has missed its deadline. The transactions are sorted depending on the scheduling method used by the system. Then, the first element in the list is broadcasted to relays nodes. Each transaction sent is removed from the active list and from the scheduler.

When the relay node receives a query, the *Deadline Controller* compares the query deadline with the current time. If it is less than the query deadline, the transaction will be transmitted to the scheduler. Otherwise, the transaction is aborted. The transactions are sorted according to the scheduling algorithm used by the system. Next, the first transaction in the list will be sent to child nodes. Once a relay node receives a query responses from its child nodes, the query will be removed from the scheduler. Otherwise, the relay node broadcasts the query again to its child nodes until the receipt of a response. When the child node receives the query, it verifies its deadline. If the transaction has missed its deadline, it will be automatically aborted. Each node manages the received queries independently of other sensor nodes.

4.3 Query Responses Collection

Once the child node has built the response message, it sends it to its parent node. The parent node receives a response message and verifies data temporal validity based on the data arrival time and the data extraction time. If the result is less than the temporal validity time fixed by users, responses are transmitted to the base station. Otherwise, the query response is not sent. When the relay node receives the same responses from child nodes, it aggregates responses into single response message, which will be transmitted to the base station. Otherwise, it transmits the query response to the base station. The response message will be verified by relays nodes, until it arrives to the base station.

5 Simulation and Experimental Evaluation

5.1 Simulation Environment

We used the Cooja network simulator for ContikiOS [4] to evaluate our system. Table 1 shows the general simulation parameters. We have performed a scenario that consists of sending multiple queries to the base station with a fixed period. The query interval defines the interval of time between two query executions. Its varied between 10 and 20 s, in order to get the ratio of successful transactions and the number of valid data, using scheduling algorithms (FIFO, RM, EDF). We have used three types of queries: The Query 1 includes a deadline clause having a value equal to the query interval. The Query 2 includes a data validity clause, having a value less than or equal to 20 s. The Query 3 includes a data validity clause having a value less than or equal to 20 s and a query deadline clause having a value equal to the query interval.

Table 1. Simulation parameters

Parameter	Values
Simulated node type	Tmote sky
Simulation area	100 m * 100 m
Number of nodes	10 nodes
Node transmission range	30 m
Maximum packet size	128 bytes
Radio transmission speed	250 kbps
MAC layer	CSMA/CA
Radio duty cycling layer	ContikiMac
Radio propagation model	UDGM distance loss

5.2 Transaction Performance Metric

We consider that the transaction is successful (committed), if it meets its deadline. The deadline is a time defined by the user in the query to be met by the transaction otherwise the transaction fails. The success ratio is given by $SRatio = CommitT/SubmittedT$ where $CommitT$ indicates the number of transactions committed by their deadlines, and $SubmittedT$ indicates all submitted transactions to the system in the sampling period.

5.3 Experimental Evaluation

Impact of Real-Time Scheduler on Success Ratio of Transactions: In this scenario, we observe the system behavior with periodic queries and the impact of the scheduling on the success ratio of transactions. We have created a transaction generator, connected to the base station using UDP socket. It can send periodic queries to the base station, according to three periods varied between 10 and 20 s. We have tested three scheduling algorithms: FIFO, RM

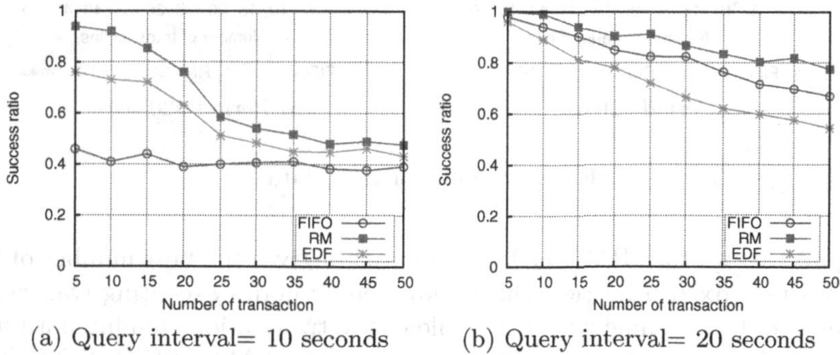

(a) Query interval= 10 seconds (b) Query interval= 20 seconds

Fig. 1. Transactions success ratio

and EDF. We have used query Q1 with a deadline equal to the query interval. We have varied the number of queries between 5 to 50 queries (cf. Fig. 1).

We note in Fig. 1a, that RM scheduling algorithm gives the best number of successful transactions. EDF scheduling algorithm is better than FIFO scheduling algorithm, but, it is inferior compared to RM scheduling algorithm. FIFO cannot handle user transactions, the first arrival is the first sent into the network. Transactions are executed regardless of their priorities, which affects the query execution time of other transactions. EDF scheduling algorithm selects the transaction having the shortest deadline to be executed firstly. It returns the best results with a small number of transactions, but, its performances decrease with a high number of transactions. It has dynamic priorities, which can affect the execution time of transactions, mainly with a high number of transactions. RM scheduling algorithm gives a high priority to transaction with small period. Transactions are executed until the end of their deadlines, which increases the number of successful transactions. We note also, that the query interval has an impact on the success ratio of transactions. The number of successful transactions using a query interval equals to 10 s (cf. Fig. 1a) is less than the results of a query interval equals to 20 s (cf. Fig. 1b).

Impact of Deadline Controllers on Number Valid Data: In this experience, we study the impact of scheduling algorithms on the number of valid data. We have sent periodic queries (query Q2) to the base station, according to three periods varied between 10, 20. We have fixed the absolute validity interval of requested data less or equal to 20 s, because we have noticed that the time required to send data from a level to another requires 10 s (cf. Fig. 2).

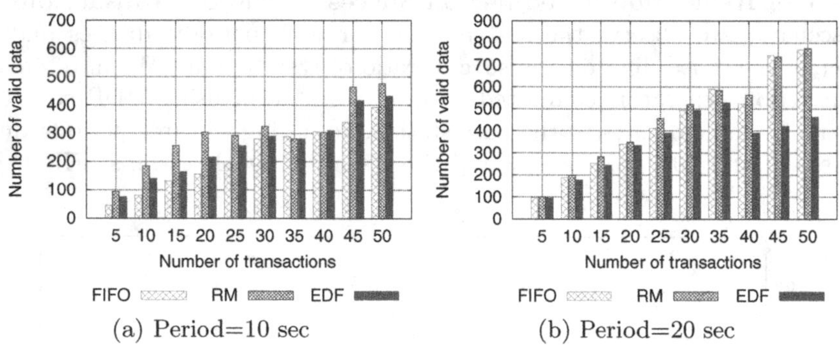

(a) Period=10 sec (b) Period=20 sec

Fig. 2. Number of valid data

Figure 2 shows that RM scheduling algorithm gives the best number of valid data. RM uses fixed priorities which allow sensor nodes executing transactions until the end. In the same way, FIFO allows the transaction running until it finishes. For this reason, we get more valid data using RM or FIFO. EDF assigns dynamic priorities, that can change during the execution. A query execution on a sensor node can be hampered by the arriving of a new query having short deadline. In result, the query cannot be executed until it finishes. We can see that EDF scheduling algorithm gives better results with a small number of transactions, and with a large sending interval. We can conclude, that RM scheduling algorithm is well suited with our system, because it gives best results through the three sending intervals, even when we increase the number of transactions.

6 Conclusions

In this paper, we have designed a real-time query processing system to improve data temporal consistency in WSN. We have proposed a new architecture of QPS including *Deadline Controllers*, to assign priorities to transactions, to reduce the base station workload and to increase the number of valid data and successful transactions. The relay selection process has improved the query dissemination time as well as the query response forwarding time to the base station. The use of *Deadline Controller* has improved the number of valid data versus the number of data received by the base station. In addition, the use of real-time schedulers in relays nodes and in the base station has minimized the workload. The RM scheduling algorithm has shown best performances in success ratio transactions. However, EDF and FIFO algorithms give best results with a small number of transactions. We plan to adapt the performance of the energy computation of our system and to improve its performances by using relaxed transactions.

Achnowledgement. This work is a part of the AMED project. This project is co-financed by the European Union with the European Regional Development Fund (ERDF) and by the Normandy Regional Council.

References

1. Belfkih, A., Duvallet, C., Amanton, L., Sadeg, B.: A new query processing model for maintaining data temporal consistency in wireless sensor networks. In: Proceedings of the International Conference on Intelligent Sensors, Sensor Networks and Information Processing, pp. 1–6. IEEE Computer Society, Singapore (2015)
2. Khoury, R., et al.: Corona: energy-efficient multi-query processing in wireless sensor networks. In: Kitagawa, H., Ishikawa, Y., Li, Q., Watanabe, C. (eds.) DASFAA 2010. LNCS, vol. 5982, pp. 416–419. Springer, Heidelberg (2010). doi:10.1007/978-3-642-12098-5_39
3. Madden, S.R., Franklin, M.J., Hellerstein, J.M., Hong, W.: TinyDB: an acquisitional query processing system for sensor networks. ACM Trans. Database Syst. **30**(1), 122–173 (2005)
4. Österlind, F., Dunkels, A., Eriksson, J., Finne, N., Voigt, T.: Cross-level sensor network simulation with COOJA. In: The Annual IEEE Conference on Local Computer Networks, Tampa, Florida, USA, pp. 641–648, November 2006
5. Winter, T., Thubert, P., Brandt, A., Hui, J., Kelsey, R., Levis, P., Pister, K., Struik, R., Vasseur, J., Alexander, R.: RPL: IPV6 routing protocol for low-power and lossy networks, March 2012
6. Yao, Y., Gehrke, J.: The cougar approach to in-network query processing in sensor networks. SIGMOD Rec. **31**, 9–18 (2002)

Centralized and Distributed Architectures: Approximation of the Response Time in a Video Surveillance System of Road Traffic by Logarithm, Power and Linear Functions

Papa Samour Diop(✉), Ahmath Bamba Mbacke, and Gervais Mendy

Laboratoire d'Informatique et de Réseaux Télécoms (LIRT), École Supérieure Polytechnique de Dakar (ESP/UCAD), Dakar, Senegal
{papasamour.diop,ahmathbamba.mbacke,gervais.mendy}@esp.sn

Abstract. In this article, we propose mathematical models that result in logarithmic, power and linear functions for the predictive evaluation of the response times of video surveillance systems in smart cities. Most of the QoS measurements and evaluations used in the current literature are hardware based and do not take into account the influence of the technical architecture. We have therefore proposed a decomposition process of video surveillance systems to obtain mathematical approximations of each treatment time. The integration of these components guided us towards generic mathematical models validated by experimentation. The measurements resulting from the experiment were approached with logarithmic, power and linear functions before adopting the most precise model between distributed and centralized architectures. The comparison between these architectures shows a much lower response time for the distributed architecture especially with the determination coefficient R^2 of the functions.

Keywords: Data transmission · Mobile · Monitoring · Forecasting perception · Distributed learning · Smart city · CCTV system

1 Introduction

In recent years, traffic monitoring systems moved many steps forward for the safety of pedestrians and drivers. Understanding behaviors by detecting and tracking the movements of traffic participants is a critical issue in urban surveillance systems, particularly in smart cities. CCTV systems are also used to maintain and manage the communication between those mobile objects and their surrounding environment while moving. Their quality depends then on the response time of the system, which is mainly based on the hardware characteristics and the architecture performances. With technology advancement, cameras, specifically for video surveillance systems, are designed to mimic as close as possible eyes and brains. These devices are capable of capturing a large amount of information and give it a meaning.

A. Puliafito et al. (Eds.): ADHOC-NOW 2017, LNCS 10517, pp. 314–327, 2017.
DOI: 10.1007/978-3-319-67910-5_26

They are integrated into video surveillance systems [1] to better manage road traffic and, supervising public or private places. Research efforts and the reduction of material costs opened the possibility of using these systems into a wider range of applications (automatic recognition and object tracking, the interpretation of the scene and the extraction or indexing special events [2], etc. ...).

There are also applications in the monitoring of industrial fields (access control or the production quality control), in the supervision of highly frequented public places (train stations, subway, [3] businesses [4]), in monitoring and analysis of elderly activities, in sport (football, golf), ...

These systems generally operate according to the following process:

- Motion detection,
- Extraction and classification of objects,
- Monitoring of objects over time,
- Behavior analysis and incident detection,
- Transmission of information (events related to objects) to decision making unit.

However, the real-time management of this mass of data is problematic. Indeed, the cameras must have a significant computing capacity in order to evaluate it, or be connected to a central collection and processing unit, engendering transmission duration issues, and this in order to identify objects for example, before they go out of their reach.

Technologies have been developed to implement smart and powerful cameras, and solutions have been proposed to better control the flow of information between intelligent cameras and a central collection and treatment unit [5]. In the case of smart cameras, solutions are now available to equip them with analysis capacities [6]. In the field of prediction, many works were done: road pedestrian path prediction in a normal environment [6], arrival time estimation of the bus considering several routes, a system of the destination and estimated future prediction according to the trajectory. In these systems, most of the evaluations are based on an assessment of an event [7] and the cameras are not equipped with learning abilities, nor collaboration.

The choice with respect to logarithmic functions is justified by the fact that in probability (the probability of passing a mobile) and in statistics, the logarithmic function is a discrete probability function, derived from Taylor's development. The function of power is a mathematical relationship between two quantities. If one of the quantities is the frequency of an event (detection of the presence of a mobile in the field of view of the camera) and the other is the size of an event, then the relation is a distribution of the power function if the frequencies decrease very slowly as the size of the event increases. Finally, for the linear function if we consider two phenomena whose intensity varies; x (the speed) denotes the intensity of the first phenomenon, and y (the number of captured images or the processing time) the intensity the second phenomenon. If x and y vary at the same time, we can estimate that the quantities are related, their variations are said to be correlated; it is then tempting to want to connect by a linear function.

Given these limitations, we propose two types of road CCTV system architectures resulting in two formal evaluation models detection performance by focusing on the time of the response of the system following a picture taking.

The first studied case is based on a forecast model using distributed learning and a shared knowledge base. The second used case evaluates distributed systems with intelligent cameras working capacity.

In both cases an assessment of the potential response time between when an image is captured, and the time that a decision is made by the system, is proposed.

This work is presented through the following 5 sections:

- Section 2 presents the state of the art in performance assessment models in a network video surveillance,
- Section 3 presents various deployment architectures,
- In Sect. 4 we divide and describe each architecture treatment via subprocesses,
- Section 5 provides for each model the logarithmic, power and linear functions associated with the coefficient of determination and makes a comparison between these models,
- Section 6 presents the conclusion and perspectives for future work.

2 State of the Art of Video Surveillance Network Performance Assessment Models

2.1 Network Performance Measurement

Much work has been mentioned on network performance in terms of throughput, delay and jitter [8].

In [8] they proposed a model to identify flow variations and time in order to optimize the system design with the correct configuration settings. A number of simulations were conducted to demonstrate the performance of the model. However, these studies were limited to the WiMax network wireless mesh. Mahasukhon et al. [9] have established a platform in which they measured the flow based on RSSI for each of the mobile nodes in a mobile WiMax network operating on the IEEE 802.16e standard. A method for improving performance in terms of throughput, end to end delay, and the jitter of a WiMax network (carrying voice calls) was proposed in [10].

Li Y et al. [11] presented a performance analysis for the frequency band allocated to the IEEE 802.16 for wireless broadband access. In this analysis, the rate was compared to the intensity of traffic. They noticed an increase in traffic intensity was associated with an increase in flow to a point of saturation.

The authors [12], among others, have led throughput tests in TCP and UDP for the downlink channels and amount of WiMax networks. Flow tests were conducted under different types of modulations and varying distances.

Reference [9] shows the work done to determine the minimum value of the signal-noise attribute to ensure acceptable levels in terms of throughput and quality of service in the WiMax networks.

2.2 Performance Measure to CCTV

Detmold et al. [13] discussed a middleware as a deployment mechanism of intelligent video surveillance systems based on the topologies described as the activity of the monitored objects. One of the contributions is the use of a distributed table to increase flexibility and scalability of the communication model. In this same line of research, IBM introduced an alternative (Tian et al.) [12] based on the intelligent management of the data collected by the safety device and open standards. For measurements of performance applied to video surveillance, [8] talks about the flow characteristics for video surveillance systems with WiMax with variations on the nodes, frame rate and size of the MSDU to check their performance. The obtained results show that, in similar circumstances, the HWW video surveillance system surpasses the legal system in terms of throughput in a factor of 1.75.

Diop P.S., Mbacké A.B., Mendy G. [14] proposed mathematical models to evaluate data processing time for a centralized and distributed architecture. Their studies have shown that the distributed architecture is much faster in terms of data processing. Compared to these mathematical models in this publication, we propose for each model a logarithmic, power and linear function associated with their coefficient of determination to validate these models in the centralized and distributed architecture.

3 Presentation of the Different Architectures

The significant problem in the design of urban traffic monitoring system is the ubiquity of detection points, the quality and reliability of the communication system and the establishment of the processing system. A widely distributed network of cameras requires infrastructure so vast that granting energy supply to the cameras and establish a communication channel between devices is problematic. The implementation of such infrastructure involves high costs and relevant policy choices that often undermine the feasibility of the entire project. Installation and network infrastructure maintenance costs for both the cameras and infrastructure are of great importance for better system performance. We proposed two generic architectures with video surveillance for each specific treatment possibility.

3.1 Centralized Architecture

In this architecture the cameras are placed at the roadside and each covers a clearly defined vision ray. A processing device is located at the remote monitoring center and cameras are connected to the latter via a wireless system. As in a centralized system, the processing unit is located at the monitoring center and handles the processing of the transmitted video stream from the cameras, followed by the treatment of video metadata. Each camera, detecting the presence of a mobile in his field of vision, sends its flow to the monitoring center where metadata are processed.

The example of prediction we consider is the anticipation of the direction of the mobile through a learning system. The goal is to predict the path that the mobile will take. Monitoring of the mobile and the prediction (activation of the next camera) will be handled at this level (see Fig. 1).

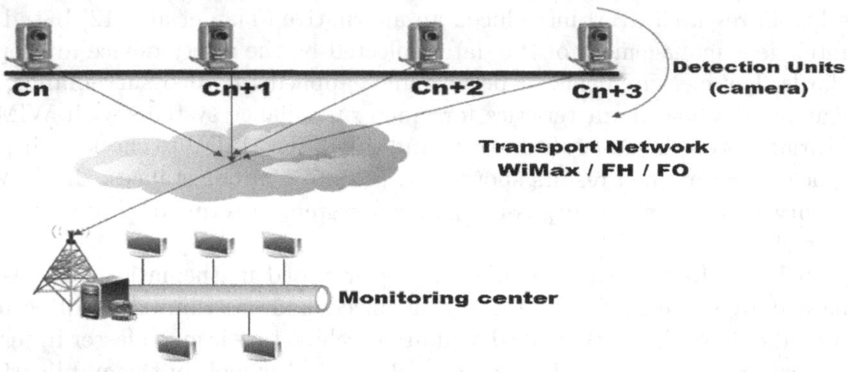

Fig. 1. Centralized architecture

3.2 Distributed Architecture

In this architecture the cameras are placed at each side of the road and have an embedded processor. A wireless link connects the cameras together and each camera is connected to the monitoring center via a wireless transmission system. Each camera that detects the presence of a mobile in its scope follows it up and analyzes it until the activation of the next camera using the onboard processor. Data processed and results will be transmitted to the monitoring center to be displayed and stored (Fig. 2). The prediction process will be triggered before the camera loses the mobile. Therefore, the current Cn camera sends a signal to the next camera C_{n+1} so that it prepares for the detection and monitoring.

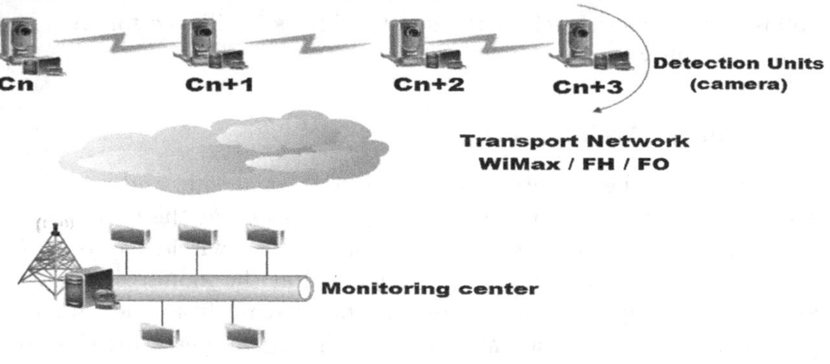

Fig. 2. Distributed architecture

Note that the average distance between the cameras is not significant because they will be positioned according to the routes and the coverage area. The description of the processing system and activation of the cameras to the prediction will be outlined in the next section.

4 Description of the Processing System

The captured images transmission delay depends on several factors such as the transmission rate, the weight of the images, the number of pictures etc. More the images weight is, the higher the transfer time is long. In our case, the transmission rate will be calculated from the initial node, i.e. since the detection of moving up to the activation of the next camera.

4.1 How Does a Camera Work?

This diagram outlines the different behaviors of a camera during the monitoring process to the activation of the next camera. Detecting a mobile for the prediction of the next camera to activate implies a set of processes between the different components of the video surveillance system. The conditions for a camera to be involved in the process are: either it detects the presence of a mobile in his field of vision, or another camera sends a signal to a mobile that is in his direction. The different states taken by a camera system are: standby activation, detection and follow-up (Fig. 3).

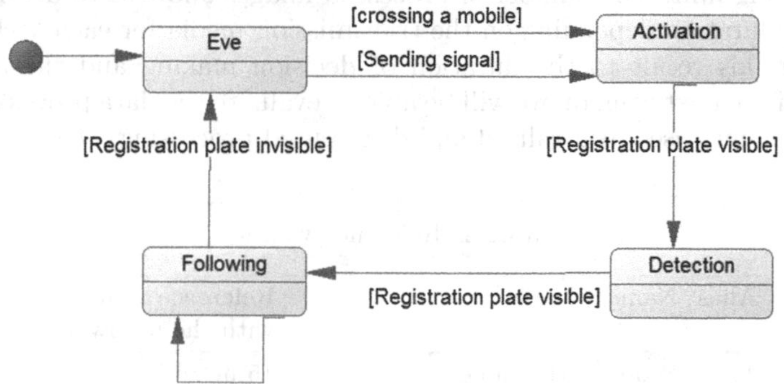

Fig. 3. State/Transition diagram of a camera

For a mobile to be detected, it must be in the field of view of a camera that is to say at a range of 15 m (experimental data taken on the basis of available material). At this level, a set of variables comes into play in the treatment process.

It is well known that the speed $= \frac{distance}{timeset}$

Let: T_{ti} = processing time of an image; T_i = size image file; P_c = camera range; D_a = next camera activation time; D_p = time decision; V_m = speed mobile; T_m = time taken; D_b = actual flow rate of the transmission medium; D_t = Duration of treatment; T_{cf} = propagation delay information; N_{it} = Number processed image.

In our case, the processing is done every 10 ms (T_{ti}). So we will have 20 ms to create a displacement vector for tracking mobile 10 ms and a centralized architecture for a distributed architecture. With an image size of 30 kb = 0.029296875 MB and a camera that covers a 15 m field of view, we have: the number of images processed by the system while a mobile is in its field of action is $N_{it} = \frac{T_m}{T_{ti}}$

With this system, we can offer the speed evaluation models of transmission at different available architectures (centralized and distributed).

After the above description, the evaluation of the treatment time between the two systems will be outlined in the next section.

5 Identification of the Logarithmic, Power and Linear Function Associated with the Coefficient of Determination for Each Mathematical Model

The evaluation consists in determining the time required for various processes to be realized. Depending on the speed of the mobile, we have the possibility to determine the time required for a mobile to be out of the camera's field of view. Therefore, we can determine the number of images to be processed by the processing unit. The number of processed images allow us to determine the processing duration depending on the transmission media for each architecture. By adding this result to the duration of decision making and the activation duration of the next camera, we will be able to evaluate the data processing times for our different cases (centralized and distributed processing) [14] (Table 1).

Table 1. Reference values

Alias	Name	Reference value with the hardware
P_c	Scope of the camera	15 m
T_{ti}	Processing time	0.01 s
Cte	Constant	2
T_i	Image File Size	0,029296875 octet
D_a	Time to next camera activation	0.01 s
D_p	Time decision making decision	0.01 s

5.1 For a Centralized Architecture

- With Time taken $=\frac{distance}{speed}$ (1) and
- Number of images processed $= \frac{Time-put}{Processing-time-of-an-image}$ (2)
- $T_m = N_{it} * T_{ti}$
- If we replace 1 in 2, we have:
- Nit$= \frac{P_c}{V_m * T_{ti} * cte}$
- $D_t = \frac{T_i}{D_b} * 8 * N_{it}$
- $T_{cf} = D_t + D_p + D_a$

Table 2. Number of image captures evolution by mobile speed in a centralized architecture

Speed (km/h)	45	50	55	60	65	70	75	80	85	90	95
Number of acquired images	60	54	49.09	45	41.54	38.57	36	33.75	31.76	30	28.42

In our model, we make a trend of the curve by an approximation of logarithmic function, linear and power. The choice of these functions is justified by the need we have to implement them in a simulator only supporting those categories of mathematical laws. Through the futur simulations, and using machine learning, we should be able to predict the vehiculars path through a distributed architecture. However, they have the following characteristics (Table 2):

- A logarithmic trend curve is the most accurate curve possible that is most useful when the frequency of data modification increases or decreases rapidly and then stabilizes. A logarithmic trend curve can use negative and/or positive values.
- A linear trend line is a weighted straight line that is used with simple linear data sets. Your data is linear if the pattern of the data points resembles a line. A linear trend line usually represents a steady increase or decrease.

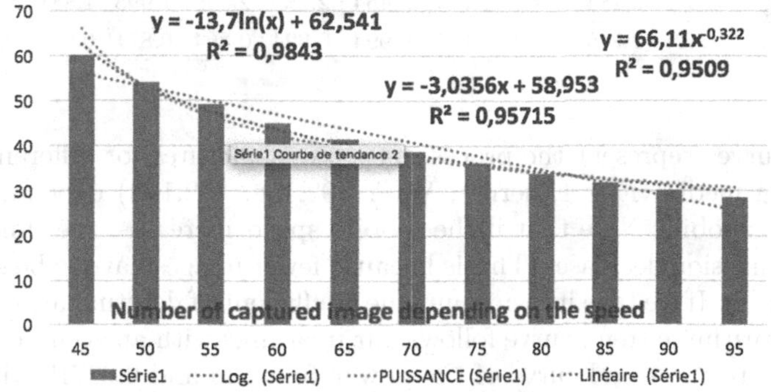

Fig. 4. Number of images processed

- A power trend curve is a curve particularly suited to data sets that compare measurements that increase at a specific speed (e.g. acceleration, every second, of a race car). You can not create such a curve if your data contains null or negative values (Fig. 4).

The analysis of this curve shows the variation in the number of images processed as a function of the speed of displacement of the mobile in a centralized architecture. We note that the higher the speed, less pictures can be processed by the camera. This is because the time required to exit the camera field decreases as the speed increases. 3 formal approximations by mathematical functions have been evaluated to get the most precise formal expression (Table 3):

Table 3. Approximation to logarithmic, power and linear functions

Logarithmic function	y = −13,7ln(x) + 62,541 and R^2 = 0,9843
Power function	y = 66,11$x^{-0.322}$ and R^2 = 0,9509
Linear function	y = −3,0356x + 58,953 and R^2 = 0,95715

The trend curve of the logarithmic function exceeds our data with a determination coefficient $R^2 = 0.9843$. With respect to the linear function ($R^2 = 0.95715$) and the power function ($R^2 = 0.9509$) on a scale of 1. This means that in relation to our results we have an accuracy of 0.02715 with respect to the trend curve of the power function and 0.0334 with respect to the linear function (Fig. 5 and Table 4).

Table 4. Transmission delay measures by physical media for a centralized architecture

RJ 45 Fast ethernet	1.426	1.286	1.171	1.075	0.994	0.924	0.864	0.811	0.764	0.723	0.686
RJ 45 gigabit	0.489	0.442	0.404	0.372	0.345	0.321	0.301	0.284	0.268	0.254	0.242
Wifi 802.11g	3.536	3.184	2.896	2.657	2.454	2.28	2.129	1.998	1.881	1.778	1.685
Wifi 802.11n	1.426	1.286	1.171	1.075	0.994	0.924	0.864	0.811	0.764	0.723	0.686

These curves represent the processing time of pictures for different types of transmission media (Fast Ethernet, WiFi 802.11g, 802.11n) depending on the speed of the mobile. Note that if the mobile speed increases, the time used for image transmission decreases. This is because fewer images have to be sent as the speed increases. If we take into account the coefficient of determination we notice that the logarithmic trend curve follows our measures with an accuracy of 0.0334 with respect to the trend curve of the power function and 0.02715 with respect to the trend curve of the function linear. The truth level of our results is shown by the fact that we have the same trends as the number of images processed

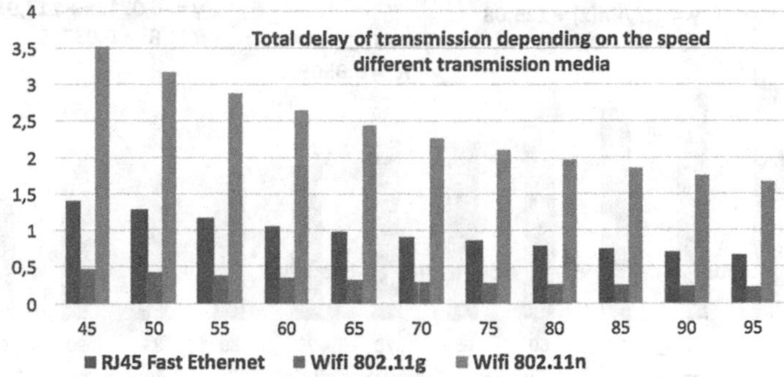

Fig. 5. Total time of transmission

that has the duration of the transmission time at the level of each transmission media. The proportionality of the trends remains the same between the number of images captured and the total duration of the transmission at the transmission media.

5.2 For a Distributed Architecture

With virtually no processing time, since treatment is done within the embedded systems in CCTV cameras, we have (Fig. 6 and Table 5):

- $N_{it} = \frac{P_c}{V_m * T_{ti} * cte}$
- $D_t = N_{it} * T_{ti} * D_p$
- $T_{cf} = D_t + D_a$

Table 5. Number of image captures and duration of treatments evolutions by mobile speed in a distributed architecture

Speed (km/h)	45	50	55	60	65	70	75	80	85	90	95
Number of acquired images	120	108	98.18	90	83.08	77.14	72	67.5	63.53	60	56.84
Duration of the treatment	0.006	0.005	0.005	0.005	0.004	0.004	0.004	0.003	0.003	0.003	0.003

Our analysis shows the variation of the number of images processed by a camera considering the mobile speed in a distributed type architecture. It should be noted that the faster the mobile moves, the number of images processed decreases by a logarithmic function. This reflects the fact that the mobile's out of view field switching time decreases as the speed increases as $= \frac{distance}{speed}$.

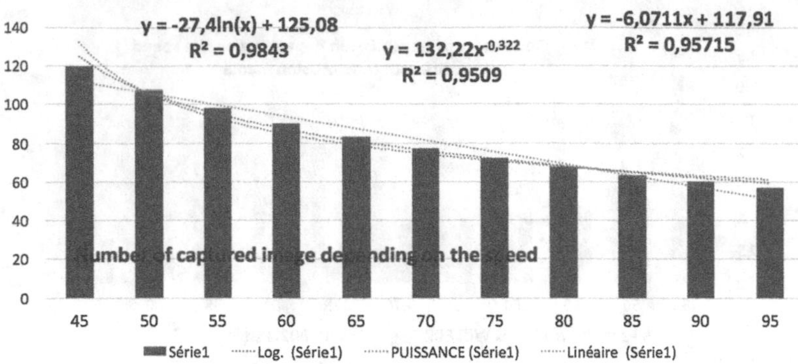

Fig. 6. Number of images processed

The choice of the logarithmic function with respect to the linear and power function is justified by the fact that, if we study the variation of the trend curves between these functions with respect to the determination coefficient R^2, the logarithmic function has a Accuracy of more than 0.0334 with respect to the power function and 0.02715 with respect to the linear function.

3 formal approximations by mathematical functions have been evaluated to get the most precise formal expression (Fig. 7 and Tables 6 and 7):

Table 6. Approximation to logarithmic, power and linear functions in a distributed architecture

Logarithmic function	y = −27,4ln(x) + 125,08 and R^2 = 0,9843
Power function	y = 132,22$x^{-0.322}$ and R^2 = 0,9509
Linear function	Y = −6,0711x + 117,91 and R^2 = 0,95715

Table 7. Transmission delay measures by physical media for a distributed architecture

RJ 45 Fast ethernet	0.016	0.015	10.015	0.015	0.014	0.014	0.014	0.013	0.013	0.013	0.013
RJ 45 gigabit	0.016	0.015	0.015	0.015	0.014	0.014	0.014	0.013	0.013	0.013	0.013
Wifi 802.11g	0.016	0.015	0.015	0.015	0.014	0.014	0.014	0.013	0.013	0.013	0.013
Wifi 802.11n	0.016	0.015	0.015	0.015	0.014	0.014	0.014	0.013	0.013	0.013	0.013

In this model, it is noted that the transmission time is low and decreases if the speed of the mobile increases. This is due to the fact that each camera is integrates a processing unit and that everything is done at this level until it no longer detects the mobile in its field of view before passing the relay to the next camera. The choice of the logarithmic function is justified by the fact

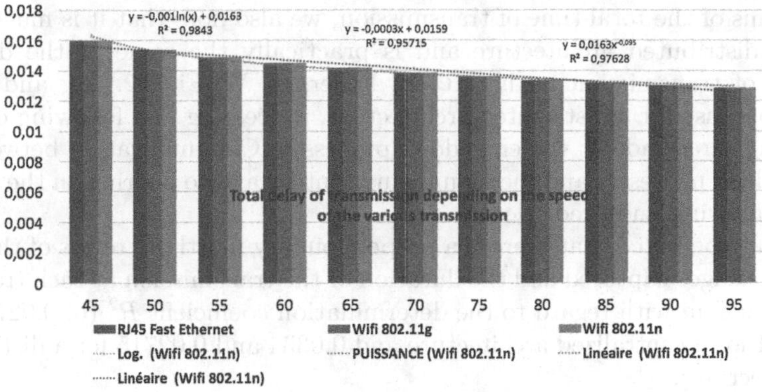

Fig. 7. Total transmission time

that it exceeds with an accuracy of 0.00802 a the function of power and 0.02715 with respect to the linear function. We have the same trends with respect to the coefficient of determination R^2 between the number of images captured and the duration of the transmission between the logarithmic function and the linear function (Fig. 8).

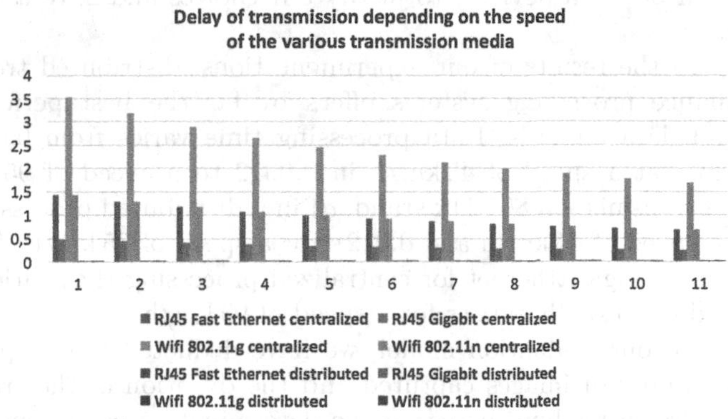

Fig. 8. Comparison between the two models

The comparative study between distributed and centralized for the different cases studied shows:

- We have more images processed in a distributed architecture when going mobile in a camera field of vision. This reflects the fact that the treatment is much faster and requires less communication between system components through the embedded processing. Therefore, we have a better decision because we have more collected information (metadata) that will allow us to have a better view and understanding of the system.

- In terms of the total time of transmission, we also note that it is much faster for a distributed architecture and is practically the same at the different types of transmission media (RJ-45 Ethernet, Wi-Fi 802.11 g and 802.11 not) because for a distributed architecture, processing and following decision makings are made at the embedded processor. Communication between the embedded processor and the monitoring center have no bearing on the process of monitoring and prediction.
- It should be noted that there is a proportionality mostly in terms of the number of images captured and the duration of the transmission at each transmission medium with regard to the determination coefficient R^2 (of 0.02715 and 0.0334 for a centralized architecture and 0.0334 and 0.02715 for a distributed architecture).

6 Contribution

The field of prediction is nowadays a major issue in research. It can be applied to several environments: medicine, transport, radar, collision detection, of road or path [1, 2, 3, 4]. But however, the current methods of prediction do not take into account the architecture of the processing system. In most cases, the architecture is fixed and is not subjected to studies for a better system performance. In our case, we did a specific study of the processing system for optimization (communication between devices) to justify our choices and have a guaranteed performance.

According to the results of our experimentations, distributed treatment in video surveillance processing systems offers by far the best performance in processing and data analysis. Data processing time varies from 0.0016 s to a mobile traveling at a speed of 45 km/h in 0.0012 to a speed of 95 km/h. We note the same transmission 801.11x standard in a distributed processing system (0.0016 s to a speed of 45 km/h and 0.012 s for a speed of 95 km/h) faster than that used in RJ45 giga Ethernet for centralized processing that varies between 0.5 s to a speed of 45 km/h in 0.4 s for a speed of 95 km/h.

In relation to our evaluation model, we have deducted the proportionality between the number of images captured and the duration of the transmission at each transmission medium based on the determination coefficient R^2 at the level of Each model (0.02715 and 0.0334 for a centralized architecture and 0.0334 and 0.02715 for a distributed architecture) between the logarithmic, linear and power functions.

7 Conclusion

This experimental and mathematical study of the assessment of the central processing time and distributed in road traffic video surveillance systems shows that the treatment is significantly faster when it is distributed. The information is processed at the same time as monitoring of the mobile. The decision to activate the next camera requires very low latency and makes this type of architecture more appropriate regarding the processing time. The type of architecture does not arise since many see after the experimental data, a distributed type architecture provides better performance in terms of speed of transmission.

We will be called to confirm these results by comparisons with various simulations using an individual centered approach (multi agents based simulations) which are closer to the distributed architecture concepts.

References

1. Gouaillier, V., Fleurant, A.E.: Intelligent video surveillance: promises and challenges. Technological and Commercial Intelligence Report (2009)
2. Chen, L., Lv, M., Ye, Q., Chen, G., Woodward, J.: A system for destination and future route prediction based on trajectory mining. Pervasive Mob. Comput. **6**(6), 657–676 (2010)
3. Krausz, B., Herpers, R.: MetroSurv: detecting events in subway stations. Multimedia Tools Appl. **50**(1), 123–147 (2010)
4. Sicre, R., Nicolas, H.: Human behaviour analysis and event recognition at a point of sale. In: 2010 Fourth Pacific-Rim Symposium on Image and Video Technology (PSIVT), pp. 127–132. IEEE (2010)
5. Communications Technology Australia (NICTA) Australian Technology Park, Bay 15 Locomotive Workshop Eveleigh, NSW 1430, Australia
6. Nasir, M., Lim, C.P., Nahavandi, S., Creighton, D.: Prediction of pedestrians routes within a built environment in normal conditions. Exp. Syst. Appl. **41**(10), 4975–4988 (2014)
7. Brulin, M.: Analyse sémantique d'un trac routier dans un contexte de vidéo-surveillance, THÈSE présentée à L'UNIVERSITÉ BORDEAUX I' ÉCOLE DOCTORALE DE MATHÉMATIQUES ET INFORMATIQUE, Soutenue le, 25 octobre 2012
8. Alabed, R.A., Mohammed, M.S.: Optimization, Improving Throughput on WiMAX Mobility. Int. J. Eng. Innovative Technol. (IJEIT) **3**(1), 6–8 (2013)
9. Mahasukhon, P., Sharif, H., Hempel, M., Zhou, T., Ma, T.: Distance and throughput measurements in mobile WiMAX test bed. In: The 2010 Military Commununication Conference, pp. 154–159, San Jose, CA, October 2010
10. Li, Y., Wang, C., You, X., Chen, H., She, W.: Delay and throughput performance of IEEE 802.16 WiMax mesh networks. IET Commun. **6**(1), 107 (2012)
11. Lubobyaa, S.C., Dlodlo, M.E., De Jager, G., Zulu, A.: Throughput characteristics of WiMAX video surveillance systems. In: International Conference on Advanced Computing Technologies and Applications, ICACTA-2015 (2015)
12. Tian, Y., Brown, L., Hampapur, A., Lu, M., Senior, A., Shu, C.: IBM smart surveillance system (S3): event based video surveillance system with an open and extensible framework. Mach. Vis. Appl. **19**(5), 315–327 (2008)
13. Detmold, H., van den Hengel, A., Dick, A., Falkner, K., Munro, D.S., Morrison, R.: Middleware for distributed video surveillance. IEEE Distrib. Syst. Online **9**(2), 1–11 (2008)
14. Diop, P.S., Mbacké, A.B., Mendy, G.: Predictive assessment of response time for road traffic video surveillance systems: the case of centralized and distributed systems. In: Hsu, C.-H., Wang, S., Zhou, A., Shawkat, A. (eds.) IOV 2016. LNCS, vol. 10036, pp. 34–48. Springer, Cham (2016). doi:10.1007/978-3-319-51969-2_4

Secure Storage as a Service in Multi-Cloud Environment

Riccardo Di Pietro[1(✉)], Marco Scarpa[2], Maurizio Giacobbe[2], and Antonio Puliafito[2]

[1] Department of Mathematics and Computer Science, University of Catania, Viale Andrea Doria, 6, 95125 Catania, CT, Italy
rdipietro@unict.it

[2] Department of Engineering, University of Messina, Contrada Di Dio, 98158 Sant'Agata, Messina, ME, Italy
{mscarpa,mgiacobbe,apuliafito}@unime.it

Abstract. Nowadays, Cloud computing is rapidly evolving due to social and cultural influences that are changing our everyday life. Innovative research in security-enabling techniques and architectures bodes to satisfy the increased interoperability among different vendors. An increasing number of service providers use Cloud computing to adapt their products to customer needs by addressing storage requirements in a secure way. The goal of our proposal is to provide a secure storage service able to store data in Multi-Cloud environment. The proposed secure storage architecture is oriented to guarantee confidentiality and integrity issues concerning information and data which are disseminated and stored in worldwide distributed Cloud environments. In this work, the SSME architecture is presented and discussed. Moreover, in order to evaluate our proposal, we conducted several experiments by considering the implementation of the SSME application as a Service in a real scenario.

Keywords: Cloud storage · Multi-Cloud · Confidentiality · Integrity · Security · AES-256 · RSA · OpenStack

1 Introduction

The growing number of requests to store and share files in Cloud environments results in an also growing number of user-friendly Cloud services in order to satisfy the above requests.

Cloud storage allows to store data in multiple remote sites usually owned by "top" companies and running their owner solutions, e.g., Google Drive, Dropbox, Amazon Simple Cloud Storage Service (S3). Besides the above-mentioned proprietary solutions, other open source solutions exist for providing Cloud storage services. For example, users can use the Swift OpenStack Object Storage to store lots of data efficiently and cheaply.

The development of the Cloud storage services marketplace particularly depends on the ability to build economies of scale. Within the *Digital Single*

A. Puliafito et al. (Eds.): ADHOC-NOW 2017, LNCS 10517, pp. 328–341, 2017.
DOI: 10.1007/978-3-319-67910-5_27

Market Strategy, the *European Union* establishes a free flow of data in Europe, by facilitating data portability and switching of Cloud service providers. The study *"SMART 2013/0043 - Uptake of Cloud in Europe"* [4] indicates that Cloud developments could lead to the growth of the European Cloud market from €9.5bn in 2013 to €44.8bn by 2020 (i.e., almost five times the market size in 2013). Approximately the 19% of EU enterprises used Cloud computing in 2014, mostly did it for hosting their e-mail systems and storing files in electronic form. Moreover, further estimates of this study highlight that four out of ten companies (39%) using the Cloud reported the risk of a security breach was the main limiting factor in the use of Cloud computing services.

In such a scenario, by moving data to the Cloud, users can avoid (i) costs (i.e., money) to build and maintain a private storage infrastructure, (ii) unauthorized access by third parties, and at the same time it is important (iii) to reduce legal uncertainly. In fact, in view of easy file storage and sharing services widely accessible by several typologies of customers and contexts (e.g., businesses, academies, and many others), more and more personal and confidential data can be exposed to privacy and security vulnerabilities.

In this paper we present the **S**ecure **S**torage in **M**ulti-Cloud **E**nvironment **(SSME)** architecture which addresses the confidentiality and integrity issues concerning data store and dissemination in worldwide distributed Cloud environments. The architecture implements a novel solution which mainly uses encryption at client-side, and a worldwide distributed middleware for data splitting, dissemination and retrieval.

The reminder of this paper is organized as follow: in Sect. 3 we give a brief overview of existing work about the use of multi-Cloud storage using cryptographic data splitting. In Sect. 2, we describe the motivation about the solution we propose. The proposed architecture is described in Sect. 4, and a detailed explanation of the implementation is in Sect. 5. In Sect. 6, we describe the results of the performance analysis. Section 7 concludes the presented work, with a discussion of some improvements planned as future work.

2 Motivations

Cloud Computing Services usually provide resource management's capabilities which ensure security on accessing services and on communicating data, but they often lack of data protection from direct malicious accesses at system level. It is mandatory to guarantee data protection mechanisms in order to deny data intelligibility to unauthorized users, even when they are (local) system administrators.

The goal of our work is to provide a software mechanism capable to store data in multi-Cloud environment in a secure way, by giving an answer to the confidentiality and integrity issues of information and data disseminated and stored in remote distributed machines all around the world. With SSME, we intend to introduce the above-mentioned data protection mechanism by combining both symmetric (AES256) and asymmetric encryption (RSA). Moreover, users can use a dynamic management of both the fragmentation schema and the pool of Cloud storage services to use to create their own multi-Cloud environment whenever they want to store a file on the Cloud. In our proposal, all client-side configurations are totally invisible on the server-side. Each run of the SSME application is tracked nowhere on the system, neither on the physical disk of servers. By using the *Trusted Control Service* (TCS, see the Sect. 4.1), users are able to *integrate* Cloud services they trust for the significant Cloud computing functionalities of *authentication* and *backup-storage*. This greatly increases their confidence in using the SSME application.

Many works in literature deal with Cloud and multi-Cloud storage systems using cryptographic data splitting. In the next Section we present the background and how the already existing approaches and solutions differ from the SSME.

3 Related Work

In [2] the authors present an architecture for a cryptographic storage service. It consists of four components: a server which processes and encrypts (AES256) data before it is sent to Cloud, a private Cloud that holds the meta-data information, and two Clouds that respectively archive one half of each user's file. The authors assume that the remote server is trusted without specifying any information on "how" this trustiness is implemented. The meta-data information (e.g., passwords, secret keys of each files, encrypted access paths) are securely stored in the private Cloud. If compared with our dynamic approach which allows to specify the size of the split fragments, here data splitting is statically fixed on half of each user's file.

In [1] the authors propose a new method for securing the user's data using the multi-Clouds in an untrusted mobile Cloud environment. This method splits data into segments that are successively encrypted, compressed and distributed via multi-Clouds while keeping one segment on the mobile device memory. Keeping one segment in the user's device will prevent any attempt to recover the distributed data, thus to avoid the grabbing of all the segments together with the key by possible unauthorized users. In contrast to our approach, their solution requires a Mobile Cloud Computing (MCC) architecture and keeps a segment in each device.

In [9] the authors presents an architecture that makes the data splitting of the uploaded file size by three. Their architecture consists of a System Database and a Middleware for data slicing and merging, and for data encryption and decryption. Also in this work data splitting is statically fixed on a third of each user's file. Another difference is they use the System Database to store the information necessary for the Middleware operations, without to specify its content (i.e., the type of information).

In [8] the authors propose a reliable storage system, *TrustyDrive*, that takes care of both the document anonymity and the user anonymity. The system architecture consists of three layers: the end users which use a storage system as a service that splits and encodes the files; the dispatcher, which provides an entry point for both the end users and the Cloud providers; Cloud providers which provide different storage space in terms of costs and performances. At client side (the end user) documents are split in blocks and for each block meta-data are associated (user meta-data). The user breaks its meta-data into blocks to be sent to the dispatcher which, in turn, computes missing information to store meta-data blocks on Cloud providers. However, the authors do not provide a description about the secure communication between end user-dispatcher and dispatcher-Cloud provider. Moreover, differently from their architecture, our mechanism of splitting and dissemination of the fragments is configurable by the user, which is able to specify the size of the fragments and the pool of Cloud providers.

In [5] the authors present three different approaches for the parallel exploitation of different Cloud storage providers using Redundant Residue Number System (RRNS) in order to enforce long-term availability, obfuscation and encryption. They conduct several experiments considering a real testbed in which a client interacts with three different providers (i.e., Google Drive, Copy, Dropbox). In contrast to our approach, their solution permits user to retrieve files even if one of the Cloud storage providers previously used is not temporarily or permanently available. On the other hand, similarly to our approach, providers cannot access the files stored within them.

The TwinCloud client-side encryption solution presented in [3] focuses on the *secure sharing* on Clouds without explicit key management. To this end, they highlight the Public Key Infrastructure (PKI)-based solution problems, i.e., costs due to get a certificate from a Certification Authority (CA) and PKI-infrastructure maintenance. Differently from the TwinCloud solution, our architecture uses both symmetric and asymmetric cryptography.

All the papers do not give specific information on how the keys are managed, at least it is not described where the key are stored and how and who can access them. Moreover, in the above-mentioned contributions, none of virtual access points (e.g., gateway, dispatcher, server) to the various multi-Cloud environments which represent the main nodes of the processing offer a scheme easily and dynamically configurable by users. Neither adopts and/or describes the use of a secure communication protocol for its communications nor works only in volatile memory. It means that the processed data are not protected against insider attacks [6].

4 Proposed SSME Architecture

Figure 1 shows a general scheme of the SSME architecture. It consists of a JAVA client-server application which uses a stateless RESTfull approach for its communication, where a client cooperates with the middleware where most of the computation is done. The *SSME middleware* is, in turn, made up two main components: a "Trusted Cloud Service" (TCS) implementing all the fundamental interfacing functions with SSME clients, and a "Server" where all the middleware file manipulations are provided.

An SSME compliant application can work in two different modes: *Upload Mode* and *Download Mode*.

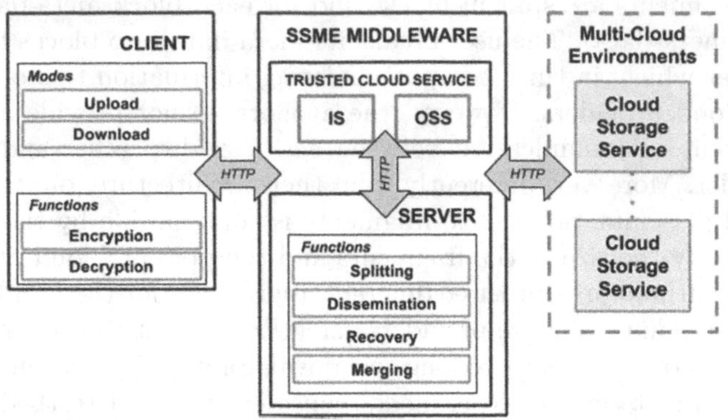

Fig. 1. SSME architecture general scheme.

4.1 SSME Middleware

Trusted Cloud Service. The **"Trusted Cloud Service" (TCS)** is the architectural component that provides some significant Cloud computing functionalities to the SSME client-server application. It is composed in turn into a two different sub-services: the **"Identity Service" (IS)** and the **"Object Storage Service" (OSS)**. As the name implies, the IS identifies the own trusted Identity Service the user may want to adopt by running the SSME client-server application. The IS, by exploiting its own token service mechanism, provides authentication and authorization on Cloud for the SSME service requests. In the same way, the OSS identifies the own trusted Object Storage Service that the user may want to adopt by using the SSME Application. The OSS provides temporary backup needed during the SSME internal operations. The OSS can also be used to store the encrypted JSON file that results as output of the SSME application. This output file represents the only *guiding light* to retrieve that particular file from the Cloud. We adopt the approach of externalizing the main function of Identity and Object Storage because each SSME user could have his/her own trusted Cloud services, for example personal or provided by his company, and may want to use these in the SSME scenario. In our testbed scenario, we adopt an OpenStack [7] compliant TCS.

SSME Server. All processing of the server are carried out without leaving any trace on local storage (hard disk) because it works only in volatile memory. In brief, the functions performed by the server are the following:

- uploading and downloading files to and from the TCS;
- fragmentation, recovery and merging of fragments to and from the Cloud services which are part of the multi-Cloud environment used at the moment;
- decryption of the information contained in the Headers of the HTTP requests received from the client;
- creation and encryption of the *json-encrypted-file* file which represents the output of the *Upload Mode.* This file represents the only way to recover the file from the Cloud.

4.2 SSME Client

In order to work, the client needs the presence of a mandatory JSON configuration file called *json-conf-file* file. For smooth functioning of our service, the *json-conf-file* file must be filled in the proper way. As the name implies, the *json-conf-file* file contains the information about the configuration of the client, in particular:

- its internal settings (including the symmetric encryption key used);
- the communication with the Server (including the public key of the Server);
- the services that make up the TCS;
- the Cloud storage service providers which are part of the multi-Cloud environment the user want to use.

In brief, the functions performed by the client, when working in either *Upload Mode* or *Download Mode*, are the following:

- encryption of the file you want to send to the multi-Cloud environment;
- decryption of the file you want to receive from the multi-Cloud environment;
- encryption of the information contained in the Headers of the HTTP requests sent from both the server and the TCS;
- decryption of the information contained in the Headers of the HTTP requests received from both the server and the TCS.

Upload Mode. During the *Upload Mode*, the client reads the information inside the *json-conf-file* file in order to instruct itself on how to contact the service. According to the reported information, the client starts the processing. The client sends to the server some HTTP requests according to the rule described in Sect. 4.3. Once the server received all the needed information, it elaborates them and returns to the client an AES256 encrypted JSON file, we called *json-encrypted-file* file. The *json-encrypted-file* file contains all the relevant information in order to retrieve and rebuild all the fragments scattered among the Cloud storage services. Referring to the classical mythology, this file serves as a modern *Ariadne's thread.* In the absence of the *json-encrypted-file* file, or if it is damaged, it is not possible to recover and rebuild the fragments stored in the Cloud. The *Upload Mode* will return a *json-encrypted-file* file for each file uploaded in the Cloud. The template structure of the *json-encrypted-file* file is shown in Fig. 2.

```
{ "File": { "FileName": "", "DirName": "", "SliceSize": "" },
  "Fragments": [ { "FragmentName": "", "FragmentNumber": "", "FragmentMD5": "", "ServiceType": "dropbox",
                   "DropboxToken": "" },
                 { "FragmentName": "", "FragmentNumber": "", "FragmentMD5": "", "ServiceType": "openstack",
                   "OpenStackUser": "", "OpenStackPassword": "", "OpenStackTenant": "", "OpenStackUrl": "" },
                 { "FragmentName": "", "FragmentNumber": "", "FragmentMD5": "", "ServiceType": "gdrive",
                   "GDriveJsonFile": "", "GDriveUrlFile": "", "GDriveIdFile": "" }
               ]
}
```

Fig. 2. The *json-encrypted-file* file template structure.

The structure is composed of two main parts: the field "File" and "Fragments". The field "File" contains the name of the original file (i.e. FileName), the name of the container/directory where the fragments are stored inside the Cloud storage services (i.e. DirName) and the dimension of the single fragment (i.e. SliceSize). The field "Fragments" contains all the information needed to retrieve all the fragments. This field is structured as a JSON vector. In the actual implementation, the elements of this vector can assume three different typology: *dropbox*, *openstack*, *gdrive*. Each of these, represents the typology of the Cloud storage services used by our SSME testbed.

Download Mode. During the *Download Mode*, the client reads the information inside the *json-encrypted-file* file to follow the *Ariadne's thread*. The *json-encrypted-file* file is related to the original file that the client wants to retrieve from the Cloud. The client decrypts *json-encrypted-file* file by using the symmetric key specified in the *json-conf-file* file and extracts the information contained to process it. The client sends to the server all the information contained in the *json-encrypted-file* file according to the rule described in Sect. 4.3. These information are needed to the server in order to retrieve and rebuild all the fragments of the file the user want to download from its own multi-Cloud environment. These fragments were previously scattered among the Cloud storage services during the *Upload Mode*.

After the *Download Mode*, the client will obtain its original file (still encrypted). At this time, firstly the client deletes the *json-encrypted-file* file related to the file just receive, then in order to remove all the fragments stored in the multi-Cloud environment, the client sends some `delete-fragment` requests to the server. In the end the client will decrypt the original file.

4.3 Communication Between the Client and the Server

All the HTTP communications between the client and the server are authenticated using the token service provided by the chosen IS. In our testbed, we used the token service provided by the OpenStack Identity Service v2.0 (Keystone). Each HTTP request sent by the client embeds some fields that contain the information necessary to carry out a given task at the server side. All these fields are symmetrically encrypted by using the AES256 algorithm. The key to decrypt these fields is also embedded in each request inside a field called "key".

The client encrypts this field by using the RSA asymmetric algorithm with a key of 2048 bytes, by using the server's public key. The server, through its own private key, can decrypt the "key" field and using it to decrypt the content of all the other fields stored in the HTTP request body. The server uses the symmetric key received from the client to encrypt all the fragments before sending them to the Cloud storage services. It is worth noting that the public key of the server is known and is freely downloadable from the web.

5 Implementation Details

In this section deeper details on how client and server interact in the proposed architecture are presented. Using the communication protocol here described it is possible to design a client compliant with a public service based on SSME.

5.1 Client-Server Communication Protocol in "Upload Mode"

The communication phase between the client and the server is done in seven different steps: (see Fig. 3)

1. First of all, in order to verify if the server is alive, the client sends to it an HTTP request (**1.A**). If the test is successfully done (**1.B**), then the client can move to the next step (**2.A**).
2. In order to be authenticated by the "Trusted Control Service" (TCS), the client sends an authentication request to the "Identity Service" (IS) (**2.A**). If the IS returns the token (**2.B**), which means that it is successfully authenticated, then the client can move to the next step (**3.A**).

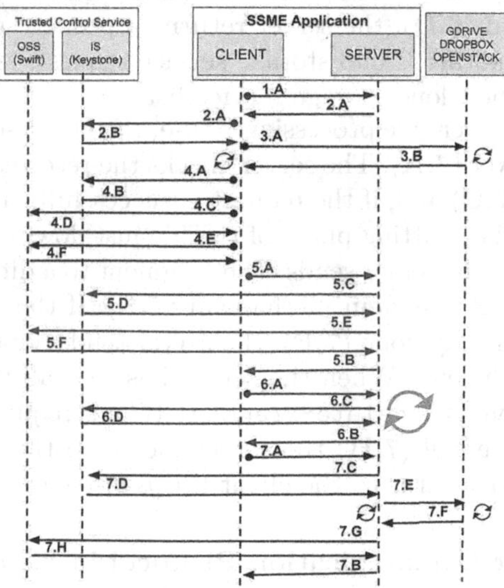

Fig. 3. Client-server communication protocol in "Upload Mode".

3. In order to test the validity of the Cloud storage services described in the *json-conf-file* file, the client checks each of them. The client sends an HTTP request to each of the Cloud storage services included in the list (**3.A**). If a Cloud storage service returns a positive response (**3.B**), the client sends another Cloud storage service test request (**3.A**), repeating the checking until the end of the storage service list.
4. In order to put the file to store in our multi-Cloud environment on the OSS:
   (a) firstly, the client sends an authentication request to the IS (**4.A**). If the IS returns the token (**4.B**), then the client can move to the next step (**4.C**);
   (b) then, the client sends an HTTP request to the OSS (**4.C**). If the container creation is successfully done (**4.D**), then the client can move to the next step (**4.E**);
   (c) finally, the client sends an HTTP request to the OSS (**4.E**). If the the create-object request was successfully done (**4.F**), and the client can move to the next step (**5.A**).
5. To start the server-side processing phase, the client sends an HTTP request to the server (**5.A**). The server checks the received request by interrogating the TCS (**5.C**), and if the request is successfully authenticated (**5.D**), then the server starts the download of the file the user want to put on the multi-Cloud environment (**5.E**) from the OSS. When the server completes the download with success (**5.F**), it returns a response (**5.B**), and the client moves to the next step (**6.A**).
6. In order to transfer all the information about the Cloud storage services to the server, the client uses an HTTP request for each Cloud storage service (**6.A**). Each request contains the information about a particular Cloud storage service the client wants to transfer at that moment. The server checks the received request by interrogating the TCS (**6.C**) and, if the request is successfully authenticated (**6.D**), the server returns a positive response (**6.B**), and the client sends another Cloud storage service request, repeating the process until the end of the Cloud storage service list.
7. To complete the server side processing phase, the client sends a `split` HTTP request to the server (**7.A**). The server checks the received request by interrogating the TCS (**7.C**) and, if the request is successfully authenticated (**7.D**), the server starts the splitting phase of the file just downloaded from the OSS. After the splitting, the server sends each fragment to a different Cloud storage service by following a random mechanism (**7.E**). If the upload-request for a fragment is successfully done (**7.F**), the server sends another upload-request, and so on, until the end. When the server has sent all the fragments to the Cloud storage services, a `delete-fragment` HTTP request is sent to the OSS (**7.G**) and, if successful (**7.H**) the server moves to the next step (**7.B**). At the end, the server returns to the client the *json-encrypted-file* file (**7.B**).

5.2 Client-Server Communication Protocol in "Download Mode"

The communication phase between the client and the server is done in nine different steps (see Fig. 4):

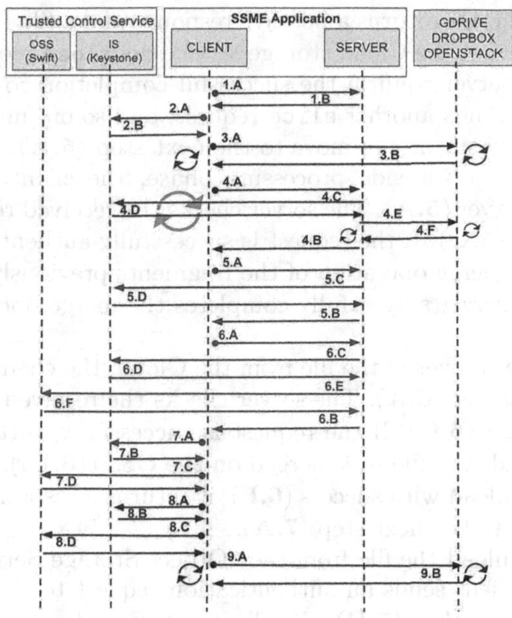

Fig. 4. Client-server communication protocol in "Download Mode".

1. First of all, in order to verify if the server is alive, the client sends to it an HTTP request (**1.A**). If the test is successfully done (**1.B**), then the client can move to the next step (**2.A**) otherwise it completes.
2. In order to be authenticated by the "Trusted Control Service" (TCS), the client sends an authentication request to the "Identity Service" (IS). The client sends to IS an HTTP request with a JSON file containing its credentials (**2.A**). If the IS returns the token (**2.B**) the client is successfully authenticated and it can move to the next step (**3.A**).
3. In order to test if the Cloud storage services described in the *json-encrypted-file* file are active and available, the client checks each of them. It sends an HTTP request to each of the Cloud storage services included in the list. All these requests (**3.A**) contain the information about a particular Cloud storage service that the client wants to check. If the Cloud storage service returns a positive response (**3.B**), which means that the Cloud `storage-service-test` request was successfully done, the client sends another Cloud `storage-service-test` request, and so on, until the end of the list.
4. To start the server side processing phase, the client transfers to the server all the information about the fragments described in the *json-encrypted-file* file. The client uses one HTTP request for each fragment. All these requests contain encrypted parameters inside the headers (**4.A**). The server checks the received request by interrogating the TCS (**4.C**). If the request is successfully authenticated (**4.D**), the server uses the information contained in the request to download fragment from the related Cloud storage service (**4.E**). If the

Cloud storage service returns a positive response and a file (i.e. the fragment) (**4.F**), meaning that the Cloud storage service `download` request was successfully done, the server confirm the successful completion to the client (**4.B**). Then the client sends another `slice` request, and so on, until the end of the fragments and the client can move to the next step (**5.A**).

5. To continue the server side processing phase, the client sends an HTTP request to the server (**5.A**). The server checks the received request by interrogating the TCS (**5.C**). If the request is successfully authenticated (**5.D**), the server start the merge operation of the fragments previously gathered server side. When the server successfully completes the merge operation, it returns a response (**5.B**).
6. To complete the recover of the file from the Cloud, the client sends an HTTP request to the server (**6.A**). The server checks the received request by interrogating the TCS (**6.C**). If the request is successfully authenticated (**6.D**), the server uploads the file just merged on the OSS ((**6.E**)). When the server completes the upload with success (**6.F**), it returns a response (**6.B**), and the client can move to the next step (**7.A**).
7. In order to download the file from the "Object Storage Service" (OSS):
   (a) firstly, the client sends an authentication request to the IS (**7.A**). If the IS returns the token (**7.B**), the client can move to the next step (**7.C**);
   (b) then, the client sends an HTTP request to the OSS (**7.C**). If the the `download-object` request is successfully done (**7.D**), the client can move to the next step (**8.A**).
8. Once the client receives the file from the OSS:
   (a) firstly, the client sends an authentication request to the IS (**8.A**). If the IS returns the token (**8.B**), the client is successfully authenticated;
   (b) then, the client sends an HTTP request to the OSS (**8.C**). If the the `delete-object` request is successfully done (**8.D**), the server moves to the next step (**9.A**).
9. At the end, the client takes care to delete all the fragment scattered on the Cloud storage service constituting the multi-Cloud environment. The client sends an HTTP request to delete each of the fragments which are present in the *json-encrypted-file* file. All these requests (**9.A**) contain the information about a particular fragment the client wants to delete at that moment. If the Cloud storage service returns a positive response (**9.B**), the client sends another `delete-fragment` request, and so on, until the end. Then the client can finish.

6 Performance Analysis

In order to evaluate the proposed system, we conducted several experiments by considering the implementation of the SSME application as a Service in a real scenario. The system was developed using the JAVA programming language for both client and server sides. We started the SSME application server on a virtual machine equipped with Ubuntu Server 14.04 and hosted on an IBM BladeCenter LS21. Instead, the client machine used in our experiments is equipped with the following hardware configuration: a CPU Intel(R) Core(TM) i7-4700MQ 2.4 GHz Dual-Core, 16 GB of central memory, Linux Ubuntu server 14.04.5 LTS 64 bit operating system and a SATA HD with 1 TB of disk storage. The middleware interacts with eight different Cloud storage services among three different Cloud storage providers: Google Drive (one instance), Dropbox (four instances) and OpenStack Swift (three instances). In our experiments, we considered different file sizes in each performed test: specifically, we used files from 10 MB to 2 GB. Therefore, we split each of them in fragments with two different size: 10 MB and 100 MB (this latter with large files at least 500 MB).

In *Upload Mode*, performance analysis consists in evaluating: the system response time (*Overall*), the encryption time (*Encryption*) and the time due to splitting and fragment dissemination (*Split*). In *Download Mode*, performance analysis consists in to evaluate: the system response time (*Overall*), the time to receive and merge all the fragments to recompose the original file (*Merge*), and the decryption time of the file just recomposed (*Decryption*).

We repeat each experiment 30 times and analyzed the collected data considering 95% as confidence level.

Figures 5 and 6 show a graphical representation of the monitored times, from 500 MB to 1 GB as file size, in *Upload Mode* and the *Download Mode* respectively. We did not depict the results obtained with lower file size only to make clearer the graphs without introducing any further information. As can be noted the system response time linearly increases with the file size both in *Upload Mode* and *Download Mode* and independently on the fragment size used for splitting the files. Both in *Upload Mode* and *Download Mode*, the impact of encryption and decryption, respectively, turn out to be negligible with respect the overall response time. When in *Download Mode* also the merge phase does not affect the performance, instead the splitting phase has an impact on the overall time in *Upload Mode*. In particular, in *Upload Mode* (Fig. 5)

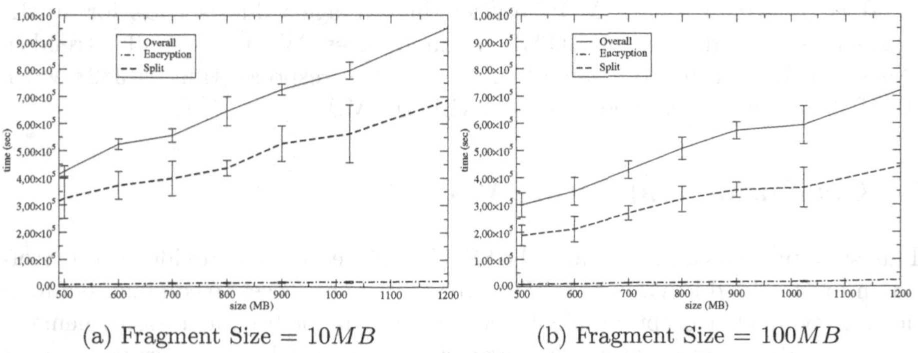

(a) Fragment Size = $10MB$ (b) Fragment Size = $100MB$

Fig. 5. Measured times in *Upload Mode*.

- encryption assumes a value between 1.5% and 5% of the overall time depending on the file size,
- Splitting and Fragments Dissemination is between 49% and 74% of the overall time depending on the file size.

In *Download Mode* (Fig. 6),

- Receive and Merging affects the Overall Time between 2.7% and 11% depending on the original file size.
- Decryption affects the Overall Time between 3.8% and 8% depending on the original file size.

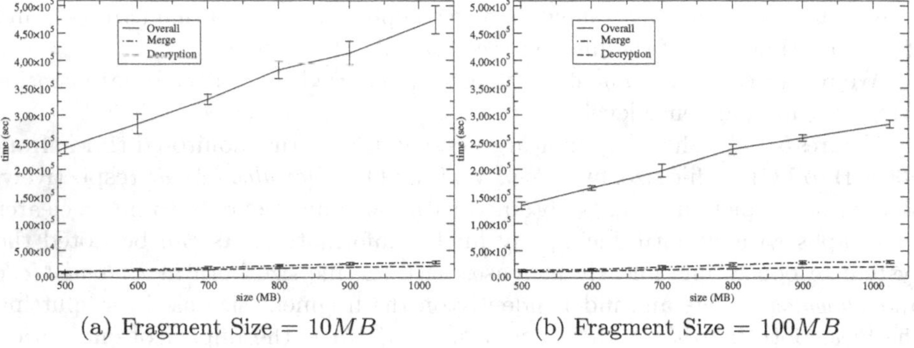

(a) Fragment Size = $10MB$ (b) Fragment Size = $100MB$

Fig. 6. Measured times in *Download Mode.*

In both cases, the results show that the Overall Time improves in performance when greater fragment sizes are used. For example, in *Upload Mode*, this trend is confirmed for 800 MB file size: the average value is 646 s for 10 MB fragment sizes and 505 s for 100 MB fragment sizes. We had a similar trend in *Download Mode* using the same file: the average response time is 382 s when 10 MB fragment size is used and 237 s with 100 MB.

7 Conclusion and Future Work

This work presents and describes the SSME architecture to provide Secure Storage in multi-Cloud Environment. The architecture proposes to combine symmetric and asymmetric cryptography by offering a dynamic fragmentation schema to users which guarantees protection against insider attacks. In order to generate encrypted data while storing on multi-Cloud Environment, the symmetric cryptography is directly applied to the data on the client-side. The asymmetric cryptography (server public key) is used to encrypt sensitive information of HTTP requests Headers, which is being exchanged in the various steps described in the

Sect. 4.3. The SSME architecture has been developed and evaluated by considering the implementation of the SSME application as a Service in a real scenario. In future, we plan to release the SSME application as a public service and, based on it, to realize an Android Mobile Application and desktop application as SSME clients in order to show the effectiveness and the flexibility of our approach.

References

1. Alqahtani, H.S., Sant, P.: A multi-cloud approach for secure data storage on smart device. In: 2016 Sixth International Conference on Digital Information and Communication Technology and its Applications (DICTAP), pp. 63–69, July 2016
2. Balasaraswathi, V.R., Manikandan, S.: Enhanced security for multi-cloud storage using cryptographic data splitting with dynamic approach. In: 2014 IEEE International Conference on Advanced Communications, Control and Computing Technologies, pp. 1190–1194, May 2014
3. Bicakci, K., Yavuz, D.D., Gurkan, S.: Twincloud: a client-side encryption solution for secure sharing on clouds without explicit key management. CoRR abs/1606.04705 (2016)
4. Bradshaw, D., Cattaneo, G., Lifonti, R., Simcox, J.: Smart 2013/0043 - uptake of cloud in Europe. Technical report, IDC EMEA, June 2015
5. Celesti, A., Fazio, M., Villari, M., Puliafito, A.: Adding long-term availability, obfuscation, and encryption to multi-cloud storage systems. J. Netw. Comput. Appl. **59**, 208–218 (2016)
6. Gunasekhar, T., Rao, K.T., Basu, M.T.: Understanding insider attack problem and scope in cloud. In: 2015 International Conference on Circuits, Power and Computing Technologies (ICCPCT-2015), pp. 1–6, March 2015
7. OpenStack: Official web site. https://www.openstack.org/. Accessed 28 June 2017
8. Pottier, R., Menaud, J.M.: Trustydrive, a multi-cloud storage service that protects your privacy. In: 2016 IEEE 9th International Conference on Cloud Computing (CLOUD), pp. 937–940, June 2016
9. Vaidya, M.B., Nehe, S.: Data security using data slicing over storage clouds. In: 2015 International Conference on Information Processing (ICIP), pp. 322–325, December 2015

Policy Management and Enforcement Using OWL and SWRL for the Internet of Things

Rustem Dautov[1,2(✉)], Symeon Veloudis[2], Iraklis Paraskakis[2], and Salvatore Distefano[1,3]

[1] Higher Institute of Information Technology and Information Systems (ITIS), Kazan Federal University (KFU), Kazan, Russia
{rdautov,s_distefano}@it.kfu.ru
[2] South-East European Research Centre (SEERC), The University of Sheffield, International Faculty CITY College, Thessaloniki, Greece
{rdautov,sveloudis,iparaskakis}@seerc.org
[3] University of Messina, Messina, Italy
sdistefano@unime.it

Abstract. As the number of connected devices is exponentially growing, the IoT community is investigating potential ways of overcoming the resulting heterogeneity to enable device compatibility, interoperability and integration. The Semantic Web technologies, frequently used to address these issues, have been employed to develop a number of ontological frameworks, aiming to provide a common vocabulary of terms for the IoT domain. Defined in Web Ontology Language – a language based on the Description Logics, and thus equipped with the 'off-the-shelf' support for formal reasoning – these ontologies, however, seem to neglect the built-in automated reasoning capabilities. Accordingly, this paper discusses the possibility of leveraging this idle potential for automated analysis in the context of defining and enforcing policies for the IoT. As a first step towards a proof of concept, the paper focuses on a simple use case and, using the existing IoT-Lite ontology, demonstrates different types of semantic classification to enable policy enforcement. As a result, it becomes possible to detect a critical situation, when a dangerous temperature threshold has been exceeded. With the proposed approach, IoT practitioners are offered an already existing, reliable and optimised policy enforcement mechanism. Moreover, they are also expected to benefit from support for policy governance, separation of concerns, a declarative approach to knowledge engineering, and an extensible architecture.

Keywords: Internet of Things · Semantic Web · Policy management · Policy enforcement · Web Ontology Language · Semantic Web Rule Language · Reasoning

1 Introduction

The unparalleled development of Information and Communication Technologies (ICT) in recent years has triggered the digital revolution, which is characterised by ubiquitous connectivity, leading to a convergence of technologies across

A. Puliafito et al. (Eds.): ADHOC-NOW 2017, LNCS 10517, pp. 342–355, 2017.
DOI: 10.1007/978-3-319-67910-5_28

different domains [8]. Rapid advances in embedded systems (especially mobile devices), wireless sensors, and mobile communications have led to the development of more and more powerful and capable personal devices. Modern tablets and even smartphones can be compared to 4–5 years old computers in terms of processing capabilities, so they can be also considered as effective computing systems. Indeed, they are more and more often taken into account as a laptop (and sometime desktop) replacement. This also contributed to their exponential growth in number and wide adoption; recent statistics [1] report on more than 2 billion smartphone users already connected to the Internet (i.e. every third smartphone owner), and quite soon, by 2020, this number is expected to reach 3 billion (i.e. every second smartphone owner). At the same time, similar statistics refers to the decreasing number of computers and indicates a reverse trend – i.e. there were almost 2 billion PC users in 2014, whereas nowadays their number has decreased to 1.5 billion, and is expected to drop to 1 billion by 2020.

Furthermore, new emerging ICT paradigms such as the Internet of Things (IoT), Cloud and Fog/Edge computing, enable novel, previously-unseen application scenarios for these devices. Smart Cities, Mobile Crowdsourcing, Digital Democracy, Citizen Science are just few simple examples among the wide range of possible scenarios, where a user can be directly and actively involved in everyday socio-technical activities by, for example, interacting through his/her personal device with the surrounding environment (e.g. smart lights, cameras, billboards, appliances, etc.) or with a remote entity (e.g. public administration).

These personal devices and their underlying infrastructure represent a 'digital image' of a user, acting as an interface between the user and the surrounding physical world, by measuring geo-localised physical data, such as weight, speed, acceleration, humidity, illumination, etc. (depending on the sensors available on-board). The massive amount of data captured in this context has laid the groundwork for the Internet of Services (IoS), which makes use of the context-aware data to provide Web services (typically in the form of Software-as-a-Service offerings) in order to generate value. The fusion between devices and data is characterised by complex networked collaboration between information sources, software, mechanical and electronic components interacting with physical entities. Today, such complex cyber-physical systems provide the technological foundation for diverse domain- and sector-specific applications, ranging from automotive and civil infrastructure, to healthcare, manufacturing, and transportation.

As a result, a heterogeneous ecosystem, constituted by embedded devices, sensors, actuators, mobile phones, and other smart connected objects, has been established through the Internet, as envisioned by the IoT. This technological fusion starts resembling a global 'melting pot', in which complex heterogeneous technologies provide solutions to 'little-local' problems. More specifically, existing solutions adopted a 'vertical' approach, in which ICT 'silos' are based on ad-hoc infrastructures, services and applications, narrow-tailored to specific problems at hand. This unscalable pattern relies on dedicated infrastructure, underpinned by specific hardware and software stacks, which will render unsustainable if, as expected, the demand for the IoT will keep increasing in the near future.

Taken together, these considerations identify the IoT as a rather fragmented 'archipelago' of isolated, local 'islands' of sensors and actuators, (static and mobile) devices (e.g. cameras, environmental sensors, personal portable devices, vehicles, etc.), network facilities (e.g. WiFi, 3 G/4 G, fiber optic, routers, base stations, cells, etc.), basic mechanisms and services for data management (i.e. collection, storage, aggregation, fusion, processing) – all related to and mainly conceived for a corresponding 'vertical' application domain. In this light, existing IoT systems seem to exist in isolation from each other – i.e. with little (or none) potential for integration, they are hardly compatible, interoperable or reusable.

1.1 Research Context

While challenges associated with timely processing of sensor data have been relatively successfully tackled by the advances in networking and hardware technologies [3,14], the challenge of properly handling data representation and semantics of IoT descriptions and sensor observations is still pressing. In the presence of multiple organisations for standardisation, as well as various hardware and software vendors, overcoming the resulting heterogeneity remains one of the major concerns for the IoT community. Moreover, apart from the syntactic heterogeneity (i.e. heterogeneity in the data representation, such as, for example, differences in data formats/encodings), heterogeneity in the semantics of the data can also be distinguished [6]. For example, different units of measurement, metric systems, or human languages are common causes for incompatibility and integrity problems in ICT.

As a potential solution to the IoT heterogeneity problem and yet-to-come standards, the research community started investigating potential ways of creating descriptive languages, which would provide an overarching modelling framework and bridge formerly-disjoint heterogeneous IoT systems at the semantic level. Such modelling languages can be seen as common vocabularies of terms, which are expected to be used by IoT practitioners to enable compatibility and interoperability when integrating IoT solutions. Simply put, two disjoint system would be able to 'communicate' to each other by expressing their data and interfaces, using a common suitable modelling language.

A representative example of how this lack of unified data representation has been addressed in the context of the IoT is the *Semantic Sensor Web* – a promising combination of two research areas, the Semantic Web and the Sensor Web [19]. Using the Semantic Web technology stack to represent data in a uniform and homogeneous manner, it provides enhanced meaning for sensor descriptions and observations so as to facilitate data analysis and situation awareness [7]. One of the main outcomes of this initiative was the Semantic Sensor Network (SSN) ontology – a thorough vocabulary, modelling the domain of physical sensor networks and observations, jointly developed by a wide group of researchers.

A number of other similar ontologies, modelling the IoT domain, have emerged in the recent years (for example, [2,4,9,11–13,15,16,18,23,25,26], to name a few). On the one hand, these ontologies are expected to provide common semantic foundation for describing various aspects of the IoT domain at different granularity levels to address the IoT heterogeneity problem. On the other hand, however, the fact that there are more and more competing, disjoint ontologies serves as yet another evidence of the isolation of existing IoT systems from each other.

Moreover, from this perspective, Semantic Web ontologies do not differ much from other modelling approaches, such as, for example, Unified Modelling Language (UML) or Extensible Mark-up Language (XML) – i.e. they all may serve to provide taxonomies of terms and relationships, to be used as 'building blocks' when describing the IoT-related application domains. A major advantage of the Semantic Web languages, which is frequently neglected in this context, is the support for *automated formal analysis*, underpinned by the built-in reasoning capabilities of Web Ontology Language (OWL) and Semantic Web Rule Language (SWRL), which are the key enabling technologies for the Semantic Web [10]. By representing data in terms of OWL classes and properties, one can perform reasoning tasks over these data and benefit from an already existing, highly-optimised and reliable analysis mechanism.

In this light, this paper is trying to tap into this idle potential for automated reasoning, and presents an approach to policy management and enforcement in the IoT context, using existing IoT ontologies and corresponding reasoning support. As it will be further described below, the proposed approach is expected to benefit from separation of concerns, extensibility, as well as increased opportunities for reuse, automation and reliability.

Accordingly, the rest of the paper is organised as follows. Section 2 contains relevant background information and briefs the reader on the foundations of the Semantic Web languages, as well as presents the target IoT-Lite ontology in more details. Section 3 presents and explains the proposed approach with a sample use case scenario, based on the IoT-Lite ontology. Section 4 summarises and discusses the main potential benefits of the approach. Section 5 concludes the paper.

2 Background: Semantic Web Languages

The Semantic Web, introduced by Berners-Lee [5] in 2001, is the extension of the World Wide Web that enables people to share content beyond the boundaries of applications and websites [10]. This is typically achieved through the inclusion of semantic content in web pages, which thereby converts the existing Web, dominated by unstructured and semi-structured documents, into a web of meaningful machine-readable information. Accordingly, the Semantic Web can be seen as a giant mesh of information linked up in such a way as to be easily readable by machines, on a global scale. It can be understood as an efficient way of representing data on the World Wide Web, or as a globally linked database. As shown in Fig. 1, the Semantic Web is realised through the combination of certain key technologies [10]. These technologies from the bottom of the stack up to the level of XML have been part of the Web standardised technology stack even before the emergence of the Semantic Web, whereas the upper, relatively new technologies – i.e. Turtle and N3, Resource Description Framework

(RDF), RDF Schema (RDFS), SPARQL Protocol and RDF Query Language (SPARQL), OWL, and SWRL – are intrinsic to the Semantic Web domain. All of these components have already been standardised by the World Wide Web Consortium (W3C) and are widely applied in the development of Semantic Web applications. The presented research specifically focuses on OWL and SWRL as the two potential ways of defining and enforcing policies in the context of the IoT, as it will be further explained in Sect. 3.

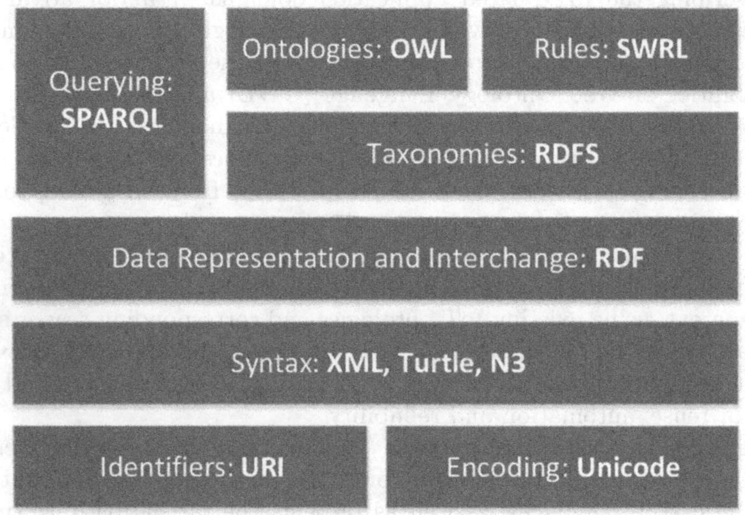

Fig. 1. The semantic web technology stack.

2.1 Web Ontology Language

OWL is a family of knowledge representation languages used to formally define an ontology – "a formal, explicit specification of a shared conceptualisation" [22]. Typically, an ontology is seen as a combination of a terminology component (i.e. TBox) and an assertion component (i.e. ABox), which are used to describe two different types of statements in ontologies. The TBox contains definitions of classes and properties, whereas the ABox contains definitions of instances of those classes. Together, the TBox and the ABox constitute the knowledge base of an ontology.

OWL provides advanced constructs to describe resources on the Semantic Web. By means of OWL it is possible to explicitly and formally define knowledge (i.e. concepts, relations, properties, instances, etc.) and basic rules in order to reason about this knowledge. OWL allows stating additional constraints, such as cardinality, restrictions of values, or characteristics of properties such as transitivity. The OWL languages are characterised by formal semantics – they are based on *Description Logics* (DLs) and thus bring reasoning power to the Semantic Web. There exists a prominent visual editor for designing OWL ontologies, called Protege,[1] and several automated reasoners written in multiple programming languages, such as Pellet,[2] FaCT++,[3] and HermiT.[4] Depending on the expressive power, the OWL family of languages can be classified into *OWL-Lite* (the most light-weight, but least expressive), *OWL-DL* (more expressive, but still with automated reasoning support), and *OWL-Full* (most expressive, but undecidable, and therefore does not have reasoning support).

2.2 Semantic Web Rule Language

SWRL extends OWL with even more expressiveness, as it allows defining rules in the form of implication between an antecedent (i.e. body) and consequent (i.e. head). It means that whenever the conditions specified in the body of a rule hold, then the conditions specified in the head must also hold. It is worth noting, that fully compatible with OWL-DL, the SWRL syntax is quite expressive, which may have certain negative impacts on its decidability and computability. The sample code in Listing 1.1 contains a rule, expressed in a human-readable syntax, and illustrates the functionality of SWRL. The following sample states that if a city is located in England, then it is also located in the United Kingdom.[5]

Listing 1.1. Example of a SWRL rule.

```
City(?city) AND
    hasLocation(?city, http://en.wikipedia.org/wiki/England) THEN
hasLocation(?city, https://en.wikipedia.org/wiki/United_Kingdom)
```

Note that with OWL and SWRL, there is typically more than one way of defining knowledge to deduce same facts. For example, the above inference, drawn by reasoning over the SWRL rule, can also be achieved by defining `hasLocation` as a transitive property between `city` and `England`, as well as between `England` and the `United Kingdom`. Even though it is not explicitly stated that a city is located in the United Kingdom, the reasoner will deduce this fact, based on the knowledge that England is located in the United Kingdom by following the transitive property. In general, the SWRL reasoning is more computationally expensive than the OWL reasoning [20], making the latter a more preferable option in most cases.

1 http://protege.stanford.edu/.
2 https://github.com/complexible/pellet.
3 https://code.google.com/p/factplusplus/.
4 http://hermit-reasoner.com/.
5 In this example, England and the United Kingdom are uniquely represented by their respective Wikipedia URLs.

2.3 An Existing Ontology: IoT-Lite Ontology

IoT-Lite ontology[6] [4] is a light-weight instantiation of the SSN ontology, actively developed by the W3C. It describes the key IoT concepts to allow interoperability and discovery of devices, sensors and sensor data in heterogeneous IoT platforms. This ontology reduces the complexity of other IoT models by describing only the main concepts of the IoT domain. This means that the IoT-Lite ontology can be extended by different models to increase its expressiveness and provide more focused modelling concepts if/when needed. That is, by following the Semantic Web principles of linking and reusing existing ontologies and datasets, it is possible to extend the core vocabulary with other relevant concepts, defined in other ontologies. This way, ontology engineers can simply import an existing, established, and trusted ontology, instead of 're-inventing the wheel' and developing their own, yet another, ontology from scratch.

3 Proposed Approach

Taking into consideration the presented features of the Semantic Web, this paper proposes leveraging reasoning capabilities of OWL and SWRL and utilise existing ontologies, describing the IoT domain, to enable creation and modification of policies to address a wide range of analysis activities in the IoT. This way, IoT practitioners can benefit from an already existing, optimised and reliable analysis engine, based on the declarative approach to defining the knowledge base.

The proposed vision will be now demonstrated thorough a sample use case scenario, based on the existing IoT-Lite ontology. Please note that the main goal of this sample scenario is to demonstrate how built-in reasoning capabilities can be used to perform analysis in the context of the IoT in a generic manner, rather than to focus on specific aspects of the IoT-Lite ontology. The proposed approach is expected to be universal and could be implemented using any other available IoT ontology. The sample pseudo-code snippets below are correspondingly simplified to make them easy to read and understand.[7]

The use case scenario focuses on a complex IoT system composed of multiple sensing devices, deployed both indoors and outdoors. Some of these devices are temperature sensors installed in rooms within a building. It is assumed that whenever any of these temperature sensors indicates a value exceeding a dangerous level of 50 degrees, the situation has to be classified as critical, and thus needs taking reactive actions. A possible way to handle this scenario would be

---

[6] https://www.w3.org/Submission/iot-lite/.

[7] The notations `ssn`, `iot`, and `dul` are established shortcuts for imported OWL ontologies, where corresponding concepts are defined.
SSN ontology (`ssn`): http://purl.oclc.org/NET/ssnx/ssn
IoT-Lite ontology (`iot`): http://purl.oclc.org/NET/UNIS/fiware/iot-lite
DOLCE Upper Level Ontology (`dul`): http://www.loa.istc.cnr.it/ontologies/DOLCE-Lite.owl.

to define explicit policies for every single temperature sensor within the building. In the worst case, such policies would be either 'hard-coded' with numerous `if/then` operators (i.e. any modifications would lead to the source code recompilation), or defined declaratively (i.e. stored in some kind of configuration file to be dynamically fetched by the analysis component). In both cases, however, the resulting knowledge base would be saturated by the excessive number of hardly manageable and possibly conflicting policies.

An alternative solution is based on using the built-in reasoning capabilities of OWL and SWRL to classify observed IoT data as instances of specific classes. More specifically, this use case demonstrates three different types of automated classification:

1. Defined Classes in OWL. Underpinned by the DLs, OWL allows creating so-called *defined classes* via a set of necessary and sufficient conditions. This means that the reasoner will classify any entity with a required set of sufficient properties as an instance of a specific class, even if this class membership was not defined explicitly. The defined class `RoomDevice` is demonstrated in Listing 1.2, which should be read as "if `sd` is a sensing device and has a rectangular coverage area, then `sd` is a device installed in a room".

Listing 1.2. Defined OWL class `RoomDevice`.

```
ssn:SensingDevice(?sd) AND
    iot:Rectangle(?r) AND
    iot:hasCoverage(?sd,?r) ≡
iot:RoomDevice(?sd)
```

2. Subclass Relationships in OWL. OWL also provides a simpler and more explicit way of defining classes and subclass relationships. It supports multiple and transitive inheritance, and, as in many other programming languages, subclasses inherit all the properties of their parent classes. The code snippet in Listing 1.3 contains two definitions. The first definition simply states that any temperature sensor is a sensing device. The second definition illustrates the transitive inheritance through a subclass hierarchy that states – in simple words – that if a device is installed in a room, then it is automatically assumed to be installed in a building as well, which in turn means it is an indoor device.

Listing 1.3. Defining OWL subclass relationships.

```
iot:TemperatureSensor IS A ssn:SensingDevice
iot:RoomDevice IS A iot:BuildingDevice IS A iot:IndoorDevice
```

3. Class Definition in SWRL. SWRL allows defining more expressive rules and takes the form of Horn-like rules, as illustrated by Listing 1.4. In simple words, the code snippet reads that if there is an indoor device `id`, indicating that its measured value has exceeded 50 degrees, the current observation has to be classified as critical.

Listing 1.4. Defining the class `CriticalObservation` using the SWRL.

```
iot:IndoorDevice(?id) AND
   ssn:Observation(?o) AND
   dul:Value(?v) AND
   iot:observes(?id,?o) AND
   ssn:hasValue(?o, ?v) AND
   swrl:greaterThan(?v, 50) THEN
iot:CriticalObservation(?o)
```

Next, the presented use case scenario assumes that there is a temperature sensor `ts` reporting a temperature level of 60 degrees in its covered rectangular area. Taking together all three definitions above, the automated reasoner will take the following steps when resolving this situation[8]:

1. Since `TemperatureSensor` is a subclass of `SensingDevice`, `ts` is classified as `SensingDevice` (according to Listing 1.3).
2. Since `ts` is a `SensingDevice` and has a rectangular coverage area, it is classified as `RoomDevice` (according to Listing 1.2).
3. Since `RoomDevice` is a subclass of `BuildingDevice`, which is a subclass if `IndoorDevice`, `ts` is classified as an instance of `IndoorDevice` (according to Listing 1.3).
4. Finally, since `ts` is an `IndoorDevice` and its measured observation is greater than 50 degrees, this observation is classified as `CriticalObservation` (according to Listing 1.4).

This way, the reasoning engine is able to identify a situation, when a dangerous level of 50 degrees has been exceeded – i.e. to detect a critical situation by inferring implicit information from the limited, explicitly provided facts. Moreover, it is worth explaining that using generic rules for a wide range of devices, as in the example above, does not affect the flexibility of the proposed approach and its ability to define fine-grained, targeted policies for individual devices. Apart from inheritance, the OWL also supports overwriting parent properties by subclasses. This means that it is possible to enforce device-specific policies, which will overwrite the default behaviour and apply only to those specific devices. This way, flexible policy enforcement at various granularity levels can be achieved.

4 Discussing the Potential Benefits

When defining monitoring and analysis policies with OWL and SWRL, IoT developers are expected to benefit from the following [6]:

[8] Please note that there are two main reasoning methods, which define the order, in which axioms are considered for evaluation – namely, *forward* and *backward* chaining [21]. In the presented use case, forward chaining is assumed to be in place.

Policy Governance. Ideally, a policy enforcement mechanism is expected to enable stakeholders to perform changes to policies and policy sets 'on the fly', i.e. in a dynamic manner that does not require re-compiling, re-deploying and restarting the entire software system. Such changes may include, for example, the introduction of new policies, or the update and retirement of existing ones. In contrast to other current approaches that tend to rely on hard-coded analysis logic, the semantic approach sufficiently addresses this requirement. In particular, by promoting a semantic representation of policies, that ontologically captures the various knowledge artefacts that are encoded in the policies, the semantic approach promotes a *separation of concerns* that disentangles the expression of policies from the actual code of the applications that enforce them. This not only enables the performance of 'on the fly' changes to policy sets, but also crucially paves the way for the construction of a novel *policy governance framework* that is capable of determining the consequences of such changes on the overall *effectiveness* of the policies through the provision of the following seminal capabilities:

- By enabling automated reasoning about the *correctness* of a newly created or updated policy by harnessing the various knowledge artefacts that it embodies;
- By enabling automated reasoning about potential *inter-policy relations*.
- By enabling the performance of *rule-based policy lifecycle management.*

With respect to the first capability, an *iterative process* that aims at defining suitable *ontological templates* for the expression of policies is advocated [24]. This process consists of a number of iterative refinement steps, where each step introduces new concepts and properties that reify the high-level concepts, and the properties thereof, that appear in the IoT-Lite ontology; the new concepts typically take the form of sub-concepts of the existing higher-level ones. The ontological templates that are ultimately defined through this process articulate all those concepts that *must*, *may* or *must not* be embodied in a policy, as well as the allowable values that these concepts may attain in a particular context of use. For instance, the iterative refinement process may reify the `geo:Point` concept of the IoT-Lite ontology with such concepts as `Building`, `Floor` and `Room`. An ontological template may then be defined to insist that a policy must invariably incorporate readings from a temperature sensor that is located in a particular room of a specific building or, alternatively, from one of the sensors that are located in a specific floor of the same building. Any newly created or updated policy that is derived as an instantiation of this template is guaranteed to satisfy this requirement and incorporate these readings. Evidently, the ontological templates enable stakeholders to influence the allowable form, or structure, that complex policies in the IoT domain may attain in a particular context of use. In this respect, they enable stakeholders to infuse into the policies their particular business logic.

With respect to the second capability, inter-policy relations such as subsumption and contradiction that are potentially brought about by changes in the policies (e.g. introduction of new policies or updates of existing ones) and which may affect the effectiveness of the policies, are determined through the use of 'off-the-shelf' DLs reasoners. Finally, with respect to the third capability, a framework may be constructed that generically determines the conditions under which *policy lifecycle actions*, such as policy updates and retirements, may be performed.

Human Readability and Ease of Use. The Semantic Web research targets at making information on the Web to be both human- and machine-readable, with languages that are characterised by an easy-to-understand syntax, as well as the visual editors for effortless and straight-forward knowledge engineering. OWL ontologies are known to be used in a wide range of scientific domains (for example, see [17] for an overview of biomedical ontologies), which are not necessarily closely connected to Computer Science, and allows even for non-professional programmers (i.e. domain specialists) to be involved in the process of policy engineering.

Extensibility. IoT systems may be composed of an extreme number of smart devices spread over a large area (e.g. traffic sensors distributed across a city-wide road network) and have the capacity to be easily extended (as modern cities continue to grow in size, more and more sensors are being deployed to support their associated traffic surveillance requirements). To keep up with this rapid growth and address the scalability issue, the proposed semantic approach, using the declarative definition, can extend the knowledge base to integrate newly-added devices in a seamless, transparent, and non-blocking manner. The same applies to the reverse process – once old services are retired and do not need to be considered anymore, the corresponding policies can be seamlessly removed from the knowledge base, so as not to overload the reasoning processes.

Increase in Reuse, Automation and Reliability. Policy enforcement mechanisms already exist in the form of automated reasoners for the OWL/SWRL languages, and the proposed approach aims to build on these capabilities. Since the reasoning process is automated and performed by an existing reasoning engine, it is expected to be free from so-called 'human factors' and more reliable, assuming, of course, the validity of ontologies and policies. Arguably, as the policy base grows in size and complexity, its accurate and prompt maintenance becomes a pressing concern so as to avoid potential policy conflicts. It is also important to keep in mind that formal reasoning based on DLs is a relatively expensive computational task, typically requiring an increased amount of computational resources (especially in the presence of numerous IoT devices). In this light, the proposed approach assumes that the policy enforcement is supposed to take place on a central (cloud-based) server, responsible for monitoring the IoT observation streams and take reactive actions, if required.

5 Conclusion

This paper discusses the possibility of utilising the idle potential for automated formal reasoning of existing IoT ontologies in the context of policy enforcement. While multiple IoT ontologies have been proposed both by the industry and the academia, there seems to be no evidence of a policy enforcement mechanism developed on top of these existing ontologies. As it was explained, the Semantic Web languages are underpinned by the Description Logics, which offer reasoning support, and enables automated classification of IoT observations. This means that IoT engineers can use existing ontological classes and properties to define policies, and, as a result, benefit from the built-in reasoning-based policy enforcement mechanisms.

As demonstrated by the sample use case scenario, with the proposed approach it is possible to define a set of policies, using various expressive OWL and SWRL constructs. The resulting policies can be generic (i.e. apply to a wide range of IoT devices) or more fine-grained (i.e. apply to specific individual devices). Moreover, IoT practitioners are expected to benefit from separation of concerns and support for policy governance, extensibility, and increased opportunities for reuse, automation and reliability.

Acknowledgements. The work presented in this paper was partially supported by the ERASMUS+ Key Action 2 (Strategic Partnership) project IOT-OPEN.EU (Innovative Open Education on IoT: improving higher education for European digital global competitiveness), reference no. 2016-1-PL01-KA203-026471. The European Commission support for the production of this publication does not constitute endorsement of the contents which reflects the views only of the authors, and the Commission cannot be held responsible for any use which may be made of the information contained therein.

References

1. Number of smartphone users worldwide from 2014 to 2020 (2017). https://www.statista.com/statistics/330695/number-of-smartphone-users-worldwide/. Accessed 14 July 2017
2. Agarwal, R., Fernandez, D.G., Elsaleh, T., Gyrard, A., Lanza, J., Sanchez, L., Georgantas, N., Issarny, V.: Unified IoT ontology to enable interoperability and federation of testbeds. In: Proceedings of 2016 IEEE 3rd World Forum on Internet of Things (WF-IoT), pp. 70–75. IEEE (2016)
3. Akyildiz, I.F., Su, W., Sankarasubramaniam, Y., Cayirci, E.: Wireless sensor networks: a survey. Comput. Netw. **38**(4), 393–422 (2002)
4. Bermudez-Edo, M., Elsaleh, T., Barnaghi, P., Taylor, K.: IoT-Lite: a lightweight semantic model for the Internet of Things. In: Proceedings of 2016 International IEEE Conferences on Ubiquitous Intelligence & Computing, Advanced and Trusted Computing, Scalable Computing and Communications, Cloud and Big Data Computing, Internet of People, and Smart World Congress, pp. 90–97. IEEE (2016)
5. Berners-Lee, T., Hendler, J., Lassila, O., et al.: The semantic web. Sci. Am. **284**(5), 28–37 (2001)

6. Dautov, R., Kourtesis, D., Paraskakis, I., Stannett, M.: Addressing self-management in cloud platforms: a semantic sensor web approach. In: Proceedings of the 2013 International Workshop on Hot topics in Cloud Services, pp. 11–18. ACM (2013)
7. Dautov, R., Paraskakis, I., Stannett, M.: Towards a framework for monitoring cloud application platforms as sensor networks. Cluster Comput. **17**(4), 1203–1213 (2014)
8. Gubbi, J., Buyya, R., Marusic, S., Palaniswami, M.: Internet of Things (IoT): a vision, architectural elements, and future directions. Future Gener. Comput. Syst. **29**(7), 1645–1660 (2013)
9. Gyrard, A., Serrano, M., Atemezing, G.A.: Semantic web methodologies, best practices and ontology engineering applied to Internet of Things. In: Proceedings of 2015 IEEE 2nd World Forum on Internet of Things (WF-IoT), pp. 412–417. IEEE (2015)
10. Hitzler, P., Krötzsch, M., Rudolph, S.: Foundations of Semantic Web Technologies. Chapman & Hall/CRC, Boca Raton (2009)
11. Hu, S., Wang, H., She, C., Wang, J.: AgOnt: ontology for agriculture Internet of Things. In: Li, D., Liu, Y., Chen, Y. (eds.) CCTA 2010. IAICT, vol. 344, pp. 131–137. Springer, Heidelberg (2011). doi:10.1007/978-3-642-18333-1_18
12. Kinkar, S., Hennessy, M., Ray, S.: An ontology and integration framework for smart communities. J. Comput. Inf. Sci. Eng. **16**(1), 011003 (2016)
13. Kotis, K., Katasonov, A.: Semantic interoperability on the web of things: the semantic smart gateway framework. In: Proceedings of 2012 Sixth International Conference on Complex, Intelligent and Software Intensive Systems (CISIS), pp. 630–635. IEEE (2012)
14. Liang, S., Croitoru, A., Tao, V.: A distributed geospatial infrastructure for sensor web. Comput. Geosci. **31**(2), 221–231 (2005)
15. Nambi, A.U., Sarkar, C., Prasad, V., Rahim, A.: A unified semantic knowledge base for IoT. In: Proceedings of 2014 IEEE World Forum on Internet of Things (WF-IoT), pp. 575–580. IEEE (2014)
16. Perera, C., Zaslavsky, A., Christen, P., Georgakopoulos, D.: Context aware computing for the Internet of Things: a survey. IEEE Commun. Surv. Tutorials **16**(1), 414–454 (2014)
17. Rubin, D.L., Shah, N.H., Noy, N.F.: Biomedical ontologies: a functional perspective. Briefings in Bioinform. **9**(1), 75–90 (2008)
18. Seydoux, N., Drira, K., Hernandez, N., Monteil, T.: IoT-O, a core-domain IoT ontology to represent connected devices networks. In: Blomqvist, E., Ciancarini, P., Poggi, F., Vitali, F. (eds.) EKAW 2016. LNCS (LNAI), vol. 10024, pp. 561–576. Springer, Cham (2016). doi:10.1007/978-3-319-49004-5_36
19. Sheth, A., Henson, C., Sahoo, S.S.: Semantic sensor web. IEEE Internet Comput. **12**(4), 78–83 (2008)
20. Sirin, E., Parsia, B., Grau, B.C., Kalyanpur, A., Katz, Y.: Pellet: a practical OWL-DL reasoner. Web Semant. Sci. Serv. Agents World Wide Web **5**(2), 51–53 (2007)
21. Sowa, J.F.: Knowledge Representation: Logical, Philosophical, and Computational Foundations, vol. 13. MIT Press, Cambridge (2000)
22. Studer, R., Benjamins, V.R., Fensel, D.: Knowledge engineering: principles and methods. Data Knowl. Eng. **25**(1–2), 161–197 (1998)
23. Toma, I., Simperl, E., Hench, G.: A joint roadmap for semantic technologies and the Internet of Things. In: Proceedings of the Third STI Roadmapping Workshop, vol. 1 (2009)

24. Veloudis, S., Paraskakis, I.: Defining an ontological framework for modelling policies in cloud environments. In: Proceedings of 2016 IEEE International Conference on Cloud Computing Technology and Science (CloudCom), pp. 277–284. IEEE (2016)
25. Wang, W., De, S., Toenjes, R., Reetz, E., Moessner, K.: A comprehensive ontology for knowledge representation in the Internet of Things. In: Proceedings of 2012 IEEE 11th International Conference on Trust, Security and Privacy in Computing and Communications (TrustCom), pp. 1793–1798. IEEE (2012)
26. Zhao, S., Zhang, Y., Chen, J.: An ontology-based IoT resource model for resources evolution and reverse evolution. In: Liu, C., Ludwig, H., Toumani, F., Yu, Q. (eds.) ICSOC 2012. LNCS, vol. 7636, pp. 779–789. Springer, Heidelberg (2012). doi:10.1007/978-3-642-34321-6_62

Wireless Systems

BSSA_{CH}: A Big Slot Scheduling Algorithm with Channel Hopping for Dynamic Wireless Sensor Networks

Chi Trung Ngo, Quy Lam Hoang, and Hoon Oh(✉)

Ubicom Lab, Department of Computer Science, University of Ulsan, Ulsan, South Korea
chitrung218@gmail.com, quylam925@gmail.com,
hoonoh@mail.ulsan.ac.kr

Abstract. To cope with the internal and external interference induced by other ambient signals, physical obstacles and the dynamic change of topology in wireless sensor networks, we propose an efficient and reliable MAC protocol combining TDMA and CSMA with channel hopping. The principle operation of the proposed protocol relies on the Big Slot Scheduling Algorithm that uses the tree routing and the notion of a sharable slot. A sharable slot of an appropriate size is allocated to each tree level and shared by all the nodes at the same level for data transmission. To adapt the dynamic change of topology, a node performs a slot scheduling in a distributed manner depending on its level. Data transmission is performed progressively, starting with the highest tree level. To mitigate external interference, the proposed protocol uses a blind channel hopping mechanism in which in channel switching, each node simply follows the selected channel sequence based on its ID every data transmission period. A sender follows the channel hopping sequence of a receiver or its parent for data transmission. We implemented the proposed protocol in Contiki OS and evaluated it in a 25-node testbed under various interference sources.

Keywords: Channel hopping · Wireless sensor networks · Internal and external interference · Sharable slot

1 Introduction

The technology of wireless sensor networks (WSNs) has evolved rapidly with its wide applicability to different Internet of Thing (IoT) applications such as smart city, smart home, and industrial monitoring and control applications, etc. The two most crucial considerations of WSNs for monitoring and control applications in industrial fields are the *energy efficiency* and *transmission timeliness and reliability* due to the difficulty of battery recharge and the need of the correct monitoring and the timely control of target place.

This research was supported by Basic Research Program through the National Research Foundation of Korea funded by the Ministry of Education (NRF-2016R1D1A1A09919658).

A. Puliafito et al. (Eds.): ADHOC-NOW 2017, LNCS 10517, pp. 359–366, 2017.
DOI: 10.1007/978-3-319-67910-5_29

Energy efficiency needs to be considered carefully because sensor nodes have the limited battery capacity. Meanwhile, the reliability of data transmission is affected by a number of following factors. First, wireless communication in WSNs is often unreliable owing to the effect of the inevitable nature of signal such as fading and multi-path propagation. Second, many WSNs deployed for different applications tend to increase the level of inter-WSN interference [1]. Third, since a WSN shares its radio medium with other communication technologies, such as WiFi, Bluetooth, and even microwave ovens, it is more likely that data transmission in a WSN can be affected by data transmissions on other networks using some overlapped channels. Lastly, wireless interference may be caused deliberately by a malicious node that performs an active jamming attack in which the malicious node transmits some garbage data continuously to prevent other nodes from accessing the channel [2].

In the literature, there have been a lot of the proposed MAC protocols that use a single channel for energy efficiency and data reliability such as TreeMAC [3], IMAC [4], ContiMAC [5]. However, the use of a single channel tends to limit the performance significantly, especially with the increase of external interference. Therefore, some multi-channel MAC protocols were proposed as a promising solution like MC-LMAC [6], WirelessHART [7] and 802.15.4e [8]. These approaches need time synchronization and a distributed/centralized slot schedule in which each data transmission is assigned a dedicated time slot or a shared slot but a distinct channel different from other data transmissions. In addition, WirelessHART and 802.15.4e utilize a channel hopping to reduce the effect of external interference. However, those approaches do not respond well to the dynamic change of topology. As the combination of TDMA and CSMA/CA, the Big Slot Scheduling Algorithm (thereafter, referred to as BSSA) [9] was proposed to ensure the reliability and timeliness of data transmission. The operation of BSSA relies on the tree routing topology and the notion of a sharable slot. All the nodes at the same level share the same sharable slot allocated to that level in transmitting data and then they employ CSMA/CA for reliable data transmission within the sharable slot. Each node performs its slot scheduling independently of the other nodes in a distributed manner, providing the high robustness against the dynamic change of network topology. However, BSSA seems to be insufficient in dealing with external interference.

In this paper we aim at extending and elaborating our previous work BSSA by using a channel hopping to cope with external interference and by enabling multiple simultaneous transmissions on different channels, named $BSSA_{CH}$. It is worth noting that BSSA_{CH} does not employ any channel negotiation mechanism, instead it generates the pseudo-random channel numbers using a Linear Congruential Generator (LCG) [10], and then each node simply selects a channel hopping sequence based on its Identification number (ID). Thus, once all children determine their parents in the tree, they can follow the channel hopping sequences of their parents easily in data transmissions.

We implemented BSSA_{CH} and the original protocol BSSA in Contiki OS and evaluated them in a 25-nodes testbed under variation of interference sources; namely, inter-WSN interference induced by the coexistence of WSNs, a jamming attack by a malicious node, and external interference caused by data transmissions of some devices using WiFi in an overlapped channel with the current WSN. The empirical results showed that

$BSSA_{CH}$ far outperforms its ancestor BSSA when working in the severe interference environment in terms of packet delivery ratio and energy consumption.

The rest of this paper is organized as follows. In Sect. 2, we overview the Big Slot Scheduling Algorithm (BSSA). In Sect. 3 we formally describe the design of the proposed approach. In Sect. 4, experimental studies are presented. Finally, we give concluding remarks in Sect. 5.

2 Overview of Big Slot Scheduling Algorithm (BSSA)

BSSA [9] combines TDMA and CSMA/CA for slot scheduling and channel competition. A big sharable slot is allocated to each tree level and its size increases exponentially when the tree level decreases. All the nodes at the same level compete each other using CSMA/CA to transmit data. Nodes at two adjacent levels have to wake up at the same time such that nodes at the higher level are in transmitting state and nodes at lower level are in the receiving state. To fulfil this purpose, BSSA borrows the wait time function proposed in [11] to generate the slot scheduling. The wait time function is expressed as follows.

$$WTime(l) = W_1 \times a^{l-1} \quad (1)$$

where, a is in the range (0,1], $W_1 = SF$ is the maximum wait time of the sink at level 1 ($l = 1$). W_1 is the given time constraint during which all data packets have to be delivered to the sink.

Within a sharable slot, a node, say x, at level l performs slot scheduling in a distributed manner based on its tree level to obtain three scheduling points, start of receiving time, $RxTime(l_x)$, start of transmitting time, $TxTime(l_x)$, and start of sleeping time, $SleepTime(l_x)$, as follows.

$$RxTime(l_x) = sTime + WTime(l_x + 1) \quad (2)$$

$$TxTime(l_x) = sTime + WTime(l_x) \quad (3)$$

$$SleepTime(l_x) = sTime + WTime(l_x - 1) \quad (4)$$

where, $sTime$ is the synchronized time point of all nodes.

When $TxTime$ starts, all the nodes at the same level competes the channel for sending data using CSMA/CA and go to the sleep mode when finishing data transmission or when $SleepTime$ starts.

3 Big Slot Scheduling Algorithm with Channel Hopping ($BSSA_{CH}$)

3.1 Time Synchronization

Since BSSA was implemented and evaluated using simulator, time synchronization is not necessary. However, for the real implementation we need a protocol to perform network-wide time synchronization. In $BSSA_{CH}$, all nodes only perform the network-wide time synchronization using the flooding time synchronization protocol (FTSP) [12] once when starting the network. A sink floods the network with a synchronization message (SYNC). Upon receiving SYNC from its neighbors, a node uses a linear regression technique to compensate for clock skew and clock offset. However, it is very costly to maintain the global time synchronization in terms of bandwidth since a sink and every node has to send out one SYNC periodically. To cope with this issue, $BSSA_{CH}$ takes advantage of MAC control messages (i.e. RTR, CTS or ACK) during data transmission to include the time synchronization and each node locally updates its global time synchronization when it performs data transmission with its parent.

3.2 Channel Hopping

A channel hopping mechanism is employed to alleviate external interference. In our approach, every node generates a table of channel sequences using the identical seed provided by a sink. Then, every node can have and maintain the same set of different channel sequences and selects its own channel sequence from the table using its node *ID* (identification number). Then, a node hops its channel from the selected channel sequence every superframe. If a node finds a bad channel at some time, it reports the channel number to a sink. Then, the sink disseminates it to all nodes so that they can remove the bad channel from every hopping sequence in the table of channel sequences.

For every superframe, when *RxTime* determined by Eq. (2) starts, each receiving node switches its channel following its own pseudo-random channel sequence, while when *TxTime* determined by Eq. (3) starts, a sending node has to switch its channel following its parent's channel hopping sequence.

We use a *Linear Congruential Generator* (LCG) [10] to generate the pseudo-random channel numbers due to its simple computation and the uniformly distributed generation sequences. The pseudo-random sequence *X* generated by LCG is computed as follows.

$$X_{n+1} = \left(aX_n + c\right) modN, n \geq 0 \tag{5}$$

where, *N* (the total number of available channels) is the modulus, X_0 is the seed ($0 \leq X_0 < N$), *a* is the multiplier ($0 \leq a < N$), and *c* is the increment ($0 \leq c < N$).

3.3 Data Transmission

When *TxTime* starts, a sender has to switch its channel corresponding to its parent's channel hopping sequence every data transmission period (=*RDTP*), while a receiver switches its channel following the selected channel hopping sequence when *RxTime* starts. Let us take Fig. 1 as an example. Node 1 acts as a sink at level 1 with the selected channel hopping sequence (26, 25, 12, 15, 20), node 2 and node 3 at level 2 have the selected channel sequence as (25, 20, 12, 26, 15) and (12, 26, 15, 25, 20), respectively. Therefore, node 2 and 3 follow the channel sequence of node 1 for data transmission, while nodes 4, 5, and 6 at level 3 follow the channel sequence of node 2 and nodes 7, 8, and 9 at level 3 follow the channel sequence of node 3.

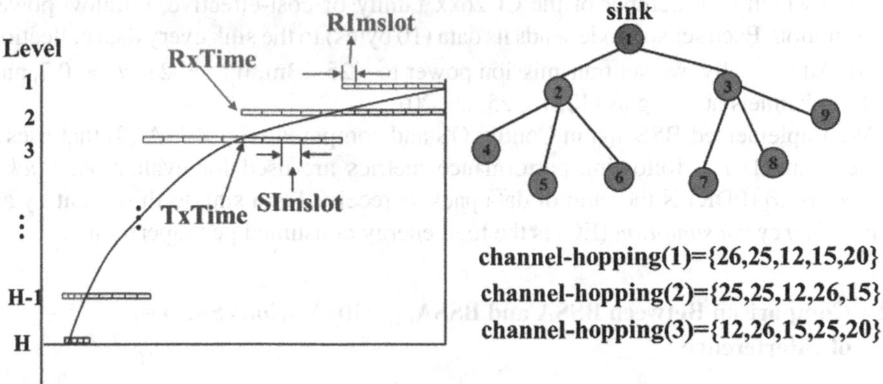

Fig. 1. Data transmission with channel hopping

There are two data transmission modes in BSSA$_{CH}$. For the nodes at level $l = 2$, we employ a *receiver-initiated (RI) transmission method* in which a sink appoints a child by sending a *ready to receive* (*RTR*) control message and asks it to send a packet every *receiver-initiated mini slot* (*RImslot*) (=7 ms in this paper). This process is repeated until the sink receives all packets from its children. In case the sink does not receive data from a child, it tries to send RTR to that child in the next *RImslot* and then changes to another child. Other nodes overhearing RTR go to sleep and wake up in the next *RImslot*. The sink acknowledges the data reception from a node by adding the ID of that node in the next RTR.

For the nodes at level $l \geq 3$, we employ a *sender-initiated (SI) transmission method* in which a sending sharable slot (SS^{Tx}) is divided into multiple *sender-initiated mini slots* (*SImslot's*) (=15 ms in this paper). Data transmission is performed at every *SImslot*. Every node uses CSMA/CA with RTS/CTS to compete with its neighbors at the same level for channel after a small random delay. A node that fails to access channel goes to sleep and retries in the next *SImslot*. Whereas, a node that succeeds in data transmission (e.g. receiving ACK from its parent) goes to sleep for the remaining time of the current superframe.

4 Experimental Evaluation

4.1 Experimental Setup

For the protocol evaluation, we deploy a testbed of 25 *umotes* (developed in the University of Ulsan) in a 5 × 5 grid topology, making an equal distance of 2 m vertically and horizontally, and a sink is placed at the top center. The umote uses the TI integrated chip CC2630 which is a member of the CC26xx family of cost-effective, ultralow power consumption. Each sensor node sends its data (10 bytes) to the sink every data collection round. Additionally, we set transmission power to −25 dBm, $W_1 = 2$ s, $a = 0.7$, and a list of channels hopping as (12, 15, 25, 26, 20).

We implemented $BSSA_{CH}$ in Contiki OS and compared with BSSA [9] that uses a single channel. The following performance metrics are used for evaluation. *Packet delivery ratio* (PDR) is the ratio of data packets received at a sink to those sent by all nodes. *Energy consumption* (EC) is the total energy consumed per superframe.

4.2 Comparison Between BSSA and $BSSA_{CH}$ with Various Sources of Interference

In this experiment, we evaluate the effectiveness of channel hopping by comparing BSSA and $BSSA_{CH}$ under three interference sources as proposed in [2]: (1) *Interference with other transmissions from nearby WSNs* (*inter-interference*): We use some additional nodes to transmit a packet of 100 bytes of random content every 20 ms without performing CCA on channel 12; (2) *Jamming case:* A WSN node transmits back-to-back packets on channel 12 without performing CCA to monopolize the wireless channel; (3) *WiFi interference:* We use one laptop to send WiFi traffic to another laptop connected to a WiFi router using the *iperf* network testing tool. The WiFi channel is set to 1 which overlaps with the channel 12 used in BSSA and $BSSA_{CH}$.

Figures 2(a) and (b) show the PDR and EC of BSSA and $BSSA_{CH}$ under various sources of interference. As for PDR in Fig. 2(a), we can see that both protocols achieve about 96% of PDR when there is no interference; however, they both show a decreasing PDR depending on different interference sources. Particularly, with inter-interference BSSA only obtains around 60% of PDR, while $BSSA_{CH}$ does 85% thanks to the channel-hopping mechanism that can reduce internal and external interference. As for the jamming case, BSSA mostly cannot work since nodes cannot access the channel, whereas $BSSA_{CH}$ can avoid this issue by switching to different channels every superframe, thereby achieving 77% of PDR. While with WiFi interference, the effect is very negligible in $BSSA_{CH}$.

Regarding energy consumption shown in Fig. 2(b), nodes consume the least energy in both protocols when there is no interference, about 2.8 mJ. It is obvious that both protocols consume more energy in the jamming case compared with other interferences; however, in this case BSSA consumes energy twofold higher than $BSSA_{CH}$. It makes sense since in BSSA nodes mostly stay active to resend data or find another parent, while $BSSA_{CH}$ uses channel hopping to deal with this problem. That also explains why

BSSA$_{CH}$ achieves better energy consumption than BSSA with inter-interference and WiFi interference.

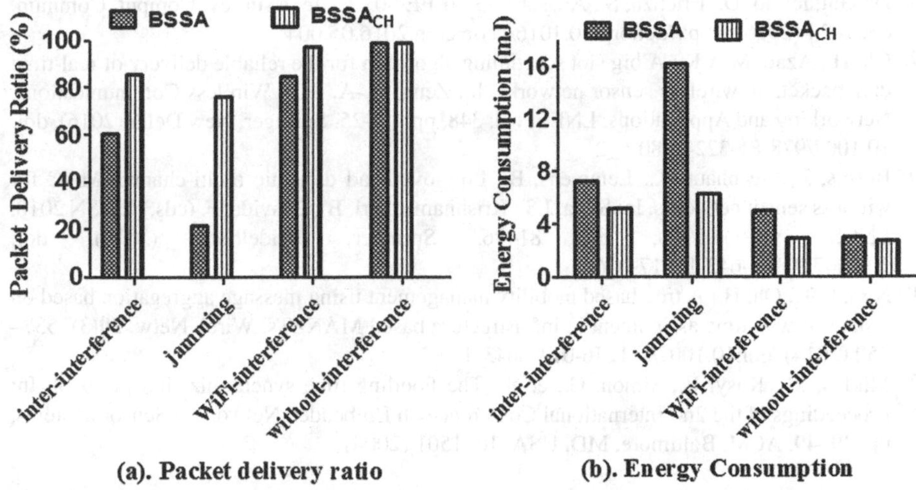

Fig. 2. Packet delivery ratio and energy consumption with various sources of interference

5 Conclusion

We proposed a channel hoping protocol based on the operation of Big Slot Scheduling Algorithm (BSSA). We implemented the proposed protocol using Contiki OS and experimented it under various interference sources. It was shown that BSSA$_{CH}$ far outperforms its ancestor BSSA in packet delivery ratio and energy consumption.

References

1. Phung, K.-H., Lemmens, B., Goossens, M., Nowe, A., Tran, L., Steenhaut, K.: Schedule-based multi-channel communication in wireless sensor networks: a complete design and performance evaluation. Ad Hoc Netw. **26**, 88–102 (2015)
2. Tang, L., Sun, Y., Gurewitz, O., Johnson, D.B.: EM-MAC: a dynamic multichannel energy-efficient MAC protocol for wireless sensor networks. In: Proceedings of the Twelfth ACM International Symposium on Mobile Ad Hoc Networking and Computing, pp. 1–11. ACM, Paris, France. 2107533 (2011)
3. Wen-Zhan, S., Renjie, H., Shirazi, B., LaHusen, R. (eds.): TreeMAC: localized TDMA MAC protocol for real-time high-data-rate sensor networks. In: IEEE International Conference on Pervasive Computing and Communications. PerCom 2009, 9–13 March 2009
4. Oh, H., Van Vinh, P.: Design and implementation of a MAC protocol for timely and reliable delivery of command and data in dynamic wireless sensor networks. Sensors **13**(10), 13228–13257 (2013)
5. Dunkels, A.: The ContikiMAC Radio Duty Cycle Protocol. Swedish Institute of Computer Science (2011)

6. Incel, O.D., van Hoesel, L., Jansen, P., Havinga, P.: MC-LMAC: a multi-channel MAC protocol for wireless sensor networks. Ad Hoc Netw. **9**(1), 73–94 (2011)
7. WirelessHART specification (2007)
8. De Guglielmo, D., Brienza, S., Anastasi, G.: IEEE 802.15.4e: a survey. Comput. Commun. **88**, 1–24 (2016). http://doi.org/10.1016/j.comcom.2016.05.004
9. Oh, H., Azad, M.A.K.: A big slot scheduling algorithm for the reliable delivery of real-time data packets in wireless sensor networks. In: Zeng, Q.-A. (ed.) Wireless Communications, Networking and Applications. LNEE, vol. 348, pp. 13–25. Springer, New Delhi (2016). doi:10.1007/978-81-322-2580-5_2
10. Borms, J., Steenhaut, K., Lemmens, B.: Low-overhead dynamic multi-channel MAC for wireless sensor networks. In: Silva, J.S., Krishnamachari, B., Boavida, F. (eds.) EWSN 2010. LNCS, vol. 5970, pp. 81–96. Springer, Heidelberg (2010). doi:10.1007/978-3-642-11917-0_6
11. Ngo, C.T., Oh, H.: A tree-based mobility management using message aggregation based on a skewed wait time assignment in infrastructure based MANETs. Wirel. Netw. **20**(3), 537–552 (2014). doi:10.1007/s11276-013-0643-4
12. Miklos, M., Kusy, B., Simon, G., et al.: The flooding time synchronization protocol. In: Proceedings of the 2nd International Conference on Embedded Networked Sensor Systems, pp. 39–49. ACM, Baltimore, MD, USA. 1031501 (2004)

A Hybrid Ant-Genetic Algorithm to Solve a Real Deployment Problem: A Case Study with Experimental Validation

Sami Mnasri[1(✉)], Nejah Nasri[2], Adrien Van Den Bossche[1], and Thierry Val[1]

[1] University of Toulouse, UT2J, CNRS-IRIT-IRT, Toulouse, France
{Sami.Mnasri,vandenbo,val}@irit.fr
[2] ENIS, LETI, University of Sfax, Sfax, Tunisia
nejah.nasri@isecs.rnu.tn

Abstract. In this paper, we investigate the problem of deploying 3D nodes in a wireless sensor network. The aim is to choose the ideal 3D locations to add new nodes to an initial configuration of nodes, while optimizing a set of objectives. In this regard, our study proposes a new hybrid algorithm which stems from the ant foraging behavior and the genetics. It is based on a recent variant of the genetic algorithms (NSGA-III) and the Ant Colony Optimization algorithm. The obtained numerical results and the simulations compared with experiments prove the effectiveness of the proposed approach.

Keywords: 3D indoor redeployment · DL-IoT · Prototyping · Testbed · ACO · NSGA-III

1 The Problem of 3D Indoor Deployment of Nodes (in WSN)/Things (in IoT)

The effectiveness of WSN (wireless sensor networks) is strongly influenced by the deployment process and the positioning of sensor nodes. In a WSN, deploying nodes is a strategy that aims to define the number of nodes, their positions and the network topology. A WSN is said to be three-dimensional if the variation in the height of the deployed sensors is not negligible with respect to the length and the width of the deployment zone. Indeed, the 2D model loses its relevance for terrestrial networks, as for marine or airborne deployments, where the 2D model does not reflect the real topography of the zone. Many research challenges in three-dimensional spaces have not been explored yet as much as for two-dimensional networks. In this paper, we are interested in improving the initial deployment by adding a set of nodes in a deterministic way. Coverage is the most important objective to be met when deploying a WSN. It is an essential subject in the design of a WSN. Coverage is generally interpreted as how a network of sensors will supervise the area of interest. The increase of the number of nodes cannot always provide a total degree of coverage, and it is also costly to maintain high-density_networks. As a result, other methods need to be applied to avoid these problems and to improve coverage after the initial random deployment. The sensor

A. Puliafito et al. (Eds.): ADHOC-NOW 2017, LNCS 10517, pp. 367–381, 2017.
DOI: 10.1007/978-3-319-67910-5_30

localization is the most relevant factor in connection with the coverage. Localization is relevant if there is uncertainty about the nodes exact location. Indeed, in WSN, the information about location is crucial especially when there is an unusual event. In this case, the sensor node that detects this event needs to locate it and then reports the location of that event to the sink node.

Another concept is closely related to our problem, it is the IoT (Internet of Things). The IoT is a scenario in which entities (devices, people or robots) are connected and have a unique identifier for each. They are able to transfer data over the network without human or automatic intervention. These objects communicate using protocols such as the Bluetooth or the 802.15.4 as it is the case in our experiments. WSN is the bridge which links the real world to the digital one. It is responsible for the hardware communication to convey the real world values detected by the wireless connected things (sensor nodes) to the Internet. While the IoT is responsible for data processing, manipulation and decision making. In our study we are interested in the DL-IoT (DeviceLayer-IoT) [1] which is a network of collection used to collect data from the sensors nodes disseminated in the network environment. Thus, our approach are valid for both WSN and IoT contexts.

For most deployment modellings, the problem of optimal placement of sensor nodes is proved to be NP-hard [2]. Therefore, this problem cannot be solved by deterministic methods, even for little number of nodes. Thus, we propose a new hybrid algorithm based on NSGA-III [3] and ACO (Ant Colony Optimization) algorithm [4] to solve it.

2 Related Works, Main Motivations and Contributions

In [5, 6], the authors propose a state-of-the-art on multi-objective approaches resolving the 3D deployment problem. Various methodologies are proposed, among others, genetic algorithms (GAs) and particle swarms (PSO). Nevertheless, no comparison using simulations or experiments between these approaches is provided in these works. In [7], Qu et al. suggest a redeployment approach based on GA and PSO. The objective is to maximize the range of detection, and to minimize the number of mobile nodes and energy consumption. However, no mathematical modeling is given for the problem. Besides, no modifications are proposed for the tested algorithms. Also, only the 2D case is considered (not the 3D one). In [8], Matsuo et al. propose a 3D deployment with a radio propagation model in disaster areas using on a hybrid algorithm based on genetic algorithm and local search. However, the behavior of the tested algorithms is not evaluated with known evaluation metrics such as Inverted Generational Distance (IGD) and Hypervolume (HV). In order to minimize the number of thermal sensors, the authors in [9] solve the problem of placing thermal sensors in the smart grid using a gappy proper orthogonal decomposition (GPOD-GA) algorithm. Nevertheless, the proposed approach allows only the optimization of a single objective and its evaluation is not based on metrics such as IGD and HV. In [10], the authors propose three strategies to resolve the deployment in terrain-aware by maximizing the coverage and minimizing the mobility cost. These strategies are based on a normalized genetic algorithm (NGA), and artificial immune system (AIS) algorithm and a particle swarm optimization (PSO) algorithm.

On the other hand, their approaches suffer from a disadvantage that concerns the high execution time of AIS and NGA.

These approaches suffer from several shortcomings. Among others, the inadequacy and inefficiency in the case of many-objective problems and real-world ones, both known for their complexity. In this work, the suggested contributions mainly concern:

- The proposition of a new algorithm with a justified hybridization which benefits from the advantages of the two algorithms ACO and NSGA-III.
- The performance of the proposed algorithm is compared with those of the ACO and NSGA-III algorithms using the HV metric, followed by a comparison between the simulation results and the real prototyping experiments. The interpretation of the obtained results is also provided.
- The use of the empirical context and the real measurements constitute a validation of the findings of the authors of the original algorithms (ACO and NSGA-III) which are validated only by tests on theoretical problems (DTLZ and ZDT).

The rest of the paper is organized as follows: in Sect. 2, the related works and the main contributions are presented. In Sect. 3, the new hybridization scheme is detailed. In Sect. 4, the performance of the AcNSGAIII is evaluated using numerical results. In Sect. 5, the simulation tests are compared to the experiments. Finally, in Sect. 6, a conclusion and different perspectives are listed.

3 Adapting the ACO and GAs for the 3D Indoor Deployment Problem

3.1 NSGA-III Algorithm

NSGA-III [3] is a recent algorithm, proposed as an extension of NSGA-II [11]. It uses a reference point based approach to solve many objective problems (MaOPs). NSGA-III use the same concept of weight vector generation in MOEA/D [12] to determine a set of reference points scattered over the objective space. At every generation of each solution, the values of the objective function are normalized to [0, 1]. Then, a reference point is associated with each solution based on its perpendicular distance to the reference line. Assigning a reference point to each solution ensures the uniform repartition of the reference points across the normalized hyper-plane. The generated offspring is combined with the parent to create a hybrid population. Afterwards, the hybrid population is divided into a set of non-domination levels according to a non-dominated sorting procedure. The next parents as composed of the solutions in the first level so on and so forth. A niche-preservation operator is used to select solutions in the last acceptable level where the solutions associated with a less crowded reference line are more likely to be to be selected. For the majority of test problems, NSGAIII which is proposed especially for many-objective optimization shows a superior performance compared to other methods such as MOEA/D and NSGA-II. Algorithm 1 illustrates the NSGA-III algorithm.

Algorithm 1 : The NSGA-III algorithm

Input: P_0(Initial Population), N_{Pop} size of population,
    t (iteration) = 0, It_{max} (Maximum iteration).
Output : P_t
While $t<It_{max}$ **do**
    Create Offspring Q_t
    Mutation and recombination on Q_t
    Set $R_t=P_t \cup Q_t$
    Apply non-dominated sorting on R_t and find $F1, F2, \ldots$
    $S_t=\{\}$, $i=1$
    While $|S_t| \leq N_{Pop}$ **do**
        $S_t=S_t \cup F_i$
        $i=i+1$
    End While
    If $|St| <> N_{Pop}$ **do**
        $P_{t+1}=\cup_{j=1}^{l-1} F_j$
        Normalize S_t using min and intercept points of each objective
        Associate each member of S_t to a reference point
        Choose $N_{Pop}-|P_t+1|$ members from F_l by niche-preserving operator
    Else $P_{t+1}=S_t$
    End if
$t=t+1$
End While

3.2 Ant Colony Optimization Algorithm

The ACO algorithm was introduced by [4] in order to resolve the problems of hard CO (combinatorial optimization). It is a bio-inspired approximate algorithm which aim to obtain good solutions with reasonable computational cost (time) when resolving CO problems. Moreover, it is a meta-heuristic which is considered as probabilistic and population-based. The ACO stems from the foraging behavior of the real ants. Indeed, the ants aim at finding the shortest path between its nest and the source of food. Instead of using visual information's, ants use a chemical substance named 'pheromone' which is left behind their trails. So, in the ACO, the artificial ants (called agents) imitate their natural counterparts in order to resolve the problems by finding the optimal solutions. The ACO algorithm is illustrated in Algorithm 2. Indeed, firstly, when the collection begins, the shortest route leading to food is not known. Thus, each ant pursues randomly a route and place pheromone. As the collection progress, ants continue putting the pheromone. As a consequence, all traveled routes contain this substance. Then, because the pheromone evaporates overtime, if an ant wants to travel, it chooses the route with the highest rate of pheromone, which corresponds to the shortest among all routes. Therefore, overtime, only one route will remain (the shortest one).

Algorithm 2 : **The ACO algorithm**

```
  Initialize the ACO parameters (pheromone, ..)
  Each_ant_builds_a_solution()
  Evaluate_the_solutions
  Initialize the number of travels per ant, t = 1
  While t < lt_max do
    Update_the_pheromone()
    Each_ant_builds_a_ new_solution()
    Evaluate_the_solutions
t = t + 1
End while
```

3.3 The Proposed AcNSGA-III: A Hybrid Framework for NSGA-III and Ant System

Despite its efficiency, the NSGA-III has some difficulties when solving mono-objective and two-objective optimization problems. These difficulties concern the low selection pressure that NSGA-III introduce to non-dominated solutions of a population when resolving two-objective problems. Moreover, they concern the small population size and the random selection process when resolving mono-objective problems [13]. Also, the ACO algorithm has a main drawback which concerns the convergence into the local optima [14].

Therefore, the idea is to propose a well-justified hybridization scheme using the two algorithms to take advantage of their strengths and to remedy their drawbacks. When hybridizing these two algorithms, most of the studies [15, 16] use the standard and basic version of the genetic algorithm. Besides, most of these studies sequentially execute the two algorithms (the standard GA then the ACO or the opposite). Thus, the final solution of one of the two algorithms is the initial solution of the other. Although this basic scheme of integration enhances the ACO convergence rate, this latter remains converging excessively which makes the problem of local optimum unsolvable. In our study, we propose a platform where the two algorithms (NSGA-III and ACO) run at the same time and interact with the same population. Thus, the ant algorithm steps are injected into the implementation of the original NSGA-III with incorporation of several modifications of the original NSGA-III. Among these modifications, the initial population of the NSGAIII which becomes the population built by the ants in the initial phase of the ACO. It should be mentioned that this is first time NSGA-III and ACO are integrated into a hybrid platform. Moreover, this is the first platform using a hybrid genetic algorithm and ACO to resolve the problem of 3D indoor deployment of nodes. The proposed algorithm, called AcNSGA-III is illustrated in Algorithm 3. It is a hybrid ant-Genetic algorithm which performs an ACO optimal selection of nodes, a dynamic pheromone updating and a mutation strategy. It accelerates the global convergence in order to speed up the local search which allows finding faster the suitable solutions of the 3D deployment problem. The global search ability and the randomness of the genetic operators are guaranteed which ensure conducting the operation of the genetic operator into generating routes by the ants if the ACO converges quickly. This allows to the latter finding the closing conditions and exits. Finally, since there is a low probability that ants and the NSGAIII process produce the same individual in the same iteration: the individual is added to the population unless it does not exist into it.

Algorithm 3 : The proposed AcNSGA-III algorithm

Input: NS (=2*N_{Pop}) size of population, t (iteration= the number of travels per ant) = 1, It_{max} (Maximum iteration= the maximum number of travels per ant).
Output : P_t
Initialize the parameters of the deployment problem (***CvD*, *F*, Ct_g, *mx*, Ng^i_n, *Expi***)
Initialize the ACO parameters (***NbA*, *ExP*, *ExV*, *EvP*, *MaxTP*, *MinTP***)
P_0(Initial Population)= Each_ant_builds_a_solution ()
V_t =Evaluate the solutions and choose $NPop$ feasible solutions
While $t<=It_{max}$ **do**
    Create Offspring Q_t
    Mutation and recombination on Q_t
    Set $R_t=P_t \cup Q_t$
    Apply non-dominated sorting on R_t and find $F1$, $F2$, …
    $S_t=\{\}$, $i=1$;
    While $|S_t| \leq N_{Pop}$ **do**
        $S_t=S_t \cup F_i$
        $i=i+1$
    End While
    If $|S_t|<>N_{Pop}$ **do**
        $P_{t+1}=\cup_{j=1}^{l-1} F_j$
        Normalize S_t using min and intercept points of each objective
        Associate each member of S_t to a reference point
        Choose $N_{Pop}-|P_t+1|$ members from F_l by niche-preserving operator
    Else $P_{t+1}=S_t$
    End if
    Update_pheromone()
    Each_ant_builds_a _new_solution()
    V_t = Evaluate the solutions and choose N_{Pop} feasible solutions
    Set $P_{t+1}=P_{t+1} \cup V_t$
    $t= t+1$;
End While

The procedure ***Each_ant_builds_a_solution()*** allows to construct the candidate solutions using a model of pheromone which is a tunable probability of distribution over the space the solution. In our case a solution is a feasible repartition of nodes in the 3D space. The procedure ***Update_pheromone()*** allows the use of the candidate solutions to update the values of the pheromone in order to ensure moving towards future better solutions. As an optimization, the operation of these two procedures can be summarized and replaced by the algorithmic sequence illustrated in Algorithm 4.

Algorithm 4 : Operation of ***Each_ant_builds_a_solution()*** and ***Update_pheromone()***

```
For all the iterations i in 1:I do
  For all the construction steps j in 1:J do
    For all the ants k in 1:K do
      Choose and move to the next possible position of node
      Update local pheromone
    endFor
  endFor
  Update global pheromone values on visited possible position of node
endFor
```

4 Numerical Results of the Algorithms

In this section, we present the parameters setting of the algorithms and the results using performance indicators. Different metrics can be used. In our case, although the Hypervolume (HV) [17] is a metric with an expensive computational cost, it is recommended because it is useful when assessing real world problems without the need of a prior knowledge of the true Pareto-front. Furthermore, when resolving a particular problem, the parameters setting has a major influence on the algorithm behavior. The used parametres are as follows:

- The size of the population (***NS***): 300 (Thus, ***Npop*** = 150).
- The operators of reproduction: It would be best to perform a recombination operation using near parent solutions in the case of MaOPs. Therefore, a large distribution index SBX operator (simulated binary crossover) is used. The recombination probability is PrOx = 0.8 using a distribution index $\tau_r = 45$. The mutation probability using the bit-flip operator, is PrMt = $2 * 10^{-3}$ using a distribution index $\tau_m = 25$.
- The number of runs: in order to guarantee statistical confident results, each algorithm is executed 25 runs using a different initial population in each execution.
- The constraints number: 5.
- The number of ants (***NbA***): 350
- The pheromone minimum threshold (***MaxTP***): 1
- The pheromone maximum threshold (***MinTP***): 15
- The pheromone exponent (***ExP***): 0.4
- The pheromone evaporation coefficient (***EvP***): 0.25
- The number of objectives and the termination condition (the maximum number of generations) are as shown in Table 1. For each instance, the best performance is demonstrated using shaded background.

The AcNGSA-III outperforms the ACO and the NGSA-III in the most of the instances which proves its efficiency. Another constatation is that the ACO is better than the NSGA-III for three objectives, while the latter one is better when the number of objectives is higher than three. This is is congruent with the observation of its authors [3] which asserts that the NSGAIII is dedicated to resolve many-objectives problems. Also, due to the increasing complexity of the problem when the number of objectives increases, for all algorithms, better HV values are obtained with smaller number of objectives.

Table 1. HV (Best, average and worst values)

Obj Nbr	Max nbr of generations	ACO	NSGAIII	AcNGSA-III
3	1300	0.984682	0.902458	0.903168
		0.983561	0.901896	0.902375
		0.982327	0.898023	0.987653
4	1800	0.885236	0.974685	0.976687
		0.884381	0.974233	0.975986
		0.884003	0.973612	0.975124
5	2600	0.878847	0.972892	0.973382
		0.872324	0.972716	0.972899
		0.871456	0.972684	0.972633

5 Comparing Simulations and Experimentations

This section aims to provide a comparison between the simulation and the experimental tests of the 3D deployment scheme in indoor WSN while satisfying different objectives. We are interesting in testing the behavior the tested evolutionary optimization algorithms (ACO, NSGA-III) and the new suggested one (AcNSGA-III).

5.1 Simulation/Experiments Parameters and Working Environment

The algorithms are tested on an Intel core Intel Core i5-6600 K 3.5 GHz computer. Our model is based on the implementation of a 433 MHz physical layer, a non-coordinated CSMA/CA access method of the IEEE 802.15.4 protocol, and a routing layer based on the reactive AODV (Ad-hoc On-demand Distance Vector) protocol. The following parameters are considered in our simulations/experiments:

- Transmission power = 60 mW;
- Reception gain = 50 mA;
- Operating temperature = 25°c;
- Bit rate = 256 kbps;
- RSSI (Received Signal Strength Indicator) = by default (100);
- FER (Frame Error Rate) = by default (0.01);
- Indoor transmission range = 7m;
- Indoor sensing range = 8m;
- Maximum number of nodes = by default (65000) according to protocol 802.15.4;
- Power supply = 3.6 v;
- Dimensions (in meters) of the 3D space: Length (x=24), Width (y=14), Height (z=7);
- The execution time of the simulations is set to 4230 seconds;
- An average of 20 simulation and experiment runs are taken;

The following tools are used:

- **OMNeT++ 4.6** [18] a free C++ platform to simulate and develop network protocols.
- **JMetal 4.5.2** [19] a java platform, which aims to develop and test different meta-heuristics solving problems of multi-objective optimization.
- **OpenWiNo** [1] an open tool for emulating IoT and WSN protocols. It is used for prototyping protocols (in different layers) and evaluating their performances.
- **Arduino 1.6.1** [20] an open software/hardware platform for modules prototyping, the WiNo nodes use it in transferring programs via serial links.
- **Teensyduino 1.2.1** an Arduino add-on to run sketches.

Figure 1 demonstrates the Teensy WiNo deployed nodes.

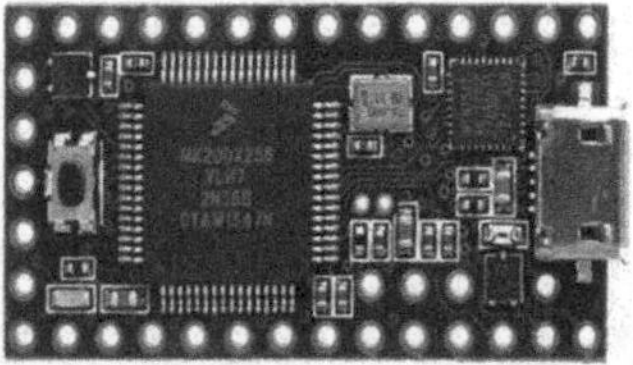

Fig. 1. The Teensy WiNo deployed nodes

5.2 Experimental Validations

The same architecture (number and type of nodes) of the 3D deployment is used both for simulations and experiments. A constant number (11) of fixed nodes with known positions is used. Also, the number of nomad nodes to be added is fixed to three. Their positions are determined with the tested algorithms. Similarly for mobile nodes, only one node is used as a trigger for the first message. In simulations, the positions of the initially deployed fixed nodes are chosen according to the distribution law used by OMNet++ which tries to uniformly distribute nodes starting from the center of the region of interest. In experiments, these positions are chosen according to the applicative objectives of the users in the building. This may lead to the non-coverage of certain zones especially at the borders if the number of fixed nodes is too small. The execution scenario is as following: In simulations, an initial message is sent from the mobile node to a random destination d; once d is found by the routing protocol, d becomes the source and a new destination is selected… etc. This cycle is repeated until a stop condition is satisfied. Among others, a maximal time for simulations. In experiments, the mobile node moves in the building and sends messages to its neighbors (nodes). Concerning the connectivity of nodes, a connectivity matrix between nodes is established. This matrix is deducted from empirical results derived from our experiments. As a consequence, the same initial connectivity links of the experiments are used. Subsequently, in order to model the dynamism of the network in simulations, these connectivity links are set to perturbations which allows modifying the initial connectivity links. Theses perturbations concern the calculation of the RSSI rates between the nodes. Indeed, a matrix of RSSI rates extracted from experiments is used initially. Then, this matrix is set to a perturbation (±30 for each value) in order to have new connectivity relations between nodes each time. Figure 2 represents the distribution of the nodes according to the OMNet++ interface in the simulations. Nodes named initial-i represents the fixed nodes and nodes named nomad-i represents the nomad nodes. Since comparing two algorithms is only possible through the use of a statistical test and due to the stochastic nature of the tested evolutionary algorithms, it is necessary to perform any test over many executions to obtain a well based judgment concerning their performance. Thus, all average values in this section, are computed based on 25 executions of the algorithms. Figure 3 shows the 3D deployment scheme of the experiments. Red nodes are the fixed nodes. Blue ones are the added nomad nodes.

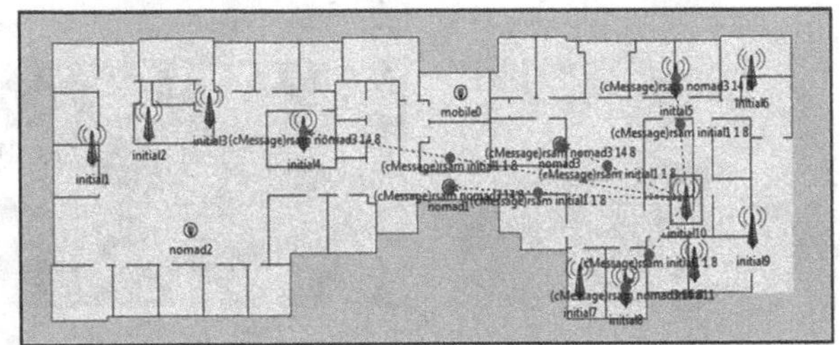

Fig. 2. The simulation scenario

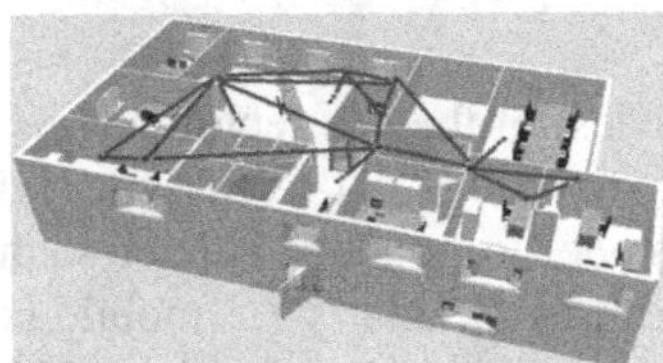 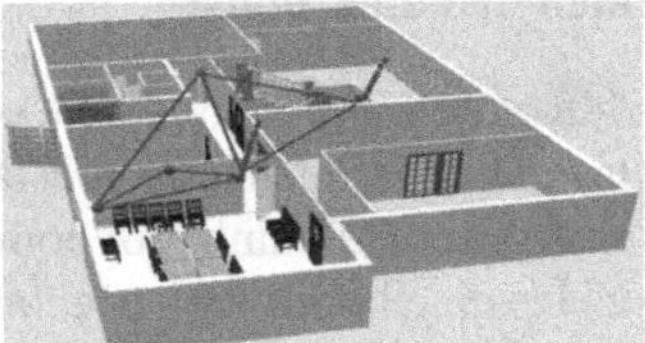

Fig. 3. 3D deployment scheme of the experiments

In the following, the different measures (RSSI, FER, etc.) are taken in both simulation and experiments.

Comparing the RSSI rates. In order to measure the localization, the RSSI metric is used since the used localization model is based on hybridization between the RSSI and the DVHop (Distance-VectorHop) localization protocol. Thus, the higher the RSSI, the better the localization is. A neighbor can be added to the neighborhood table of a node only if the RSSI value of the detected node is greater than a predefined threshold. The theoretical value of this threshold is set to 90. Initially, RSSI levels are based on our empirical experiments. Then, as mentioned before, in order to guarantee dynamism within the network, disruption of the value of RSSI is introduced via a random function. Figure 4 shows, for different number of objectives considered by the tested algorithms, the average of the RSSI rates measured for all the nodes in connection with (detected by/detecting) the mobile node.

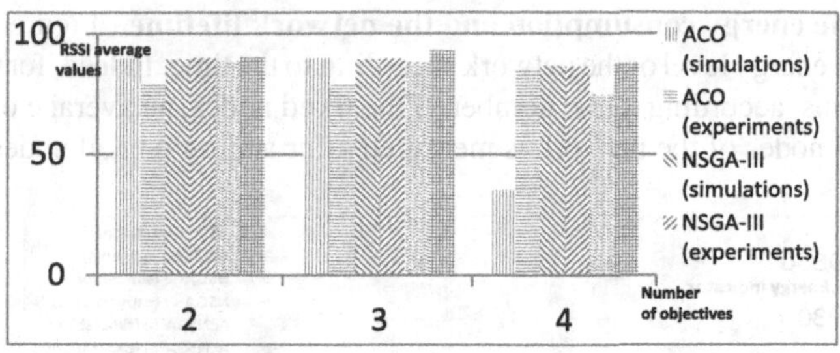

Fig. 4. Comparing the average RSSI rates

Comparing the FER rates. To measure coverage, we use the FER as a metric to assess the quality of links between nodes. Thus, the lower the FER, the better the coverage is. Although the FER values are less variable than those of the RSSI, for each node's pair {node C, node i}, an average value extracted from four values is taken with an interval of 10 s between the four values. Initially, FER rates are based on our empirical experiments. Afterwards, to guarantee dynamism within the network, disruption of the FER values is introduced via a random function (±0.04 to ±0.2). Figure 5 shows the average FER rates measured for all nodes in connection with (detected by /detecting) the mobile node.

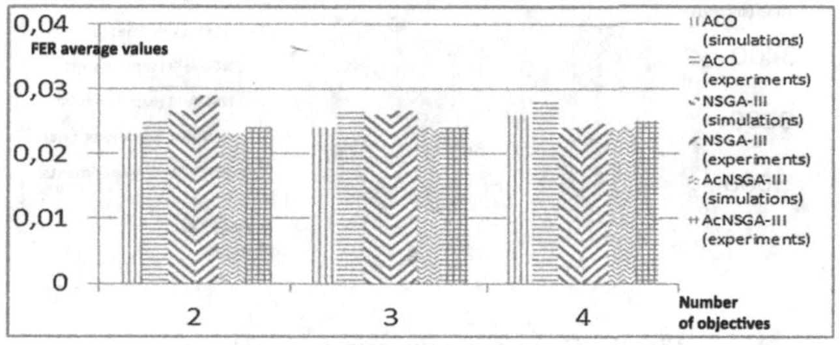

Fig. 5. Comparing the average FER rates

Comparing the number of neighbors. In order to assess the network connectivity, the number of neighbors of the target (the mobile node) is measured. Figure 6 shows the average of the number of neighbors of the mobile node per objective.

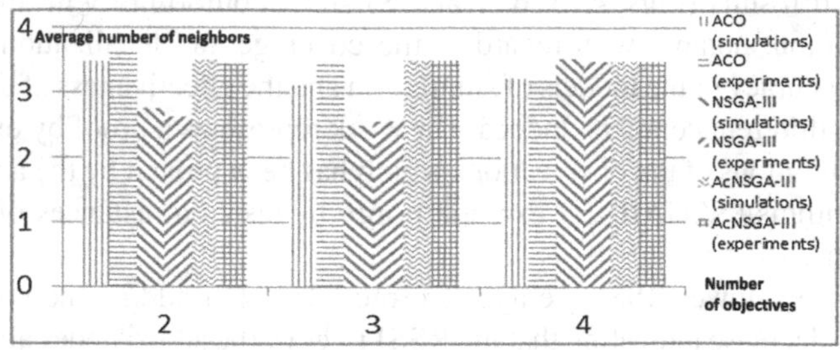

Fig. 6. Comparing the average number of neighbors

Comparing the energy consumption and the network lifetime. Figure 7 shows the variation of the energy level of the network according to the time. Indeed, for the different tested algorithms, according to the number of the fixed nodes, an average of the energy indicator of all nodes of the network is measured after adding nomad nodes.

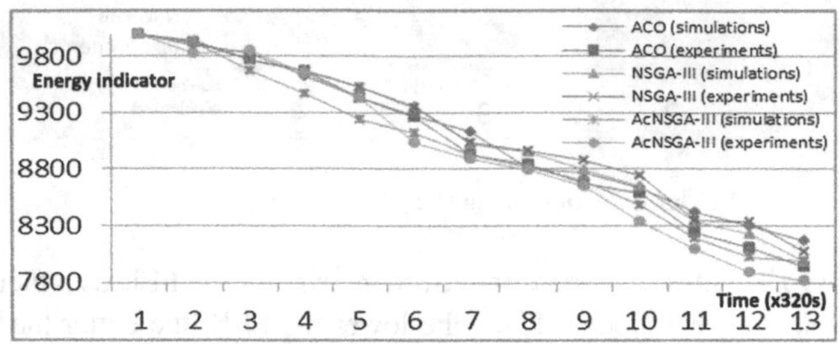

Fig. 7. Comparing the average energy consumption levels

Figure 8 illustrates the lifetime of the network. It shows for different number of objectives, the time in which the first node of the network is switched off.

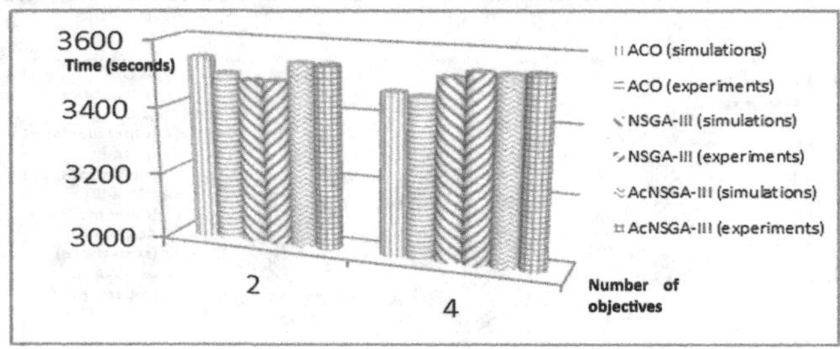

Fig. 8. Comparing the average lifetime

5.3 Discussion and Interpretations

After comparing the simulations and the experiments, different interpretations can be considered:

- The obtained results (Figs. 4, 5, 6, 7 and 8) show conformity with the results of experimentation, notably with regard to the coverage and localization rates. This proves the accuracy of the models of simulation and the effectiveness of the proposed approach in different contexts. Indeed, our work represents a proof by experimentation and simulations of the observations which has been proved by the authors of the tested algorithms (NSGAIII for example) only by tests on instances of theoretical test problems.
- In several cases, lower RSSI averages are recorded after adding the nomad nodes. Despite this decrease indicating that the RSSI rates of the added nodes are lower than the RSSI values of the fixed nodes, the localization rate, the coverage rate and the

number of neighbors are improved. Given the objectives set by our approach, this decrease in RSSI averages is understandable, since adding a node in a location ×1 so that it will be close to several nodes with a lower RSSI value will be better than adding it in a location ×2 with a higher RSSI value but smaller number of neighbors.
- The error rates (FER) are more important in experiments then in simulations. This is due to the influence of the activities of persons in the building during experiments (for example opening and closing doors) which generates the perturbation of the signal.
- By comparing the efficiency of the tested algorithms, as proved by numerical results (Sect. 4), the simulation/experimental results show that this efficiency is relative to the number of objectives to be optimized. Table 1 shows that for less than three objectives, the ACO is more efficient than NSGAIII, while the latter is more effective than ACO if the number of objectives exceeds three. This is explained by the fact that the ACO is dedicated for multi-objective problems, while the authors of NSGAIII propose this latter as an adaptation of the NSGAII for many-objective problems having more than three objectives. However, the AcNSGA-III has an almost stable behavior and is not influenced by the variation of the number of objectives.

6 Conclusion and Perspectives

In this paper, we proposed a hybrid algorithm called AcNSGA-III based on NSGA-III and ACO to resolve the problem of 3D indoor deployment in WSN. In order to prove the efficiency of the new algorithm, the performance of the above mentioned algorithms is assessed using the Hypervolume metric, then by simulations and experimental tests while considering five objectives. The proposed algorithm outperforms the traditional NSGA-III and ACO. Different improvements may be incorporated to our approach: We can further include the dynamic redeployment of nodes and the existence of obstacles in the 3D space while considering different other objectives, such as the network connectivity. Moreover, as a perspective, we seek to intensify the real deployed network by adding new nodes in order to investigate the influence of the network density on results. Also, the implementation of a more realistic energy model based on the management of the BO and SO values [21] of the 802.15.4 CSMA/CA protocol.

References

1. Van den Bossche, A., Dalce, R., Val, T.: OpenWiNo: an open hardware and software framework for fast-prototyping in the IoT. In: Proceedings 23rd International Conference on Telecommunications, Thessaloniki, Greece, pp. 1–6, 16–18 May 2016. doi:10.1109/ICT.2016.7500490
2. Cheng, X., Du, D.Z., Wang, L., Xu, B.: Relay sensor placement in wireless sensor networks. ACM/Springer J. Wirel. Netw. **14**(3), 347–355 (2007). doi:10.1007/s11276-006-0724-8
3. Deb, K., Jain, H.: An evolutionary many-objective optimization algorithm using reference-point-based nondominated sorting approach, part I: solving problems with box constraints. IEEE Trans. Evol. Comput. **18**(4), 577–601 (2013). doi:10.1109/TEVC.2013.2281535

4. Dorigo, M., Maniezzo, V., Colorni, A.: Ant system: optimization by a colony of cooperating agents. IEEE Trans. Syst. Man Cybern. Part B (Cybern.) **26**(1), 29–41 (1996). doi:10.1109/3477.484436
5. Aval, K.J., Abd Razak, S.: A review on the implementation of multiobjective algorithms in wireless sensor network. World Appl. Sci. J. **19**(6), 772–779. ISSN 1818-4952 (2012). doi:10.5829/idosi.wasj.2012.19.06.1398
6. Mnasri, S., Nasri, N., Val, T.: An overview of the deployment paradigms in the wireless sensor networks. In: Proceedings International Conference on Performance Evaluation and Modeling in Wired and Wireless Networks, Tunisie, 04–07 November 2014
7. Qu, Y.: Wireless sensor network deployment. Ph.D. dissertation, Florida International University, Miami, Florida, USA (2013)
8. Matsuo, S., Sun, W., Shibata, N., Kitani, T., Ito, M.: BalloonNet: a deploying method for a three-dimensional wireless network surrounding a building. In: Proceedings of the Eighth International Conference on Broadband and Wireless Computing, Communication and Applications, pp. 120–127 (2013). doi:10.1109/BWCCA.2013.28
9. Jiang, J.A., Wan, J.J., Zheng, X.Y., Chen, C.P., Lee, C.H., Su, L.K., Huang, W.C.: A novel weather information-based optimization algorithm for thermal sensor placement in smart grid. IEEE Trans. Smart Grid **99**, 1–11 (2016). doi:10.1109/TSG.2016.2571220
10. Sweidan, H.I., Havens, T.C.: Coverage optimization in a terrain-aware wireless sensor network. In: Proceedings of the 2016 IEEE Congress on Evolutionary Computation, Vancouver, BC, pp. 3687–3694 (2016). doi:10.1109/CEC.2016.7744256
11. Deb, K., Pratap, A., Agarwal, S., Meyarivan, T.: A fast and elitist multiobjective genetic algorithm: NSGA-II. IEEE Trans. Evol. Comput. **6**(2), 182–197 (2002). doi:10.1109/4235.996017
12. Zhang, Q., Li, H.: MOEA/D: a multiobjective evolutionary algorithm based on decomposition. IEEE Trans. Evol. Comput. **11**(6), 712–731 (2007). doi:10.1109/TEVC.2007.892759
13. Ibrahim, A., Rahnamayan, S., Martin, M.V., Deb, K.: EliteNSGA-III: an improved evolutionary many-objective optimization algorithm. In: Proceedings IEEE Congress on Evolutionary Computation, Vancouver, BC, Canada, pp. 973–982, 24–29 July 2016. doi:10.1109/CEC.2016.7743895
14. Sim, K.M., Sun, W.H.: Ant colony optimization for routing and load-balancing: survey and new directions. IEEE Trans. Syst. Man Cybern. Part A: Syst. Humans **33**(5), 560–572 (2003). doi:10.1109/TSMCA.2003.817391
15. Shen, H.: A study of welding robot path planning application based on Genetic Ant Colony Hybrid Algorithm. In: Proceedings IEEE Advanced Information Management, Communicates, Electronic and Automation Control Conference, Xi'an, China, pp. 1743–1746, 3–5 October 2016. doi:10.1109/IMCEC.2016.7867517
16. Huang, P., Chen, J.: Improved CCN routing based on the combination of genetic algorithm and ant colony optimization. In: Proceedings 3rd International Conference on Computer Science and Network Technology, Dalian, China, pp. 846–849, 12–13 October 2013. doi:10.1109/ICCSNT.2013.6967238
17. While, L., Hingston, P., Barone, L., Huband, S.: A faster algorithm for calculating hypervolume. IEEE Trans. Evol. Comput. **10**(1), 29–38 (2006). doi:10.1109/TEVC.2005.851275
18. The OMNeT platform (2016). https://omnetpp.org/omnetpp. Accessed 9 June 2016
19. The jMetal platform (2015). http://jmetal.sourceforge.net/. Accessed 2 Mar 2015

20. The Arduino platform (2017). https://www.arduino.cc/en/main/software. Accessed 5 Jan 2017
21. Farhad, A., Farid, S., Zia, Y., Hussain, F.B.: A delay mitigation dynamic scheduling algorithm for the IEEE 802.15.4 based WPANs. In: Proceedings International Conference on Industrial Informatics and Computer Systems, Sharjah, UAE, pp. 1–5, 13–15 March 2016. doi:10.1109/ICCSII.2016.7462430

Interference Analysis for Asynchronous OFDM in Multi-user Cognitive Radio Networks with Nonlinear Distortions

Hanen Lajnef(✉), Maha Cherif Dakhli, Moez Hizem, and Ridha Bouallegue

Innov'Com Lab, Sup'Com, University of Carthage, Tunis, Tunisia
hanen.lajnef@supcom.tn

Abstract. In this paper, we investigate the effect of timing synchronization errors for multicarrier techniques CP-OFDM based Cognitive Radio networks in presence of non-linear high-power amplifier (HPA). We first develop a theoretical analysis of nonlinear distortion (NLD) effects and the asynchronous interference between primary user (PU) and secondary users (SU). Then, we evaluate the performance of the CP-OFDM based CR-system, in terms of bit error rate (BER) for a Rayleigh flat fading channel. Finally, the simulation results as compared to the theoretical approach confirm the validity of our analysis.

Keywords: Cognitive radio · CP-OFDM · Interference · Timing synchronization · Nonlinear distortion

1 Introduction

Cognitive Radio (CR) is a form of wireless communication in which a transceiver can intelligently detect which communication channels are in use and which are not, and instantly move into vacant channels while avoiding occupied ones [1]. The main objective of CR is to allow the secondary user (SU) to use the available spectrum resource to the primary user (PU) by detecting the existence of spectrum holes.

Thus, the detection of PU is one of the main ultimatums in the development of CR technique. Furthermore, compared to the conventional wireless communication systems, the CR system sets new technique for resource allocation (RA) problems due to the interference from adjacent channels used by SU to PU [2].

For the CR technology, most efforts have been devoted on the Orthogonal Frequency Division Multiplexing (OFDM) based CR systems especially for perfect synchronization [3].

Using multicarrier technique of Cyclic Prefix CP-OFDM, the performance of Cognitive Radio Networks depends extensively in the manner well the orthogonality among subcarriers and which is retained at the primary user received. In an asynchronous system, this orthogonality is destroyed due to synchronization errors which include timing offsets and carrier frequency/phase offsets.

A. Puliafito et al. (Eds.): ADHOC-NOW 2017, LNCS 10517, pp. 382–393, 2017.
DOI: 10.1007/978-3-319-67910-5_31

In literature, many studies have deeply investigated for the problem of asynchronous in multicarrier systems [4–6]. In the context of CP-OFDM, the multicarrier signals are constructed by a sum of N independent streams transmitted over N orthogonal subcarriers. If we consider high values of N, according to the central limit theorem [7], the superposition of these independent signals leads to a complex Gaussian multicarrier signal. For this reason, a classical OFDM exhibits large peak-to-average power ratios (PAPR), i.e., large fluctuations in their signal envelopes. Moreover, the performance of the transceiver is very sensitive to NLD due to high-power amplifiers (HPA) [8].

The main objective of this paper is to develop a theoretical approach for the asynchronous interference in presence of the NL HPA for simple case (only one PU and one SU) and for generalized case (N PUs and N SUs). Besides, we studied the effects of the nonlinear distortion due to the HPA and the impact of timing errors on the bit error rate (BER) performance for multi-user OFDM-CR system under Rayleigh fading channels.

So, the comparison of the linear and nonlinear cases from the spectral behavior of the signal transmitted by the primary user (PU) is discussed; and lastly, by Monte-Carlo simulation method, we are interested in describing the receiver performance through its interference model for different situations of interest.

The rest of this paper is organized as follows: Sect. 2 describes the system model with brief introduction to OFDM modulation in addition to some elementary information on commonly used memory less HPA model, exhibiting AM/AM and AM/PM distortions. Section 3 presents the effects of nonlinear distortion for asynchronous interference. Section 4 provides the BER analysis. Section 5 briefly introduces the simulation parameters, and finally Sect. 6 brings the paper to a conclusion.

2 System Model

2.1 Cognitive Radio Consideration

Cognitive radio is a form of wireless communication in which a transmitter / receiver can smartly detect communication channels that are in use and those that are not, and can travel in unused channels using the spectrum Sensing technique. This optimizes the use of available radio frequencies (RF) spectrum while minimizing interference with other users [1].

The primary user is perfectly synchronized with its base station (PU-BS), but it is not necessarily synchronized with the other secondary users that cause interference from this asynchronous as shown in Fig. 1.

2.2 OFDM Transmission Chain

In OFDM based cognitive radio systems, the transmission bandwidth must be large in order to support high data rates [11]. However, a large contiguous bandwidth may not be available for opportunistic transmission of CR. In addition, in a dynamic spectrum

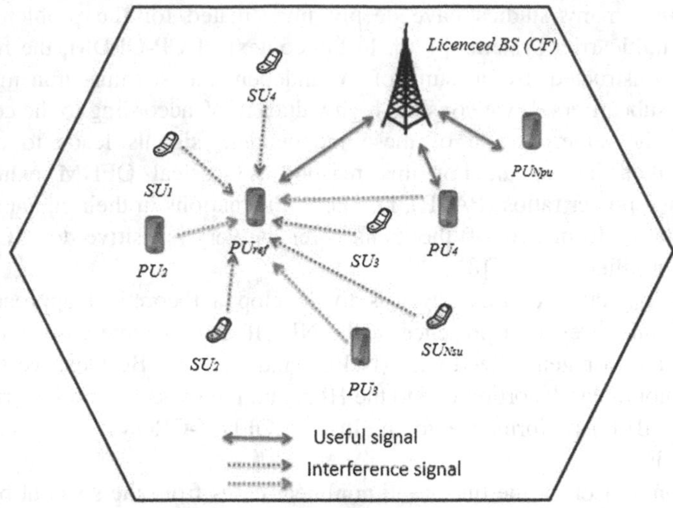

Fig. 1. Cognitive radio network

access (DSA) network, multicarrier-based cognitive radio transceivers need to deactivate some of their subcarriers to avoid causing interference to primary users.

2.3 HPA Model

Consider the presence of HPA nonlinear distortions. These distortions can be modeled using the Bussgang theorem [7–9] which states that for a Gaussian signal $s_L(t)$ which undergoes a non-linear distortion and produce an output signal $s_{NL}(t)$:

$$s_{NL}(t) = K_0 s_L(t) + d(t) \tag{1}$$

Where

$$s_L(t) = x\, g(t) e^{j\frac{2\pi}{T}kt + j\varphi} \tag{2}$$

$$s_{NL}(t) = K_0 x\, g(t) e^{j\frac{2\pi}{T}k + j\varphi} + d(t) \tag{3}$$

Where

- K_0 : The nonlinear distortion term uncorrelated with the input signal $s_L(t)$.
- x: The complex transmitted symbol.
- φ: The phase offset.
- $g(t)$: The transmit pulse shape.
- $d(t)$: is a zero mean additive noise.

Concerning the channel model, we consider a frequency selective multipath channel, whereas all the interfering signals from the CR BS experience Rayleigh fading. The impulse response of multipath channel between the CR BS and the reference mobile station (PU) is:

$$\mathrm{h}(t) = \sum\nolimits_{i=0}^{L-1} h_i \delta(t - \frac{n_i}{N} T) \tag{4}$$

Where

- T is the symbol duration
- N is the total number of subcarriers
- L is the total number of nodes
- n_i are the maximum delay spreads of the channel.

Where $n_0 < n_{k1} < \cdots < n_{k,L-1} < C$ and C is the maximum delay spreads normalized by the sampling period $\frac{T}{N}$ and $h_{k,i}$ are the complex channel path gains, which are assumed mutually independent [3].

The received signal can thus be written as follows:

$$\begin{aligned} r_L(t) &= h(t) \otimes s_L \\ &= \sum\nolimits_{i=0}^{L-1} \sum\nolimits_{k=1}^{M_k} x_k h_i(t) \delta\left(t - \frac{n_i}{N} T\right) g(t) e^{j\frac{2\pi}{T}kt} + n(t) \end{aligned} \tag{5}$$

Where

- M_K denotes the total number of sub-carriers of all spectrum holes.
- τ_{su} denotes the timing offset between the reference PU and the SU undergoing interfering.
- $n(t)$ Additive White Gaussian Noise (AWGN).

3 Interference Analysis

In [6], the interference model is calculated for linear CP-OFDM signal. In this section, we will derive the interference model of nonlinear CP-OFDM, and we will show the effect of the nonlinear distortion (NLD).

3.1 Single User Case

The interference signal corresponding to the secondary user is described by equations below:

$$\begin{aligned} Z_{su}(t) &= \sum\nolimits_{k=1}^{M_K} r_{NL,su}(t - \tau_{su}) \\ &= \sum_{i=0}^{L-1} \sum_{k=1}^{M_k} K_0 x_k h_i(t) \delta\left(t - \frac{n_i}{N} T\right) g(t - \tau_{su}) e^{j\frac{2\pi}{T}k(t-\tau_{su})} \\ &\quad + \sum\nolimits_{i=0}^{L-1} h_i(t) \delta\left(t - \frac{n_i}{N} T\right) d(t - \tau_{su}) \end{aligned} \tag{6}$$

Supposing we have only one PU and one SU, the asynchronous OFDM interference signal received on the *kth* subcarrier for a PU receiver, considering the transmission of a single complex symbol $x_{k',0}$ on the k'^{th} subcarrier by another user (PU or SU), is given by:

$$y_{k'}(\tau,\varphi) = \langle Z_{su_1}(t-\tau,\varphi), f(t-(T+\Delta))e^{-j\frac{2\pi}{T}k'(t-k(T+\Delta))}\rangle \tag{7}$$

We can deduct the Eq. (7),

$$\begin{aligned} y_{k'}(\tau,\varphi) = K_0 x_k e^{-j\frac{2\pi}{T}k\tau + j\varphi} \int_{-\infty}^{+\infty} g\left(t' - \tau\right) f\left(T+\Delta-t'\right) e^{j\frac{2\pi}{T}lt'} dt' + \\ \int_{-\infty}^{+\infty} d(t') f\left(T+\Delta-t'\right) e^{j\frac{2\pi}{T}k't'} dt' \end{aligned} \tag{8}$$

For simplicity's sake:

$$D = \int_{-\infty}^{+\infty} d(t') f\left(T+\Delta-t'\right) e^{j\frac{2\pi}{T}k't'} dt' \tag{9}$$

The Eq. (8) becomes:

$$y_{k'}(\tau,\varphi) = K_0 x_k e^{-j\frac{2\pi}{T}k\tau + j\varphi} \int_{-\infty}^{+\infty} g\left(t' - \tau\right) f\left(T+\Delta-t'\right) e^{j\frac{2\pi}{T}lt'} dt' + D \tag{10}$$

The value of τ determines the limits of the integral, then we can obtain Eq. (11).

$$y_{k'}(\tau) = u_k \begin{cases} K_0\delta(l) + D, \tau \in [0,\Delta] \\ K_0 e^{j\frac{\pi l}{T}(T+\tau+\Delta)} \frac{\sin\left(\frac{\pi l(T+\Delta-\tau)}{T}\right)}{\pi l} + D, \tau \in [\Delta, T+\Delta] \\ 0 \quad elsewhere \end{cases} \tag{11}$$

The interference power can thus be written as follows:

$$I(\tau) = \left|y_{k'}\right|^2 \tag{12}$$

We can also write it:

$$\begin{aligned} I(\tau) &= E\left(\left|K_0 x_k e^{-j\frac{2\pi}{T}k\tau + j\varphi} \int_{-\infty}^{+\infty} g(t'-\tau) f(T+\Delta-t') e^{j\frac{2\pi}{T}lt'} dt' + D\right|^2\right) \\ &= E\left(\left|K_0 x_k e^{-j\frac{2\pi}{T}k\tau + j\varphi} \int_{-\infty}^{+\infty} g(t'-\tau) f(T+\Delta-t') e^{j\frac{2\pi}{T}lt'} dt'\right|^2\right) + E\left(|D|^2\right) \end{aligned} \tag{13}$$

and we assumed that:

$$E\left(|D|^2\right) = E\left(\left|\int_{-\infty}^{+\infty} d\left(t'\right) f\left(T+\Delta-t'\right) e^{j\frac{2\pi}{T}k't'} dt'\right|^2\right) = \sigma_d^2 \tag{14}$$

and

$$I(\tau,l) = \begin{cases} K_0^2\delta(l) + \sigma_d^2, \tau \in [0,\Delta] \\ K_0^2\left|\frac{\sin^{\pi l(T+\Delta-\tau)}/T}{\pi l}\right|^2 + \sigma_d^2, \tau \in [\Delta, T+\Delta] \\ 0 \quad elsewhere \end{cases} \tag{15}$$

A. **Multiuser Case:**

In general case, the resulting interference power is the sum of interference powers coming from all other NL Secondary Users (SU) in reference to the NL Primary user PU_{ref}, we follow indeed the same steps as in the simple case.

The interference signal corresponding to the secondary user is described by equation below:

$$X_{SU} = \sum_{i=1}^{N_{SU}} x_{Su,i} = \sum_{i=1}^{N_{SU}} \sum_{k=1}^{F_k} x_k g(t-\tau_i-\Delta_{si}) e^{-j\frac{2\pi}{T}k(t-\tau_i-\Delta_{s,i})} \tag{16}$$

Where:

- x_k: The complex transmitted symbol of the *ith* secondary User.
- F_k: The number of unlicensed subcarriers.

The general interference of the entire NL secondary user referred to NL PU_{ref} can be written as follows:

Case 1: $\Delta_{si} \le \tau_{pu,i} \le \Delta$

$$I_{k_i',PU_{ref}}\left(\tau_{pu,i}, \varphi_{pu,i}, l_i\right) = \left|\sum_{i=1}^{N_{su}} (K_0 A_i \frac{\sin\left(\frac{\pi l_i}{T}\left(T+\Delta+\Delta_{s,i}-\tau_{pu,i}\right)\right)}{\pi l_i} + D)\right|^2 \tag{17}$$

With

$$A_i = x_{PU,i} e^{-j\frac{\pi}{T}l_i\left(2\Delta_{s,i}+\Delta+T\right)+j\varphi_{pu,i}}$$

Case 2: $\Delta_{si} \le \tau_{pu,i} \le \Delta - \Delta_{si}$

$$I_{k_i',PU_{ref}}\left(\tau_{pu,i}, \varphi_{pu,i}, l_i\right) = \left|\sum_{i=1}^{N_{su}} (K_0 B_i \frac{sin\left(\frac{\pi l_i}{T}\left(T+\Delta+\Delta_{s,i}-\tau_{pu,i}\right)\right)}{\pi l_i} + D)\right|^2 \tag{18}$$

With

$$B_i = x_{PU,i} e^{-j\frac{\pi}{T} l_i \left(3\Delta_{s,i} + \Delta + T + \tau_{pu,i}\right) + j\varphi_{pu,i} - j\frac{\pi}{T} k'_i \tau_{pu,i}}$$

Case 3: $\boldsymbol{\tau_{pu,i} \geq \Delta - \Delta_{si}}$

$$\mathrm{I}_{k'_i,\mathrm{PU_{ref}}}\left(\tau_{\mathrm{pu,i}}, \varphi_{\mathrm{pu,i}}, \mathrm{l_i}\right) = 0 \tag{19}$$

4 Bit Error Rate Analysis

In this section, we will investigate the theoretical Bit Error Rate (*BER*) with and without HPA, the block diagram of an OFDM based CR transceiver is depicted in Fig. 2.

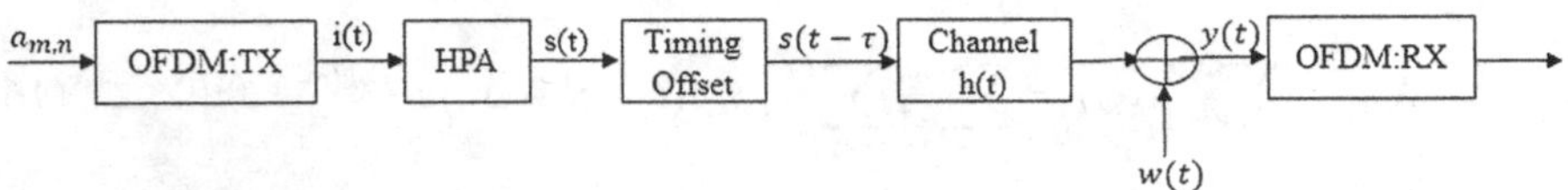

Fig. 2. Unsynchronised NL CP-OFDM System Block scheme

4.1 Linear OFDM Based CR System

In the literature the *BER* of M-QAM modulation [9–11] in Rayleigh fading channel is given by:

$$\overline{BER}^{L}_{MQAM} = \frac{(M-1)}{Mlog_2 M}\left[1 - \frac{1}{\sqrt{1 + \frac{M^2-1}{3(SINR)log_2 M}}}\right] \tag{20}$$

with *SINR* is the signal-to-interference-plus-noise ratio. It will be calculated in the single and multi-user case.

A. Single user Case: Using [4–6], the asynchronous interference power coming from the kth secondary user can be calculated using the following expression:

$$P^{k}_{SU}(\tau, l) = d_k^{-\beta} \sigma_x^2 \mathrm{I}_{\mathrm{k,PU_{ref}}}\left(\tau_{\mathrm{pu,i}}, \varphi_{\mathrm{pu,i}}, \mathrm{l_i}\right) |H_i(k)|^2 \tag{21}$$

Where d_k^{β} is the distance between the reference primary user and the *kth* secondary user interfering, β is the path loss exponent and σ_x^2 is the transmitted symbol power.

Finally, we present the following expression of the *SINR* for single user case (only one PU and one SU):

$$SINR(\tau, l) = \frac{d_0^{-\beta}|H_0|^2}{P_{SU}^k(\tau, l) + \gamma} \tag{22}$$

Where $\gamma = \frac{\sigma_n^2}{\sigma_x^2}$.

B. Multiuser Case: In the presence of multiple secondary user, the total asynchronous interference power coming from the kth Secondary user is given by:

$$P_{SU}^k(\tau, l) = \sum_{k=1}^{N_{su}} d_k^{-\beta} \sigma_x^2 I_{k,PU_{ref}}(\tau_{pu,i}, \varphi_{pu,i}, l_i)|H_i(k)|^2 \tag{23}$$

and by substituting Eq. (23) in (22), the *SINR* can be written:

$$SINR(\tau, l) = \frac{d_0^{-\beta}|H_0|^2}{\sum_{k=1}^{N_{SU}} B_{k,l} + \gamma} \tag{24}$$

where $B_{k,l} = d_k^{-\beta} I(\tau_k, l)|H_k(k)|^2$.

4.2 NonLinear OFDM Based CR System

It is known in the literature [9, 12] that the BER of M-QAM modulation running under a Rayleigh fading channel is given by Eq. 25.

$$\overline{BER}_{MQAM}^{NL} = \int_0^{SINR_{NL}} \frac{4}{log_2 M} Q\left(\sqrt{\frac{3\gamma_b log_2 M}{M-1}}\right) \frac{|K_0|^2 \sigma_x^2 \sigma_n^2}{\Omega\left(|K_0|^2 \sigma_x^2 - \gamma\sigma_d^2\right)^2} exp$$
$$- \left(\frac{\gamma\sigma_n^2}{\Omega(|K_0|^2 \sigma_x^2 - \gamma\sigma_d^2)}\right) d\gamma \tag{25}$$

Where Ω is the average fading power.

A. Single user Case: In presence of non linear HPA for PU signal reference and one SU, the SINR can be written as follows:

$$SINR_{NL}(\tau, l) = |K_0^2| \frac{d_0^{-\beta}|H_0(k)|^2}{P_{SU}^k(\tau, l) + \gamma_{NL} + \gamma} \tag{26}$$

where: $\gamma_{NL} = \frac{d_k^{-\beta} E[|D|^2]|H_i|^2}{\sigma_x^2}$ is the variance of the received nonlinear noise.

(1) **Multiuser Case**

When the number of SU exceeds one, the $SINR_{NL}$ become:

$$SINR_{NL}(\tau, l) = |K_0^2| \frac{\sigma_x^2 d_0^{-\beta} |H_0|^2}{\sum_{k=1}^{N_{SU}} B_{k,l} + \gamma_{NL} + \gamma} \quad (27)$$

5 Simulation Results and Discussion

In this section, the interference expression, the bit error rate expression, and simulation results are presented.

We will verify that nonlinear distortions and synchronization errors caused severe degradation in the performance of CP-OFDM based Cognitive Radio networks.

The CR network as shown in Fig. 1 with one primary System and one secondary system is simulated for different number of users and spectrum holes.

The system simulation parameters are showed in Table 1. We have considered the Pedestrian-A model as a Rayleigh fading propagation channel where the parameters are given in Table 2 [5, 11].

Table 1. System simulations parameters

Total bandwidth B	10 MHz
Center frequency	2.5 GHz
Symbol duration (T)	102.4e–6 s
Number of blocks/frame (Nbloc)	100
Number of subcarriers (N)	1024
Cyclic prefix duration	T/8
Modulation	QPSK/OQAM
Timing offset (τ)	[0 T]
Phase offset (φ)	[0 2π]
Input back-off (IBO)	8
Non Linear factor (K0)	0.7560 + 0.1641j
Variance of the NL distortion (σd2)	0.0033

Table 2. Channel parameters

Parameter	Value
Pedestrian multipath delays	$10^{-9} \cdot [0, 110, 190, 410]$ s
Pedestrian multipath powers	[0, −9.7, −19.2, −22.8] dB

Firstly, Fig. 3 compares the interference level of the output signal amplified by a NL HPA using Eq. (23) for each timing offset. We have used a CP-OFDM signal only 600 subcarriers (from 50 to 650) are activated for Primary user, the reminder of the subcarrier are devised for the other secondary user using the dynamic spectrum access (DSA), it can be defined as a system where the secondary users (SU) can temporarily communicate over the spectrum holes of primary User.

Figure 3 depict the interference level performance considering an IBO of 8 dB scheme. Four cases of types of Secondary User signal investigated: (a) the Linear CP-OFDM with timing offset $\tau = \frac{T}{2}$, (b) the NL CP- OFDM with timing offset $\tau = \frac{T}{2}$, (c) the Linear CP-OFDM with timing offset $\tau = \frac{T}{4}$, (d) the NL CP-OFDM with timing offset $\tau = \frac{T}{4}$. For (a) and (c) a significant degradation is present when the timing offset increase. Difference between case (a) and (b), the presence of HPA causes higher interference level.

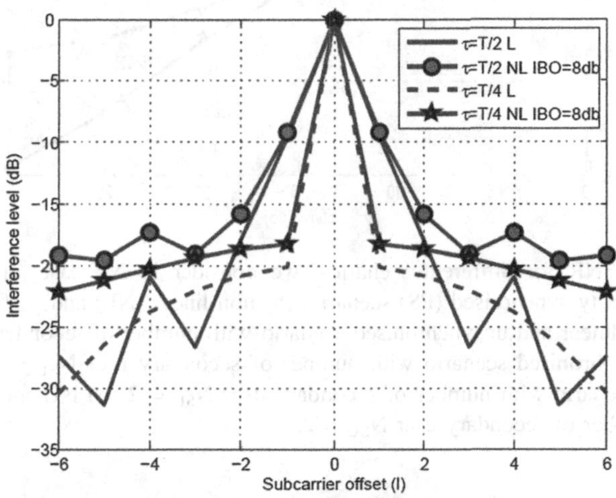

Fig. 3. The interference level in CP-OFDM Single Case with and without NL-HPA.

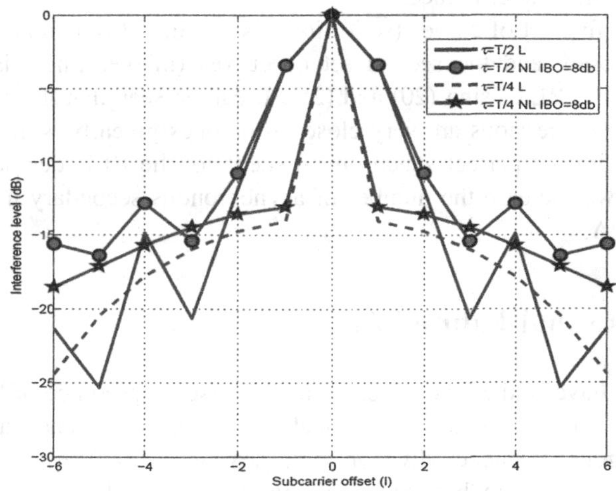

Fig. 4. The interference level in CP-OFDM multi-user case with in without NL-HPA for τ = T/2 and τ = T/4.

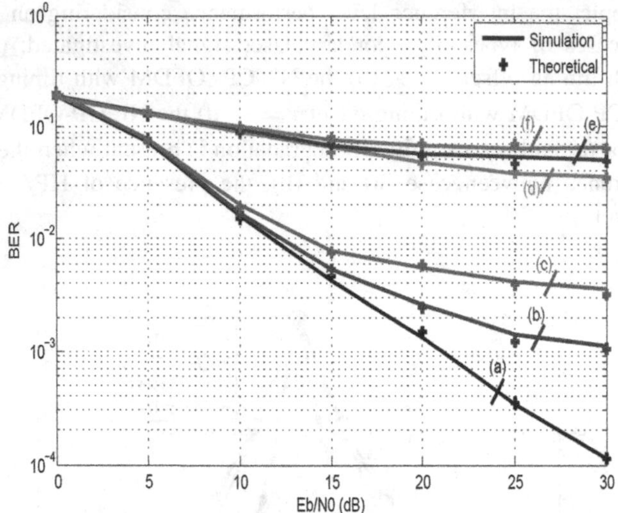

Fig. 5. BER vs. SNR for diffrerent scenarios, we consider in NL case an $IBO = 8$ dB : (a) linear and perfecty synchorised (PS) scenario, (b) non linear (NL) and perfecty synchorised (PS) scenario, (c) linear and unsynchronised scenario with number of secondary user $N_{SU} = 1$, (d) NL and unsynchronised scenario with number of secondary user $N_{SU} = 1$, (e) linear and unsynchronised scenario with number of secondary user $N_{SU} = 2$, (f) NL and unsynchronised scenario with number of secondary user $N_{SU} = 2$.

In Fig. 4, we present the asynchronous interference by changing the number of secondary users in presence of NL HPA ($IBO = 8$ dB). The interference decrease of the number of SU. It's also clear that even though the number of secondary user is increased, the use of linear CP-OFDM scheme is still better than NL CP-OFDM in terms of asynchronous interference.

In Fig. 5, the first set of curves (solid lines) is obtained from simulation results of the system proposed, and the second set of curves (marker lines) is provided by theoretical study the BER using (20) and (25). It can be seen that the results obtained from closed-form expressions are very close to the ones given by simulation.

Comparing with the perfect synchronism scenario, the BER degradation is more important when we increase the number of asynchronous secondary user and adding the nonlinear HPA.

6 Conclusions and Future Work

In this paper, we have studied the effect of timing offset in presence of HPA nonlinear distortions. And provides a new theoretical aspect of interference analysis in the context of orthogonal frequency division multiplexing.

Indeed, the combination between this asynchronism and non-linearity caused a severe degradation in the performance of cognitive Radio system, which leads us to work on introducing the compensators for eliminate the nonlinear distortion effects in our next research.

Acknowledgment. This research is supported by Innov'Com laboratory.

References

1. Mitola, J., III., Maguire, G.Q., Jr.: Cognitive radio for flexible mobile multimedia communications. In: IEEE International Workshop on Mobile Multimedia Communications (MoMuC 1999), pp. 3–10 (1999)
2. Wang, X.C.X., Chen, H.H., GuiZani, M.: Cognitive radio network management. IEEE Veh. Technol. Mag. **3**(1), 28–35 (2008)
3. Hamdi, K., Shobowale, Y.: Interference analysis in downlink OFDM considering imperfect intercell synchronization. IEEE Trans. Veh. Technol. **58**(7), 3283–3291 (2009)
4. Elmaroud, B., Abbad, M., Aboutajdine, D.: BER analysis of asynchronous and non linear FBMC based multi-cellular networks. In: IEEE 27th Annual International Symposium on Personal, Indoor, and Mobile Radio Communications (2016)
5. Medjahdi, Y.: Modelisation des interfrence et analyse des performances des systmes OFDM/FBMC pour les communications sans fil asyn-chrones. Ph.D. thesis, Le CNAM (2012)
6. Aissa, S.B., Hizem, M., Bouallegue, R.: Asynchronous OFDM interference analysis in multi-user cognitive radio networks. In: Wireless Communications and Mobile Computing Conference, International, pp. 1135–1140. IEEE (2016)
7. Dardari, D., Tralli, V., Vaccari, A.: A theoretical characterization of nonlinear distortion effects in OFDM systems. IEEE Trans. Commun. **48**(10), 1755–1764 (2000)
8. Bussgang, J.: Crosscorrelation functions of amplitude-distorted Gaussian signals. Research laboratory of electronics, Massachusetts Institute ofTechnology, Cambridge (1952)
9. Dakhli, M., Zayani, R., Belkacem, O.B., Bouallegue, R.: Theoretical analysis and compensation for the joint effects of HPA nonlinearity and RF crosstalk in VBLAST MIMO OFDM systems over Rayleigh fading channel. EURASIP J. Wirel. Commun. Networking **61**, 15 (2014)
10. Cherif, M., Zayani, R., Bouallegue, R.: A theoretical characterization and compensation of nonlinear distortion effects and performance analysis using polynomial model in MIMO OFDM systems under rayleigh fading channel. In: ISCC 2013, 18th IEEE Symposium on Computers and Communications, Split, Croatie, 07–10 July 2013
11. Bouhadda, H., Shaiek, H., Roviras, D., Zayani, R., Medjahdi, Y., Bouallegue, R.: Theoretical analysis of BER performance of nonlinearly amplified FBMC/OQAM and OFDM signals. EURASIP J. Adv. Sig. Proc. **60**, 18 (2014)
12. Saleh, A.A.M.: Frequency-independent and frequency-dependent nonlinear models of TWT amplifiers. IEEE Trans. Commun **29**, 1715–1720 (1981)

Author Index

Abassi, Ryma 77, 92
Aby, Affoua Therese 290
Aitouazzoug, Nadia 55
Amanton, Laurent 307
Anastasi, Giuseppe 63
Arena, Antonio 63
Avgeris, Marios 25

Battat, Nadia 55
Belfkih, Abderrahmen 307
Belghith, Abdelfettah 276
Bellavista, Paolo 144
Berrocal, Javier 144
Bolettieri, Simone 213
Bouallegue, Ridha 382
Brea, Victor 3
Bruneo, Dario 135

Calafate, Carlos T. 121
Campobello, Giuseppe 231
Cano, Juan-Carlos 121
Castano, Marco 231
Chaal, Dina 171
Chahboun, Asaad 171
Chalhoub, Gérard 262
Chehida, Aida Ben 77
Chikhaoui, Ons 77
Corradi, Antonio 144

Dakhli, Maha Cherif 382
Dautov, Rustem 39, 342
Dechouniotis, Dimitrios 25
Di Pietro, Riccardo 328
Dini, Gianluca 63
Diop, Papa Samour 314
Distefano, Salvatore 39, 135, 342
Duvallet, Claude 307

El Barrak, Soumaya 161
El Fatmi, Sihem Guemara 92
Esmukov, Kostya 135

Fanjul, Jacobo 221
Fatmi, Sihem Guemara El 77
Förster, Anna 107, 239
Foschini, Luca 144
Fucile, Agata 231

Garrido-Hidalgo, Celia 3
Ghrab, Dhouha 276
Giacobbe, Maurizio 328
Gonnouni, Amina El 161
Gore, Manoj Madhava 184

Hakem, Nadir 290
Hernández-Orallo, Enrique 107
Herrera-Tapia, Jorge 107
Hizem, Moez 382
Hoang, Quy Lam 359
Hortelano, Diego 3

Ibáñez, Jesús 221

Jemili, Imen 276
Jha, Bhaskar 184

Kalatzis, Nikos 25
Kheddouci, Hamamache 55
Kuppusamy, Vishnupriya 239

Lajnef, Hanen 382
Lehmann, Frédéric 171
Leite, J.R. Emiliano 199
Liu, Ju 239
Longo, Francesco 135
Lopez-Camacho, Vicente 3
Loucera, Carlos 221
Lyhyaoui, Abdelouahid 161, 171

Makhoul, Abdallah 55
Manzoni, Pietro 107, 121
Martins, Paulo S. 199
Mbacke, Ahmath Bamba 314
Medjahed, Sabrina 55
Mehrotra, Prakhar 184
Mendy, Gervais 314
Merlino, Giovanni 135
Mingozzi, Enzo 213

Misson, Michel 262, 290
Mitrakos, Dimitrios K. 255
Mnasri, Sami 18, 367
Mosbah, Mohamed 276

Nasri, Nejah 18, 367
Ngo, Chi Trung 359

Oh, Hoon 359
Olivares, Teresa 3

Pandey, Mayank 184
Papavassiliou, Symeon 25
Paraskakis, Iraklis 342
Perazzo, Pericle 63
Puliafito, Antonio 135, 161, 328

Roussaki, Ioanna 25
Ruiz, M. Carmen 3

Sadeg, Bruno 307
Sang, Liu 239
Santamaria, Ignacio 221
Scarpa, Marco 328
Segreto, Antonino 231
Serrano, Salvatore 161
Servajean, Marie-Françoise 290
Sewak, Anurag 184
Soler, David 121
Spyrou, Evangelos D. 255

Tall, Hamadoun 262
Tanganelli, Giacomo 213
Tomas, Andrés 107

Udugama, Asanga 107, 239
Ursini, Edson L. 199

Val, Thierry 18, 367
Vallati, Carlo 63, 213
Van Den Bossche, Adrien 18, 367
Veloudis, Symeon 342

Wang, Jinpeng 262

Zambrano-Martinez, Jorge Luis 121

Printed in the United States
By Bookmasters